AirPort® & Mac® Wireless Networks For D

Cheat Sheet

Finding Your Way Around the AirPort Admin Utility

Apple's AirPort Admin Utility packs an awful lot of wallop into a small space. The utility's single configuration window can present as many as seven separate tabs depending on the model base station you are configuring. Under each tab you'll find a configuration pane loaded with setting controls. Here's a summary of what's where in this essential utility.

Tab Label	***What you can set in the configuration pane***
AirPort	Base station name, administrative contact name, base station location, administrative password for the base station, AirPort network name, closed network setting, broadcast channel, and compatibility mode for older AirPort cards. The Wireless Security button produces a sheet for setting the security method, the network password, and associated security options depending on the chosen method. The Base Station Options button produces a sheet for configuring remote management settings, connected USB printers, and logging. The Wireless Options button produces a sheet for setting the station's multicast rate and transmitter power, and for enabling interference robustness.
Internet	The Internet connection method, which can include Ethernet, modem, PPPoE, America Online dial-up, and wireless distribution system. Different connection settings provide for entering dial-up phone numbers and specifying dialing options, entering passwords, and making TCP/IP settings, as appropriate for the connection type.
Network	Configure how IP addresses are distributed in the AirPort network, DHCP and NAT settings, IP address range used, network welcome message, dial-in settings, America Online parental control setting, and LDAP server list.
Port Mapping	Add, import, export, edit, or remove entries specifying which public network port number corresponds to which private port number at which local address in the AirPort network.
Access Control	List of client AirPort IDs (MAC addresses) allowed to join the AirPort network, and settings for integrating base station with an enterprise RADIUS server for client authentication.
WDS	Settings for configuring base station to form part of a wireless distribution system as a main base station, relay base station, or remote base station. Additional settings for adding or removing other base stations from the WDS as appropriate.
Music	Enable AirTunes music transmission from iTunes, AirTunes speaker name that appears in iTunes, optional password for enabling music playback, and allow AirTunes transmission through base station's Ethernet port. This tab is only available on AirPort Express base stations.

For Dummies: Bestselling Book Series for Beginners

AirPort® & Mac® Wireless Networks For Dummies®

Cheat Sheet

Connecting to an AirPort Network

Mac OS X offers several methods for connecting your Mac to an AirPort network.

- **Internet Connect:** This application lives in your Applications folder. Open it, click the AirPort icon in the toolbar, and choose a network from the Network pop-up menu.
- **AirPort Status Menu:** This optional menu appears in the right side of the menu bar. Click it to choose an AirPort network.
- **Automatic:** In your Mac's Network Preferences, choose AirPort from the Show pop-up menu and click the AirPort tab. Using the By Default, Join pop-up menu, you can specify whether your Mac joins nearby networks automatically, or whether it looks for a preferred network to join. The Options button presents a sheet on which you can control whether the Mac requests permission before automatically joining wireless networks.

Adding and Removing the AirPort Status Menu

The AirPort status menu shows the current strength of your AirPort network connection, and provides many of the AirPort-related features of the Internet Connect application in a compact, always available form.

- **To add the status menu to your menu bar:** Open Internet Connect and check Show AirPort Status in Menu Bar, or open the AirPort pane of Network preferences and check Show AirPort Status in Menu Bar.
- **To remove the status menu:** While pressing the ⌘ key, click and drag the menu off the menu bar then release the mouse.

Making a Secure Connection

Depending on the AirPort card installed in your Mac and the type of wireless network to which you connect, you may encounter a number of different security options. These include:

- **WEP *(Wired Equivalent Privacy):*** It comes in two varieties: 40 bit and 128 bit. Both versions require you to enter a password to join the network. Experts consider 40-bit WEP to be insecure, and recommend using the 128-bit version or some other security option.
- **WPA Personal *(Wi-Fi Protected Access):*** This also comes in two varieties, WPA Personal and WPA2 Personal. AirPort Extreme cards support both. WPA2 provides stronger security. You use a password to join a WPA Personal network.
- **WPA Enterprise:** Used in corporate and other large networks, this security option employs a special server to verify your login request. This server uses the *802.1X* protocol, which is designed for high-security connections. You use your Mac's Internet Connect application to set up the 802.1X connection. Ask the network administrator for the specific 802.1X connection settings you need to make, as they vary from network to network.

AirPort® & Mac® Wireless Networks FOR DUMMIES®

AirPort® & Mac® Wireless Networks FOR DUMMIES®

by Michael E. Cohen

WILEY

Wiley Publishing, Inc.

AirPort® & Mac® Wireless Networks For Dummies®

Published by
Wiley Publishing, Inc.
111 River Street
Hoboken, NJ 07030-5774

Published by Wiley Publishing, Inc., Indianapolis, Indiana

Published simultaneously in Canada

For general information on our other products and services, please contact our Customer Care Department within the U.S. at 800-762-2974, outside the U.S. at 317-572-3993, or fax 317-572-4002.

For technical support, please visit `www.wiley.com/techsupport`.

Wiley also publishes its books in a variety of electronic formats. Some content that appears in print may not be available in electronic books.

Library of Congress Control Number: 2005924601

ISBN-13: 978-0-7645-8971-3

ISBN-10: 0-7645-8971-7

Manufactured in the United States of America

10 9 8 7 6 5 4 3 2 1

1B/RY/QR/QW/IN

WILEY

About the Author

Over the years **Michael E. Cohen** has been a teacher, a programmer, a multimedia producer, a writer, and one of the world's foremost fifth-rate cartoonists. The co-author of *Teach Yourself VISUALLY iLife* and *The Mac Xcode 2 Book,* among other titles, Michael has been playing in Apple's orchard for a long, long time, and still has the memory board — a whole 512K! — from his Apple Lisa to prove it. He lives in Santa Monica with the Digital Medievalist and a closet full of ancient technology.

Dedication

For Lisa.

Author's Acknowledgments

First, thanks go to the various folk at Wiley who helped bring this book into the light of day: You can see many of their names in the credits pages somewhere in the front of this book. Go there. Look. Admire.

Special thanks to Mike Roney and my agent, Carole McClendon, for helping me out of a pickle midway through this journey.

Several people at several companies provided me with information and equipment as I wrote this book, and I owe them many, many thanks for their generosity and help. In no particular order, but with deep appreciation, I wish to express my gratitude to Jason Litchford and Sam Levin and the folk at Griffin Technology, Mike Ridenhour at Keyspan, Tricia Arana and Genea Sobel at Roku, Ryan Sommer and the people at Palm, and Patrick Cosson at Slim Devices. You folk have provided some of the best toys this middle-aged eight-year-old has ever played with.

Thanks also go to Bruce Kijewski for the timely and generous loan of his Bluetooth wireless keyboard and mouse, Bruce Dumes for his technical insights into Bluetooth modem connections, and, of course, the Digital Medievalist for her research, commentary, criticism, and blinding intelligence.

Publisher's Acknowledgments

We're proud of this book; please send us your comments through our online registration form located at www.dummies.com/register/.

Some of the people who helped bring this book to market include the following:

Acquisitions, Editorial, and Media Development

Project Editors: Maureen Spears; Tonya Cupp

Acquisitions Editor: Tom Heine

Copy Editor: Scott Tullis

Technical Editor: Maarten Reilingh

Editorial Manager: Robyn Siesky

Media Development Manager: Laura VanWinkle

Media Development Supervisor: Richard Graves

Editorial Assistant: Adrienne Porter

Cartoons: Rich Tennant (www.the5thwave.com)

Composition Services

Project Coordinator: Kathryn Shanks

Layout and Graphics: Carl Byers, Andrea Dahl, Lauren Goddard, Stephanie D. Jumper, Lynsey Osborn, Ron Terry

Proofreaders: David Faust, Dwight Ramsey, TECHBOOKS Publishing Services

Indexer: TECHBOOKS Publishing Services

Special Help: Teresa Artman

Publishing and Editorial for Technology Publishing

Richard Swadley, Vice President and Executive Group Publisher

Barry Pruett, Vice President and Publisher, Visual/Web Graphics

Andy Cummings, Vice President and Publisher, Technology Dummies

Mary Bednarek, Executive Acquisitions Director, Technology Dummies

Mary C. Corder, Editorial Director, Technology Dummies

Publishing for Consumer Dummies

Diane Graves Steele, Vice President and Publisher

Joyce Pepple, Acquisitions Director

Composition Services

Gerry Fahey, Vice President of Production Services

Debbie Stailey, Director of Composition Services

Contents at a Glance

Table of Contents

Introduction

Using my amazing and internationally renowned powers of deductive reasoning, I can tell that you either use, or are thinking about using, a Mac, and that you have some interest in using Apple's AirPort wireless network technology with it.

Good for you! You won't regret it.

Apple has done something rather remarkable with AirPort: It has turned wireless computer networking from a complex configuration ordeal that can make a seasoned engineer weep into something almost anyone can set up and use. Although network technology traditionally has come with a shipload of acronyms, jargon, numbers, and complex concepts, you won't need to know very much about any of that stuff to get going with your AirPort network. I know this is true, because my own silver-haired mother has an AirPort network that she set up herself, and she doesn't know TCP/IP from TCBY. If she can do it, you can.

Ten years ago I watched a friend struggle through the process of installing a home network: drilling holes through walls, stringing cables, wriggling through crawl-spaces, wiring cable jacks, and, eventually, paying a contractor to fix the things he got wrong. The whole process consumed the better part of a month.

Today my friend could build an even better home network by taking a quick trip to the Apple Store to pick up an AirPort base station. Once he got home, setting up the AirPort network would take less than an hour.

That's progress.

About This Book

Like most books in the *For Dummies* series, you don't have to read this one from cover to cover — although if you do, you'll certainly learn a heck of a lot about wireless networking. In fact, much of this book is like a network, with parts connecting to other parts in several different ways, and with many different paths you can follow to get from one place to another.

I've tried to make each chapter, and, usually, each main section within each chapter, stand on its own. When a topic needs some explanation that I've presented more fully elsewhere in the book, I've tried to give you enough to go on from right where you are without launching you on a page-flipping odyssey — although I also provide you with the cross-references you need should you wish to follow up on some of the details.

The book also contains more than its fair share of screen shots: after all, why should you spend time and trouble trying to imagine what you should be seeing when I can just show you?

One thing I do recommend: before you attempt any of the procedures that I've given you, read the procedure through first. This isn't a mystery novel — it's okay to know how it all turns out ahead of time.

How to Use This Book

This book uses a number of conventions for telling you how to get things done — and, after all, this book is all about getting things done.

When I tell you to select something from a menu, it often looks like this:

Choose File➪Save Configuration As.

This means that you click the File menu on your menu bar and choose the Save Configure As item on that menu.

When I want you to hold down the ⌘ key when pressing another key, it appears something like this:

Press ⌘+A.

This means, hold the ⌘ key down, tap the A key, and then release the ⌘ key. When reading the book aloud, you can pronounce ⌘ "Command."

Finally, when I refer to a Web site, the address usually looks like this:

`www.example.com`

I may or may not preface the address with `http://` — I usually do for more complex Web addresses, but, if you don't see one, you should preface the address with it anyway.

What You're Not to Read

Ignore that heading floating above this sentence. You should read everything in this book: Once you assimilate all the information that I've collected here, you'll be a better person and your hair will be shinier and more manageable.

That's not to say that there are some parts you can skip over and come back to later, at your leisure. Sidebars, for example, provide you with interesting context and background for various tasks, but you don't need to read them to perform any of the tasks I present. Similarly, paragraphs sporting the jaunty Technical Stuff icon provide useful technical details to give you a clearer picture of what a particular task involves, but you don't have to read — or understand — their contents in order to get on with the job at hand.

Foolish Assumptions

For starters, I foolishly assume that you have a Mac, and one that's of a recent enough vintage to have wireless network capabilities.

Even more foolishly, I assume you are using the latest version of Mac OS X, which is currently version 10.4, also known as "Tiger." You don't *have* to run Tiger in order to use AirPort networking on your Mac, of course. However, Apple keeps changing, and usually improving, how networking operates on the Mac with each successive version of Mac OS X. Having used each version of Mac OS X, starting with version 10.0, I can tell you that Mac OS X 10.4 offers better AirPort support than earlier versions. Furthermore, this book would be much longer, and would still not be finished, if I tried to cover how to do each task, using Mac OS X 10.3, 10.2, 10.1, and 10.0. In short, if you are taking the leap into wireless networking, you may as well upgrade your operating system as well. That way, you'll have the latest technology and we'll both be on the same page.

Finally, I assume you have an Internet connection that you want to use wirelessly. Again, you don't *need* an Internet connection to use a wireless network, but most of the advantages that a wireless network offers involve an Internet connection.

How This Book Is Organized

I've divided this book into five main parts, with each part containing two or more related chapters. Within each chapter, I've divided the material into topics and sub-topics and, occasionally, sub-sub-topics. You can read the book's chapters, and each chapter's main topics, in any order that makes sense to you: linearity is *so* last century.

Part I: Wireless Basics

After offering an introductory look at wireless networking in general, and AirPort networking in particular in Chapter 1, this part provides information about the hardware involved in wireless networking. Chapter 2 covers the various AirPort cards you can use, and provides descriptions of how you can put them in various model Macs. Chapter 3 looks at the AirPort base stations — the core of your network — that Apple has offered over the years, and how to set up and use them.

Part II: Knitting a Network

The three chapters in this part describe in detail how to set up different sorts of AirPort networks. Chapter 4 takes you on a tour of how to set up a home network. In Chapter 5 you can learn where to put your AirPort base station in order to increase signal strength and avoid interference. Chapter 6 shows you how you can set up an AirPort network even if you don't have an AirPort base station.

Part III: It's (Almost) All Fun and Games . . .

This part details all the cool things you can do with your AirPort network when you aren't working. In Chapter 7, you learn how to use AirPort to send music around your house or office. Chapter 8 looks at the collaborative tools and toys that AirPort technology makes possible. And Chapter 9 focuses on sharing media and playing games on your network.

Part IV: Taking Care of Business

In this part, you can find out the stuff that warms the cold cockles of a CEO's heart. Chapter 10 takes you into the core of a modern business network and shows you how to work with the corporate network professionals to fit your wireless Mac into the enterprise network. Chapter 11 takes you out on the road, and details how you can stay in touch with the home office from almost anywhere on the planet.

Part V: The Part of Tens

Four sets of ten items for your dining and dancing pleasure: Ten tips for troubleshooting your network, ten suggestions for making your network safer, ten pointers to places on the Web to find out more about wireless networking, and ten facts of dubious importance to make you shake your head and mutter, "What a wild, weird, wireless world we live in."

Bonus Material

But wait. . . there's more! Simply go online wirelessly using the easy techniques provided in this book and you can win a free Bonus Chapter about the other wireless technology that Apple packs into its Macs: Bluetooth. Our special exclusive one-time only bonus chapter is a solid, fact-filled entertainment thrill-ride through the wonders of Bluetooth that's guaranteed to impress your family and friends and to double your Mac's gas mileage. "Biting into Bluetooth" isn't available in stores, but only at this specially reserved toll-free URL: `www.dummies.com/go/airport`.

Icons Used in This Book

TIP

When you see this icon, it means that the paragraph beside it has a time-saving or brain-unstressing tip to make your networking experience easier.

REMEMBER

Almost everything in this book is memorable, but when you see this icon, it means that you really want to take special note of the stuff in the paragraph beside it. Think of it as the icon version of a yellow highlighter.

TECHNICAL STUFF

Paragraphs and sidebars that go into the geeky details of how something works proudly bear this icon. Although you're reading experience may be the poorer if you skip these items, you won't miss anything crucial by doing so. Personally, though, I like the geeky details.

Pay very close attention when you see this icon: It means that the step described or the feature discussed in the associated paragraph can cause trouble if you don't take care — and by "trouble" I mean losing data, or harming equipment, or otherwise putting you in the position of slapping your forehead and crying, "D'oh!!"

Where to Go from Here

Turn the page. The fun is just starting.

Part I
Wireless Basics

The 5th Wave By Rich Tennant

"Frankly, the idea of an entirely wireless future scares me to death."

In this part . . .

This part covers the basics, by which I mean the basic components you use for creating wireless networks. We're talking about hardware, ladies and gentlemen.

In the following pages, I tell you about AirPort cards and how to install them, as well as AirPort base stations and what the different models offer.

Chapter 1

Cutting the Network Cable

In This Chapter

- Hitting the ground running
- Joining wireless networks your way
- Collecting the wireless puzzle pieces
- Judging wireless pros and cons
- Knowing the limits
- Touring home and office networks
- Thinking globally

Over the last century the world has wrapped itself in wires and cables: telegraph cables, telephone cables, power cables, network cables.

Consider what lies behind the typical office desk. If you were to take a time-lapse movie camera and travel back about a hundred years, you might see something like the following: At first there'd be a lone wire for a desk lamp, possibly accompanied by a second wire in the form of a telephone cord. After a while, an office intercom cable would appear, along with its associated power cord, followed shortly by another power cord or two for an electric typewriter and an adding machine. A multi-plug electrical outlet extender would also enter the picture at around the same time, from which would sprout the growing number of power cords that by this time have overwhelmed the existing wall socket. Snaking into the picture with increasing speed as you got closer and closer to the present, a whole bunch of other cables and cords would arrive to feed both electricity and information into an expanding desktop computer system. The last frame of this movie would show a hopelessly tangled Gordian knot of dust-encrusted cables and cords, looking very much like the one behind my desk, and perhaps behind yours, too.

If you would like to become a twenty-first century Alexander the Great and take a big whack at that tangled wire-and-dust knot, this chapter introduces you to your sword: Apple's AirPort wireless network technology. Get ready to start swinging.

Arriving at the AirPort

You just took delivery of your shiny new iBook with its highly touted built-in wireless capability, and now you feel ready, and maybe even more than ready, to join the wireless generation. Only one problem remains: You don't know where to begin. For that, I can offer you the words of wisdom spoken to me by a grizzled old programmer back in the days of punch cards and tape drives, words that remain as true today as they were then: It works better when you turn it on.

Turning the AirPort on

When you start up your Mac, a lot of stuff happens behind the gray Apple logo and spinning gear that you first see on the screen. One of the many things that the Mac does as it starts is look for a network connection — any network connection — and try to set that connection up. If your Mac has AirPort installed, this startup process may include looking for nearby wireless networks that your Mac can join. That is, it can look for wireless networks *if* your Mac's AirPort is turned on.

REMEMBER

You need a device known as an *AirPort base station*, or a similar wireless device, to create a wireless network, often called an *AirPort network* when you use an AirPort base station to create it. An AirPort base station contains a radio transmitter, which provides a network signal that AirPort can receive, and a receiver, to which AirPort can transmit. You can find out more about Apple's various AirPort base stations in Chapter 3.

Here's how you can turn on your Mac's AirPort capability:

1. **Open the Internet Connect application.**

   You can find this application in your Applications folder. You can use Internet Connect to establish Internet connections various ways, as you see in several other places in this book.

2. **In the toolbar at the top of Internet Connect's window, click AirPort.**

   The window displays the current state of your Mac's AirPort, as shown in Figure 1-1. In the figure, you can see that the AirPort does not have a connection to any nearby AirPort base stations, which is not surprising, given that AirPort power is turned off.

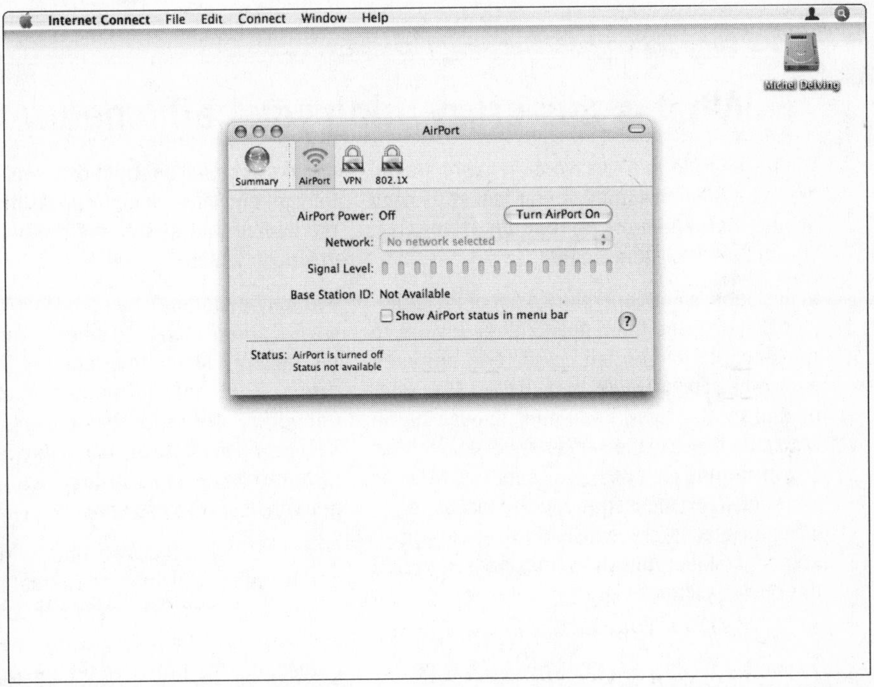

Figure 1-1: You can turn AirPort on using the Internet Connect application.

3. **Click Turn AirPort On.**

   If you compare Figure 1-2 to Figure 1-1, you might not notice the changes in the Internet Connect window, but they are significant: Figure 1-2 now lists AirPort power as being on, the Turn AirPort On button now says Turn AirPort Off, and the Status field now notes that the AirPort is not associated with any network.

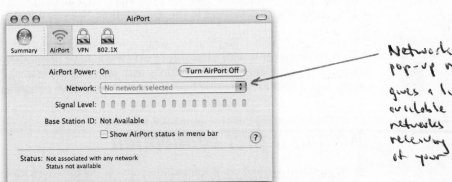

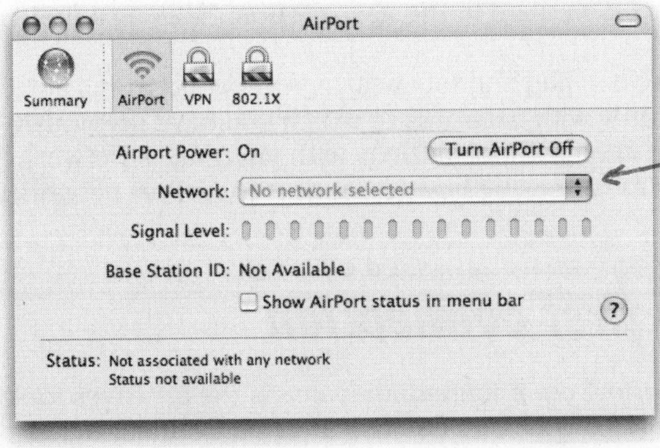

Figure 1-2: When you turn on AirPort power, the Internet Connect window subtly changes.

Network pop-up menu — gives a list of available AirPort networks within receiving range of your Mac

What is this crazy thing you call a network?

You're going to see the word *network* tossed around a lot in this book. If you feel at all hazy about what a network is, read on. Otherwise, you can skip this little sidebar.

In this book, a network refers a set of technologies that allow computers and other devices to transfer information back and forth between each other. For example, if you have a Mac connected to a network, and your spouse has a Mac connected to the same network, your Mac can exchange files with your spouse's Mac, or can send it text messages or voice messages or video messages, and can help you engage in all sorts of collaborative activities, many of which this book describes.

Every device on a network has its own *address*: a number that uniquely identifies the device. No two devices on a network can have the same address or Bad Things happen; after all, the network uses addresses to route information to the right device.

All of the information zipping around the network comes in the form of *packets*: bite-sized chunks of digital information wrapped in a shell that, among other things, carries the address of the device that sent it and the address of the device to which it is going.

You may encounter two physical kinds of networks: *wired* networks, where every device on the network sends information to every other device over cables of some sort, and *wireless* networks, where the computers and other devices on the network use radio transmissions instead of wires to carry information. An AirPort network is a wireless network.

Networks can comprise both wired portions and wireless portions at the same time. For example, you can create a network in which, say, a printer connects to the network with cables, but the Macs on the network connect wirelessly. One of an AirPort base station's many functions is to join the wired and wireless portions of a network so that it appears to be a seamless whole to the network's users.

Now that you have turned on AirPort, you can try connecting to a nearby AirPort network, as described in the next section, "Making a first connection."

Although your Mac can find and join nearby AirPort networks automatically, by default it does so only with networks to which you have previously connected your Mac. You can change these settings with your Mac's Network Preferences, as described later in "Choosing how to connect to AirPort networks."

Making a first connection

When you turn AirPort on, it immediately scans the airwaves looking for available wireless networks. You can use Internet Connect to instruct your Mac to join one of those networks.

When you join an AirPort network, your Mac can use that network's Internet connection if it has one. As you see in Chapter 4, most AirPort networks provide shared Internet access to the computers using the network.

Here's how you join an AirPort network using Internet Connect:

1. **Open Internet Connect and click Turn AirPort On if your AirPort is off.**

   If you have just finished the procedure in the previous section, Internet Connect should already be open and AirPort should already be turned on.

2. **Click the Network pop-up menu.**

   You should see a list of the available AirPort networks within receiving range of your Mac, as shown in Figure 1-3. Of course, if you are not within range of any AirPort networks, no networks appear on this menu. If that's the case, just read through the remaining steps and try them again when you find yourself in range of an AirPort network.

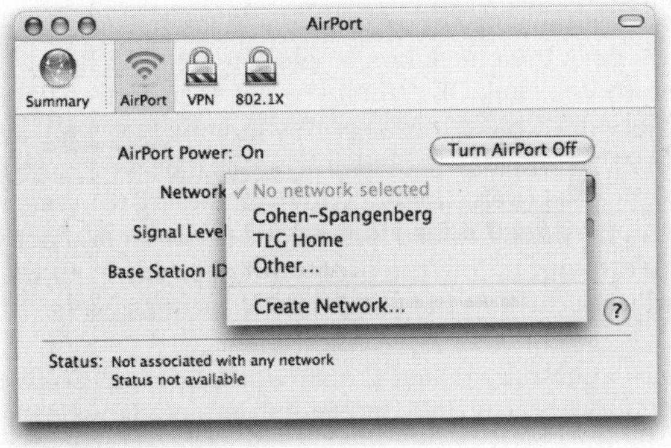

Figure 1-3: The Network pop-up menu offers nearby AirPort networks you can join.

If you have a Mac laptop with AirPort, and aren't connected to a wireless network yet, you might want to find a nearby wireless *hotspot* to use when trying out the procedures in this chapter. Chapter 11 describes what a hotspot is and provides more information about how to find and use wireless hotspots.

3. **On the Network pop-up menu, click the name of a network.**

   In many cases, after you perform this step, your Mac has joined the network, and you can begin working — or playing. Often, however, AirPort networks will require you to enter a password or other form of identity verification before your Mac can join the network. This process is called *authentication,* and it allows the network to restrict access to trusted individuals. Figure 1-4 shows a typical authentication sheet you may see when attempting to join a protected AirPort network. You can find out more about how to protect an AirPort network in Chapter 4, and how to use the kinds of authentication that business networks often require in Chapter 10.

Figure 1-4: Protected networks require you to provide some authentication.

4. **Type the network password in the Password field and click OK.**

   You can skip this step if the network you have selected does not require a password.

   TIP: You can also skip this step the next time you join the same protected network if you click the check box labeled Remember Password in My Keychain before you click OK. When you set that option, your Mac records the network's password securely in your *keychain,* a standard system feature that the Mac provides to store passwords and other sensitive information securely, freeing you from having to remember all the passwords you may need to use for various services and activities. The next time you attempt to join the protected network, your Mac retrieves the password from the keychain and uses it automatically.

That's it. If you've successfully joined the network, Internet Connect indicates that fact in the window's Status field. Internet Connect also shows you the strength of the network's broadcast signal in the window's Signal Level indicator, and gives the *hardware address* (also known as the *MAC address*) of the base station providing the network in the window's Base Station ID field. You can find out more about AirPort broadcast signals in Chapter 5 and read about hardware addresses in Chapter 4.

After you've connected your Mac to an AirPort network, you can quit Internet Connect. The application does not have to remain open in order for you to use the network connection.

TIP: You can avoid most future trips to Internet Connect by clicking the Show AirPort Status check box in Menu Bar window. This setting puts a status menu on your menu bar; the menu appears as an AirPort signal strength indicator so you can see, at a glance, how strong your network connection is. When you click the status menu, you see a list of available AirPort networks, as well as several other useful options, which are described elsewhere in this book. Figure 1-5 shows this menu in action and also shows the Internet Connect window for comparison.

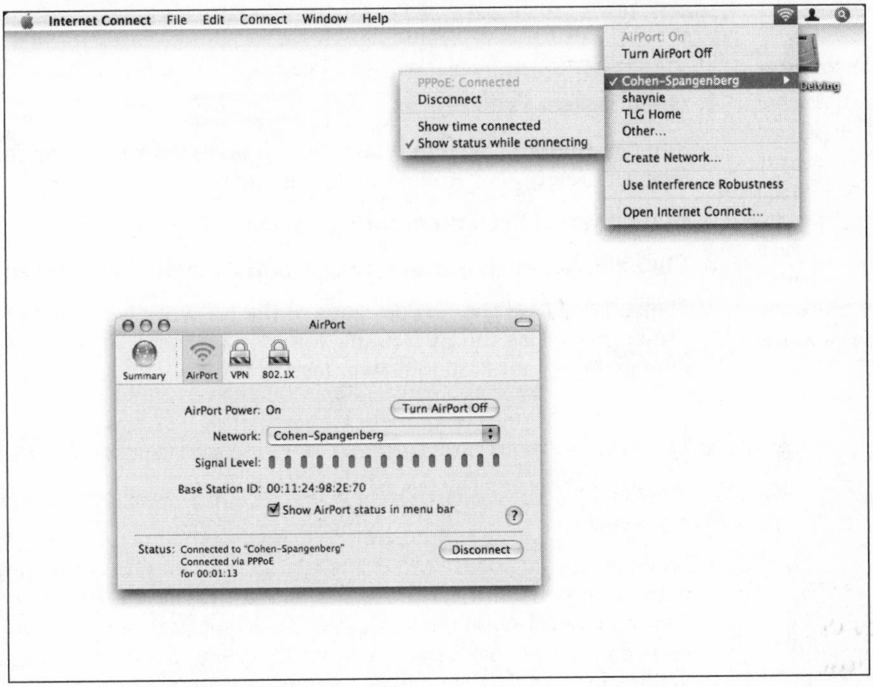

Figure 1-5: The AirPort status menu offers many of Internet Connect's functions.

You may be wondering about the Disconnect button in the lower-right corner of the Internet Connect window, shown in Figure 1-5. This button appears when your Mac has joined an AirPort network. You click the Disconnect button to instruct the AirPort base station to close its Internet connection. When the AirPort base station doesn't have an Internet connection, the button label changes to Connect, and, as you might expect, you click it to instruct the base station to create an Internet connection. You can find out more about configuring a base station to establish an Internet connection in Chapter 4.

Choosing how to connect to AirPort networks

In order to hide the uninteresting complexities of wireless networking from its users, Mac OS X employs several strategies. For example, when you join an AirPort network for the first time, Mac OS X remembers that network: The next time your Mac comes within range of that network, Mac OS X joins that network again automatically. You can use the Network preferences window in your Mac's System Preferences to change how your Mac joins wireless networks.

System Preferences → Network

Options for how your Mac joins wireless networks

System Preferences → Network → AirPort

To get to the AirPort settings of your Mac's Network preferences, follow these steps:

1. **Open System Preferences.**

   You can click the System Preferences icon in your Dock, or choose System Preferences from your Apple menu.

2. **In the System Preferences window, click the Network icon.**
3. **Click the Network window's Show pop-up menu and choose AirPort.**

   Figure 1-6 shows the AirPort pane of the Network window in System Preferences. Use the By Default, Join pop-up menu in this pane to change how your Mac joins wireless networks.

your wireless card's MAC (media access control) address Cf. 349, 84

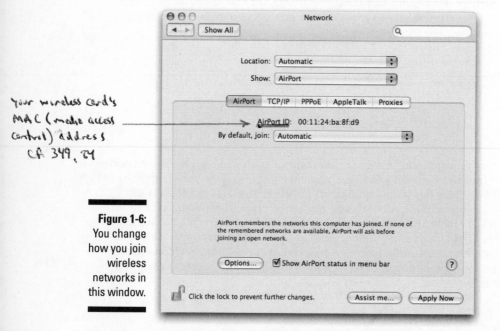

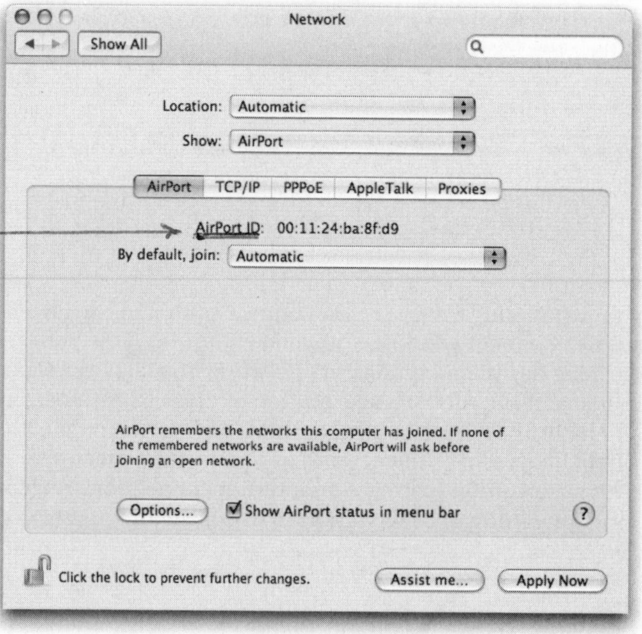

Figure 1-6: You change how you join wireless networks in this window.

TIP

You need to turn on your Mac's AirPort before you can use the By Default, Join pop-up menu. When AirPort is turned off, this pop-up menu shows the current setting but does not allow you to click it.

Here are the two options that appear on the By Default, Join pop-up menu:

- **Automatic:** With this setting, your Mac automatically joins any wireless network you've previously joined successfully whenever it detects that network. If two or more such networks are in range, your Mac joins the network that it joined most recently. If your Mac can find no previously joined wireless network in range, but does detect a nearby open network, your Mac presents a dialog asking you if you wish to join that network. Unless you or someone else has already tinkered with your Mac's AirPort settings in Network preferences, the Automatic item on this menu is selected.

Wireless networks, including AirPort networks, can be *open* or *closed*. An open network broadcasts its name, allowing your Mac to display that network name in Internet Connect and in your AirPort status menu. A closed network does not broadcast its name, so your Mac can only join it if you both know that it exists, and you know its name. You can find out more about creating closed networks in Chapter 4.

- **Preferred Networks:** Use this setting to specify the list of preferred networks your Mac will join. Figure 1-7 provides an example of how the AirPort pane of my Network preferences window looks when I choose the Preferred Networks setting: as you can see, I have quite a few preferred networks on my list.

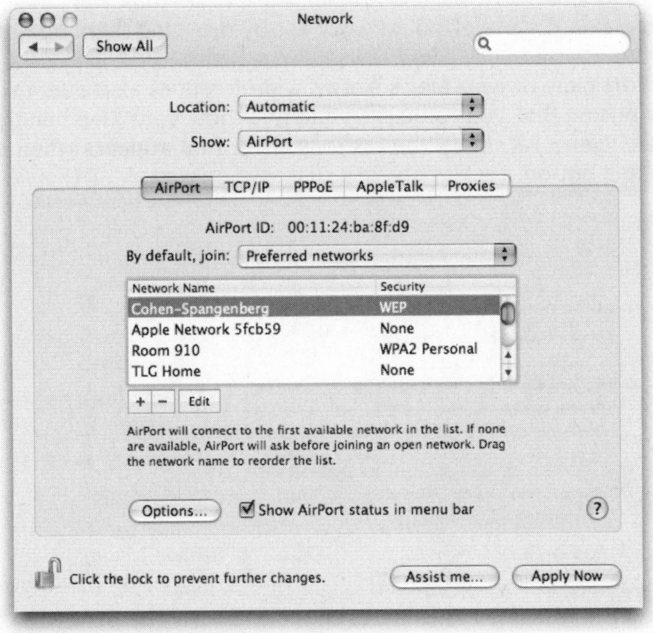

Figure 1-7: Your Mac keeps track of wireless networks it can join.

The preferred networks list gives you control over which wireless networks your Mac joins:

- **+ button:** Click this button to add networks to the preferred list manually.

- **– button:** Click this button to remove selected networks from the list. You may want to use this button to remove networks that your Mac has added to the list, which it does whenever it joins a new wireless network. If you travel a lot, and use a lot of different wireless networks, this list can become lengthy.
- **Edit:** Click this button to change the name of selected network in the list and to change the authentication method your Mac uses when it connects to that wireless network. This button can come in handy if, for example, a wireless network administrator changes the name of a network you regularly use: You can edit the network name in this list, and your Mac will connect to the renamed network the next time you come within its range.

Your Mac works its way down the preferred network list, starting at the top, when it attempts to find and join wireless networks. You may want to rearrange the list to put the network you use most use on top, so that when you find yourself in an environment where your Mac can detect two or more preferred networks at once, such as in school or an office building, your Mac will try to join the network you regularly use first. You can change the order of the items on the preferred network list by dragging them up or down.

The AirPort pane of your Mac's Network preferences also offers you an Options button that you can use to fine-tune how your Mac handles AirPort networks. Figure 1-8 shows the Options sheet that appears when you click the Options button.

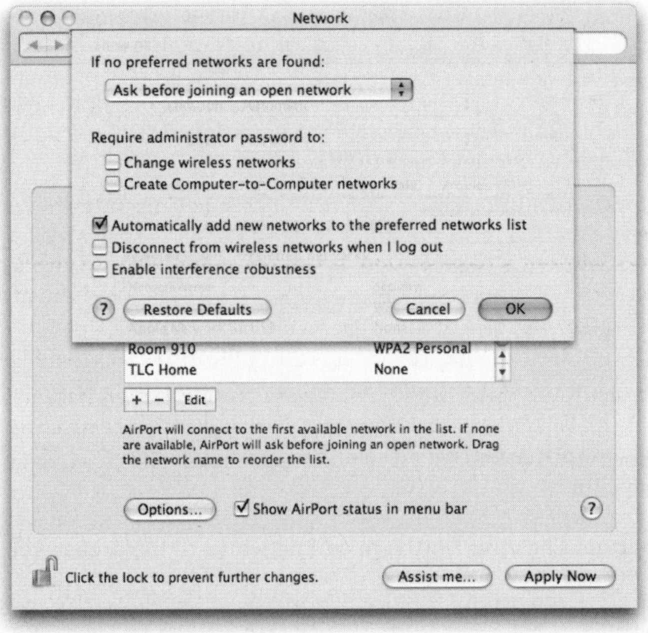

Figure 1-8: Fine-tune your Mac's AirPort network settings with this Options sheet.

The If No Preferred Networks Are Found pop-up menu at the top of the Options sheet controls how your Mac behaves when your Mac has not already joined a wireless network and none of your Mac's preferred wireless networks are in range. The menu provides three choices:

- **Ask Before Joining an Open Network:** When you choose this item, your Mac alerts you when it detects an open network. You can then choose to join the network.
- **Automatically Join an Open Network:** When you choose this item, your Mac joins the first open network it detects. If the network does not require a password or some other authentication, you may not even notice your Mac has joined the network unless you look at your AirPort status menu or open Internet Connect.

  The AirPort pane of your Network preferences has a Show AirPort Status in Menu Bar check box, just like Internet Connect does. If you use AirPort regularly, you owe it to yourself to devote the small amount of menu bar space required to show the AirPort status menu. That way, you can see when your Mac joins, or loses contact with, nearby wireless networks, avoiding networking surprises.
- **Keep Looking for Recent Networks:** When you choose this item, your Mac ignores open networks that you've never joined before, or that don't appear on your list of preferred networks, and keeps looking for a preferred network. When your Mac finds one of those, it will join it.

Here's what the other Options sheet items do:

- **Require Administrator Password to Change Wireless Network:** When you check this box, a dialog asks for an administrator's password each time you attempt to join a wireless network. A business might want to check this box on an employee's computer, for example, to keep the computer linked to a specific wireless network in a setting that might have several wireless networks present.
- **Require Administrator Password to Create Computer-to-Computer Networks:** Two or more AirPort-equipped Macs can establish small networks without a base station. A teacher might want to check this box on classroom computers, for example, to keep students from creating in-class networks and passing notes wirelessly during class. You can find out more about computer-to-computer networks in Chapter 6.
- **Automatically Add New Networks to the Preferred Networks List:** Uncheck this box to keep your Mac from adding networks to your preferred network list. By default, this box is checked.

- **Disconnect from Wireless Networks When I Log Out:** By default, after your Mac joins a wireless network, the Mac remains connected even when you log out of your user account, so that the next person who logs in can use the same wireless network. Check this box to require users to join a wireless network when they log in. Of course, if they log in when the Mac is in range of a preferred network, the Mac joins that network automatically.
- **Enable Interference Robustness:** This option applies only to Macs equipped with AirPort Extreme capability. It provides better network performance in the presence of interfering radio transmitters. The next section of this chapter explains what AirPort Extreme is, Chapter 4 describes the interference robustness option in more detail, and Chapter 5 discusses how interference can affect an AirPort network.

Any changes you make to your Mac's Network preferences don't take effect until you click Apply.

Assembling a Wireless Network

Apple has put a lot of time, thought, and money making wireless networking simple enough for "the rest of us" to use. These days, setting up a wireless AirPort network can require less expertise than it takes to set the timer on a VCR — which is a good thing, too, because who wants to see a bunch of Mac screens blinking 12:00 over and over and over? Apple's hardware and software work together to make what used to be — and for many users of other computer systems still is — a complex process into a simple matter.

Getting the hardware

Depending on how new your Mac is, getting the right hardware for using a wireless network may involve no work at all. These days, Apple builds everything you need to use wireless networking into nearly every Mac model they make, either as a build-to-order option or as a standard feature. However, if you want to do more than use a network that someone else has set up, you may also need one or two other components.

The following list comprises the various items you might need to establish your wireless network:

- **AirPort card:** This small device plugs into a socket inside your Mac desktop or portable computer and converts the Mac's networking signal into a radio transmission. It attaches to a small antenna, which, in most cases, Apple builds into the Mac.

Apple has made two different sorts of AirPort cards since it introduced AirPort networking in 1999: the original AirPort card, and the newer AirPort Extreme card. You don't need to decide which card to get, though; older Macs use the original AirPort card and newer Macs use the AirPort Extreme card. Many of the newest Macs don't use any card because Apple has incorporated the necessary equipment into the circuitry.

You can find out more about which card to get and how to install it in various Macs in Chapter 2.

- **AirPort Base Station:** Also known as a *wireless access point,* this item transmits and receives network information among the computers on the wireless network.

  Apple has made four AirPort base station models over the years, and all of them work with any Mac that has an AirPort or AirPort Express card. They also work with non-Mac wireless computers and wireless-enabled *Personal Digital Assistants (PDAs).*

  The AirPort base station provides the central hub of your network. Although you can create a wireless network without a base station (as Chapter 6 explains), most AirPort networks have at least one AirPort base station handling the traffic. In Chapter 3, you find out more about the different AirPort base station models Apple has manufactured over the years and how to use them.

- **Ethernet cable:** I know, I began this chapter waxing rhapsodic about how wireless technology helps you reduce cable clutter, but you still need *some* cables. An *Ethernet cable* is a standard cable that has a plug on each end — called an RJ-45 plug — that looks like a large version of the plug at the end of most telephone cables — which, by the way, is called an RJ-11 plug. Although some older networks may use different cables, these days most networks tend to connect devices, using Ethernet cables.

  You'll need an Ethernet cable, for example, to connect your base station to a device like a cable modem or DSL modem, which an Internet Service Provider (ISP) may provide to give you high-speed Internet access. Chapter 4 describes these devices and how to use them in your wireless network. You may also need Ethernet cables to attach non-wireless equipment, such as printers or network fax machines, to your network.

- **Telephone cable:** If you get your current Internet connection from a dial-up account, you need one of these to connect your base station to the telephone jack. Make sure you get an AirPort Extreme base station, though: Apple's compact AirPort Express base station doesn't have modem capabilities. See Chapter 3 to read more about the capabilities of the different AirPort base stations.

- **Network hub:** You use a network hub to connect multiple wired devices to a network. You require one of these only if you have several network devices that need connecting. Chapter 4 covers how to arrange a network that includes both wired and wireless devices.

You won't need all of the items in the preceding list to set up a simple wireless network. If you don't need to share Internet access, you only need the first two items on the list. In fact, as described in Chapter 6, you don't even need a base station if you can leave your AirPort-equipped Mac turned on, because the Mac itself can act as a base station and share its Internet connection with other AirPort-equipped Macs.

pp. 155-183

Plugging in

AirPort networks can, and often do, consist of both wired and wireless devices, all of which connect to each other through the AirPort base station. When you lay out your network, you should consider which devices need to connect with wires and which don't. This helps you decide where to put your various network devices, including your AirPort base station. It also helps you decide which sort of AirPort base station you need.

As discussed in Chapter 3, Apple currently sells two kinds of AirPort base station: the AirPort Extreme and the AirPort Express. If you need to have a wired segment of your network, you need an AirPort Extreme; an AirPort Express lacks the necessary physical connectors to accommodate a wired network segment.

Here is a short list of the kinds of network devices that tend to require wired connections:

- **Older Macs:** All Macs manufactured before 1999 and many manufactured as much as a year later can't have AirPort installed.
- **Macs without AirPort cards:** In most Macs that can have AirPort installed, you need to purchase and install an AirPort card. An AirPort-capable Mac that lacks such a card must use a wired connection. Chapter 2 describes the Macs that can use AirPort and the types of AirPort cards they require.
- **Network printers:** Network printers have been regular participants on computer networks ever since Apple released the first LaserWriters back in the mid-1980s. In fact, the very first Macintosh networks were designed to facilitate printer sharing. However, few network printers include wireless capability even today, possibly because these devices tend to stay in one place.
- **USB printers:** *Universal serial bus (USB)* printers, unlike network printers, are usually intended to connect directly to a single computer. However, Mac OS X allows a networked Mac to share a directly connected USB printer with other network users. Furthermore, both the AirPort Express and AirPort Extreme base stations provide a USB port and USB printer-sharing capability. That allows a USB printer to serve any computer on the AirPort network. Nonetheless, such printers connect to the Mac or the base station via cables and have no wireless capability of their own. Read more in Chapter 4 about connecting a USB printer to an AirPort base staton.

To complement this list, these network devices often use wireless connections:

- **Portable Macs:** The first Macs to have wireless capability were the first iBook models, and every portable Mac model released since has had wireless capability. Because portables are designed for mobility, it makes sense that they are on this list. See Chapter 2 for more information about the wireless options available on various Mac portables.
- **Desktop Macs with AirPort:** AirPort has also been an option on nearly all Macs since 1999, and is a standard feature on the latest iMacs. Although desktop computers tend to move around less than portables, connecting them via wireless networks often makes good sense because it reduces wiring costs and allows more location flexibility. Chapter 2 also describes the wireless options available on desktop Mac models made in the last few years.
- **PDAs:** Some recent PDAs (Portable Digital Assistants), such as the recently released Palm LifeDrive, incorporate the wireless network technology that AirPort uses. Chapter 11 discusses wireless-enabled PDAs.

When you set up your AirPort network, the wired devices all plug into a device called a *network hub*. A hub usually has four or more Ethernet connectors, called *ports,* to which network devices connect. One of those devices is your AirPort base station. See Chapter 3 and Chapter 4 for more about how and why you connect an AirPort base station to a network hub.

Because the entire network's wired devices, including your base station, connect to the network through a hub, you have to run cables through your home or office to connect those devices to the hub. Many offices already have Ethernet cable in place, but unless you want to engage in a home-wiring exercise, most people locate their printers and other wired network devices pretty close to one another. Chapter 4 discusses the factors you should consider when setting up a home network.

Touching base with the Internet

To establish a Mac wireless network, you need only an AirPort base station and a single Mac with wireless capability — but what's the fun of that? For most of us, a network without an Internet connection doesn't seem like much of a network. Most AirPort networks also provide an Internet connection that the network's users can share.

As described in Chapter 3, much of the setup process for an AirPort base station involves configuring the base station to obtain and share an Internet connection. This means, of course, that you must have an Internet connection to share in the first place.

You usually get an Internet connection in one of two ways:

- **From an Internet Service Provider (ISP):** ISPs provide most home Internet connections. Common ISPs include local telephone companies and cable television companies, which often make a good part of their income by providing Internet service to their customers. In addition, many home users get Internet access from nationally available ISPs, such as EarthLink and NetZero, not to mention the great granddaddy of them all, America Online. Smaller local ISPs also provide their fair share of Internet connections. Chapter 4 describes how to set up an Internet connection, using the most common methods available to home Internet users.

  TIP: If you need to find an ISP, one place to look is `www.thelist.com`, which for many years has compiled and made available comprehensive lists of the various ISPs available around the world.

- **From an existing Internet connection:** You tend to find such connections in schools, corporations, and other large institutions, in public venues such as libraries, and even in restaurants, hotels, book stores and coffee shops. You can find more about using a corporate Internet connection in Chapter 10. Chapter 11 discusses using available connections in hotels, cafes, and similar venues.

After you have an Internet connection source, setting up your base station or your Mac to use it does not have to be a brain-straining challenge. For setting up a new AirPort network, Mac OS X offers the AirPort Setup Assistant, a program that walks you through the process of setting up a base station to use your Internet connection. Chapter 3 describes this program and how you can use it.

Deciding to Go Wireless

Haven't decided whether to go wireless? Actually, I'm really not all that surprised: For example, if your current network configuration works for you, the time-honored principle of "If it ain't broke, don't fix it" can exert a powerful force. I assume, however, that you have picked up this book because the prospect of using wireless network technology has some appeal for you. In this section, I lay out some of the issues you need to consider when going wireless.

Weighing what you gain and lose

I had a high-school physical education teacher who regularly mouthed the platitude, "No pain, no gain." Like all platitudes, this one was overreaching, but, also like all platitudes, it contained some truth. When you switch from wired to wireless networking, you may experience a gain — in many cases, a considerable gain — but chances are that gain comes with at least a little accompanying pain. You have to decide if the gain is worth the pain to you.

Losses

Take a look at some of the things you may have to give up when you adopt wireless network technology:

- **Speed:** These days, the Ethernet port on the newest Macs can transfer information across a network at up to 1 billion bits per second — at least theoretically. By comparison, current wireless networks provide a theoretical top speed of only 56 million bits per second. Of course, most networks, wired or wireless, don't operate at anywhere near their theoretical top speeds, and many other factors, such as the number of active network users and the speed of the various devices attached to the network, can affect a network's speed far more than whether the connection is wireless. Nonetheless, if you absolutely require extremely fast network throughput, you may find that wireless networking just won't work for you.

  But put the speed issue in perspective: At a billion bits per second you can transfer the entire text of a book like *War and Peace* in well under a twentieth of a second. By comparison, Tolstoy's classic would take slightly more than half a second to transfer at the theoretical top speed of a wireless network. If you don't need to transfer gargantuan 19th-century Russian novels over a network faster than 20 or so copies a minute, the speed limits of wireless technology probably won't cause you any problems.

- **Security:** On a wired network, you have physical control over who gets to use the network: If you don't want someone to use the network, just yank that person's network cable or, even simpler, don't provide one. A wireless network, on the other hand, extends invisibly in all directions for dozens of yards, and you have to take active steps to protect unauthorized individuals from intercepting your wireless network transactions or from joining the wireless network. If you don't want to take the time and trouble to secure your wireless network, you may want to stick with cables.

- **Money:** If you have already invested in a wired network, going wireless means spending additional dollars. Adding AirPort capabilities to a nonwireless Mac, for example, can take a big chunk out of a hundred-dollar bill, and purchasing an AirPort base station costs even more. On the other hand, wired networks aren't free, either: Between the costs of network routers and Ethernet cables, along with the trouble required to run the cables to where you need them, you can spend a nontrivial amount of money or time.

- **Control and predictability:** You can see a wired network — just follow the cables — and if one wired network client seems to have a bad connection, it's pretty easy to swap cables and see if that solves the problem. A wireless network has no visual indication of its presence, and all sorts of things can make network performance unpredictable, such as nearby wireless networks that interfere with yours, or electrical devices that generate radio interference. Although Chapter 4 can help you diagnose and solve many wireless network problems, you may need to develop some deductive skills beyond those required for fixing wired network problems.
- **Hard-earned knowledge:** If you've built your own wired network, I imagine that you spent more than a little time and energy understanding the technology, figuring out how all the pieces work together, and acquiring a nice collection of troubleshooting tips. When you go wireless, you have to acquire some new knowledge and skills. On the other hand, that's why you have this book, right?

Gains

Here's a list of some of the things you gain by going wireless:

- **Mobility:** This, obviously, is the main attraction for portable Mac users, and I don't think I really need to explain why. But an AirPort network makes sense for desktop Mac users, too: You can set up your Mac wherever you desire without worrying about a network cable, and you can move the Mac more easily.
- **Expandability and flexibility:** An AirPort Express base station can serve as many as 10 simultaneous network users, and the AirPort Extreme can serve as many as 50. When you can add network users, either at home or at the office, without routing additional cables, you save both time and money.
- **Security:** I'll bet you're thinking, "Wait a minute — didn't he just say that wired networks are more secure than wireless?" Well, yes, they are, — but only if you don't secure your wireless network. If you do, however, you can protect your network transmissions with very secure encryption, using the latest wireless technology, as Chapter 10 explains. And wires do not provide complete security: Just do a Google search on the phrase **"packet sniffer"** to find out why.
- **Sheer utter coolness:** What can I say? Wireless networking is cool.

Avoiding show-stoppers

Going wireless, of course, makes sense only if you actually *can* go wireless. If you want to use an Airport network, you need to make sure that no stumbling blocks lie on your path.

Here's a short list of possible stumbling blocks you may encounter:

- **Crowded wireless network environment:** In some locations, such as apartment buildings, college dorms, or office buildings, you may find many base stations in your immediate vicinity. As you see in Chapter 4, adjacent base stations can interfere with one another. Fortunately, you can usually overcome the crowded airwaves by coordinating with your neighbors, or by adjusting your AirPort base station settings.
- **Other environmental interference:** Radio interference can adversely affect wireless networks, and the causes may be quite subtle. I recently tried to help someone set up a new wireless home network. Everything worked perfectly until the sun went down, at which point the network seemed to go dead. It seems this person's home had halogen track lights that automatically came on at sunset, and the lights were generating radio interference that drowned out the network transmissions. Turning off the track lights solved this particular problem. You can find out more about how to overcome radio interference in Chapter 4.
- **Physical obstructions and distances:** AirPort networks have relatively limited ranges and, although radio transmissions can penetrate walls to some degree, that penetrating power has its limits. Once again, Chapter 4 tells how to extend range with external antennas and wireless relay stations.
- **Corporate policy:** Many corporations have rather strict policies concerning how you can connect to the company network. You may encounter policies that discourage or even prohibit wireless network use. Chapter 10 discusses how to deal with such situations.
- **Lack of Internet access:** Okay, this could really be a show-stopper if you plan to use your network only for Internet sharing. However, a wireless network offers some advantages even if you don't have Internet access; besides, you can get Internet access almost anywhere on the planet if you are willing to pay enough money.

Generally speaking, with a bit of knowledge, some ingenuity, and, occasionally, the strategic expenditure of a few extra dollars, you can set up a wireless network almost anywhere you like.

Seeing the AirPort from 10,000 Feet

Networks come in all shapes and sizes, ranging from a simple connection between two devices to the globe-spanning Internet itself. The number of possible network configurations is staggering, but you don't need to know them all: Becoming familiar with just a few typical network situations can help you enormously when you come to set up your own AirPort network.

Checking out a typical home AirPort network

You can count on your fingers the number of computers that a typical home network connects: Such networks have to serve the needs of a small group of people.

Home networks provide most or all of the following services:

- **Internet connection sharing:** Using an AirPort base station to share an Internet connection means no more arguments about who gets online next. This connection usually comes in through the telephone line, as either a dial-up or DSL connection, or through the TV's cable connection. Chapter 4 describes those sorts of connections and how to set up an AirPort base station to work with them.
- **Shared printing:** In most homes, printers sit idle most of the time, so sharing a single printer with the household over an AirPort network makes good economic sense — usually, even when shared, a home printer still sits idle most of the time.
- **Occasional file sharing:** In a home, personal computers tend to be, well, personal. Even so, family members or housemates occasionally may need to exchange files, such as mailing lists during the holidays or financial information during tax season. File sharing over a network beats swapping floppy disks — especially because floppy disks and floppy disk drives are relatively rare items among Mac users these days.
- **Entertainment:** AirPort networks allow family members to share music and photos with one another, as discussed in both Chapter 7 and in Chapter 9.

For most home networks, a single AirPort base station suffices. That base station, as you might expect, usually sits near the source of the home's Internet connection, to which it needs to be physically connected. The computers on the network, on the other hand, can be located anywhere the AirPort base station signal can reach, which, for all but the most palatial homes, usually includes the entire residence.

REMEMBER

If you want to share a printer over the network, place the printer near the base station as well, along with any other network devices that don't use a wireless connection: Those devices connect to the network with cables. If possible, avoid running cables between rooms.

Looking at a typical office network

Businesses, like networks, can come in all shapes and sizes. A network for a small business, such as a retail store or doctor's office, might not differ much in its networking needs from the home network described previously. Even the Internet connection for a small office network might be provided the same way as for a home network. Larger businesses and organizations that include many offices and work areas, possibly spanning several sites, are another matter.

Information provides the life blood of many modern business and organizations, and the network is the circulatory system through which that blood flows. The network carries e-mail from cubicle to cubicle and instant messages among work groups; it provides employees with access to the various file servers that contain the spreadsheets and memos and reports and databases that they create and use; it transmits data generated by in-house business applications to back-end data-mining applications and business process suites — have your eyes begun to glaze over yet?

The point is simply this: In most modern enterprises, the network constitutes an absolutely essential asset, and an asset over which the enterprise, quite rightly, exercises tight control, in the form of an *information technology (IT) staff* that sets and enforces network policies.

Most enterprise network policies deal with two concerns:

- **Physical integrity:** Enterprise networks often consist of multiple networks joined in various ways into a larger network. Each of the subnetworks may have different physical characteristics, depending on when they were first set up and why they were created. The way various subnetworks interconnect in an enterprise network — which network professionals like to call the network *topology* — can be dizzyingly complex. The IT staff keeps the complex enterprise network together and running smoothly in part by controlling what kinds of devices use the network, and where those devices make their network connections.

  In short, for reasons of physical network integrity, you won't use wireless network technology in an enterprise network unless the IT staff allows you to use it and can control how you use it.

- **Security:** To keep the enterprise's valuable information safe from outsiders, the IT staff sets security policies and procedures that permit only authorized users to gain network access. Also, because some information stored on the network may need to be restricted to small groups, the IT staff often grants different users different access privileges. Security policies may also control which users have Internet access, and which users can access the network away from the office.

  In short, for security reasons, you won't use wireless technology in an enterprise network unless the IT staff allows you to use it and can control the ways in which you use it.

This doesn't mean you won't be able to set up or use a wireless network at the office. Many enterprises have noted that wireless networks, sometimes referred to as *air-gapped LANs* in IT circles, can offer economic advantages. You may run across enterprises that have either established wireless networks within the larger enterprise network, or created IT policies and guidelines for how individual work groups in the enterprise can establish their own wireless networks.

Chapter 10 describes some of the technical and political challenges you may encounter when using wireless networks in a large enterprise and provides ways to meet them.

Contemplating the big picture

I doubt that Apple's AirPort technology would even exist if the Internet had not become such a powerful economic and social force in the 1990s. That's not to say that wireless networks would not exist: Enterprises, without the Internet, would find wireless network technology useful and economical in particular cases. But the unprecedented growth of Internet use by the general public — which happens to be Apple's core market — made the creation of easy-to-use wireless networking hardware and software not just commercially feasible but wildly successful for Apple. Today, AirPort capability isn't even an option on many Macs: It's a standard feature.

Providing easy, secure Internet access to wireless network users pervades the AirPort experience. The AirPort Setup Assistant, which is described in Chapter 3, devotes a good deal of its prepackaged intelligence to helping you establish an Internet connection. As shown in Chapter 4, Apple's AirPort Admin Utility — the master-control program for your AirPort base station — not only devotes one of its seven configuration panes to setting up an Internet connection, but four of the remaining six panes also include settings that control some part of Internet connectivity.

When you establish an Internet connection with your AirPort base station, your network becomes part of the Internet itself. After all, that's what the Internet really is: a global collection of individual networks all interconnected.

Chapter 2

Picking a Wireless Card

In This Chapter

- Installing AirPort and AirPort Extreme cards
- Accessing your Mac's AirPort card slot
- Looking at alternative wireless cards for older Macs
- Opting for a USB wireless adapter

When a prestidigitator says, "Pick a card . . . any card," and, following some amusing patter, tells you exactly which card you picked, you can probably figure out that magic isn't really involved. The choice was never really yours in the first place. Nor is any magic involved when you pick an AirPort wireless card for your Mac. For most Macs — those that Apple calls "AirPort-ready" — the only choice you can make is whether to add a wireless card at all.

For AirPort-ready Macs, the Mac model that you have determines which AirPort wireless card you can use. For a handful of other Macs, you don't even have a choice: either they come with the proper AirPort card built right in, or they don't support wireless technology at all. Don't worry about this last kind of Mac. The only AirPort-unready Mac that Apple currently makes is their rack-mount Xserve G5 server, which you are unlikely to have sitting around the house. But, for the rest, whether desktop Mac or laptop, there's an AirPort card made just for it. And installing the card is often child's play.

This chapter doesn't attempt to provide the detailed step-by-step instructions for installing the appropriate AirPort card in the Airport-ready Mac models. For one thing, Apple has been producing AirPort-ready Macs since the summer of 1999, and that encompasses a *lot* of different Macs, each with its own unique AirPort card installation process. For another, Apple's Web site provides Web pages, illustrated downloadable PDF files, and, in some cases, even QuickTime movies, that describe exactly how to install AirPort cards in Airport-ready Macs.

Instead, this chapter describes how to open each of the AirPort-ready Macs to get access to the AirPort card slot — calling to your attention any quirks or curiosities in the process, as well as directions for where to obtain Apple's detailed installation instructions — so that you can decide if you want to install the card yourself.

Installing an AirPort Card

Since Airport's debut in the summer of 1999, Apple has produced just two types of Airport card: The original AirPort card and the newer AirPort Extreme card. Which card you can use depends on when your AirPort-ready Mac was made.

- All AirPort-ready Mac models released prior to the introduction of AirPort Extreme use Apple's original AirPort card.
- Most AirPort-ready Mac models released after the introduction of AirPort Extreme require AirPort Extreme cards.

Installing a card, whether the original AirPort card or the AirPort Extreme card, requires that you make two different connections: One connection attaches the card to the Mac's circuitry, and the other connection attaches the card to the Mac's built-in antenna. The shape of the connector that attaches the card to the Mac's circuitry differs between the two cards, which is one reason why you can't plug an AirPort Extreme card into a Mac designed for the original AirPort card.

Figure 2-1 shows a diagram of an AirPort Extreme card. Note the protruding connector on the card's right side, which fits into a slot inside the Mac. Figure 2-2 shows a diagram of an original AirPort card. It has no connector extending from it; instead, the end of the card contains two rows of pinholes that connect to two rows of pins in the AirPort card socket in older AirPort-ready Macs.

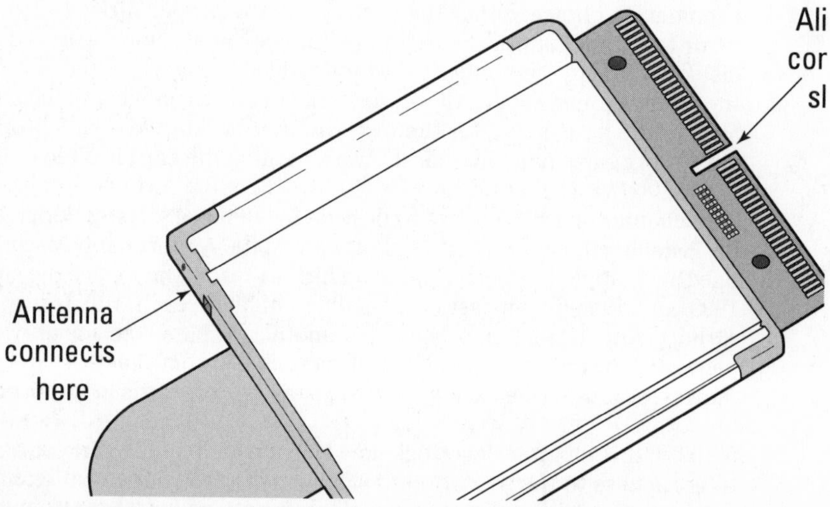

Figure 2-1: An AirPort Extreme card has a protruding connector.

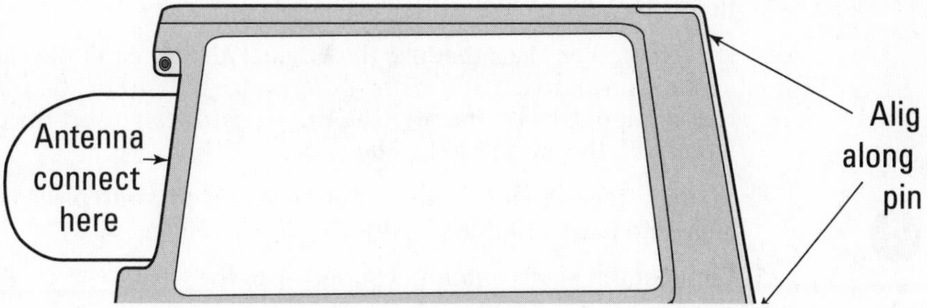

Figure 2-2: An original AirPort card employs pinhole connectors along its bottom.

The basic installation procedure goes like this:

1. **Go to Apple's Web site to download and print a copy of Apple's installation instructions before you begin working.**

   In case you are wondering where to go on Apple's site, don't worry: I have provided links to Apple's installation instructions for each of the AirPort-ready Macs described in this chapter.

2. **Turn off your Mac and remove all connecting cables, including the power cable. For laptops, also remove the battery.**

   Apple's instructions often recommend that you let a laptop sit for about a half hour after you remove the battery. This allows any electrical charges inside your laptop to dissipate and reduces the chance of electrical damage while you are poking about inside your Mac.

3. **Open your Mac to reveal the AirPort slot.**

   Not surprisingly, Apple places this slot in different places in different Macs, but getting to it usually involves removing a covering of some sort. For example, on a dual USB iBook, the slot is beneath the easily removable keyboard; on a flat-screen iMac G4, the slot is behind a small cover on the machine's bottom.

   TIP: If you need to remove any screws to get at the AirPort slot, have a small dish or cup handy into which you can place the screws when you open your Mac. Such screws are often small and hard to replace.

4. **Align the card's connector with the corresponding connector in the Mac and gently slide the card into place.**

   When you install the AirPort card, pay close attention to how the card is oriented in the illustrations in Apple's instructions. The fronts of both types of card have an AirPort label, as shown in Figures 2-1 and 2-2. The backs of the cards have a bar code label on them. The diagrams in Apple's instructions show you whether you should insert the card face up or face down. Usually, the correct orientation is face down so you can easily see the bar code when looking inside your Mac. You want to get the orientation right, because the cards plug in only one way.

For a few older Macs that use the original AirPort card, the card itself actually fits onto a card carrier — known as the AirPort Card Adapter — that is supplied with the original AirPort card. You insert the carrier along with the card into the Mac.

Do not attempt to force the card (or card carrier) into place if it doesn't slide into place relatively easily.

5. **Connect the Mac's antenna connector to the card.**
6. **Latch the card down, if necessary.**

   Some Macs have a strong wire latch that folds down over the card and locks the card into place.
7. **Close the Mac.**
8. **Reconnect the cables, and, for laptops, reinsert the battery.**

Apple's instructions explain how you can ground yourself electrically before installing an AirPort card. Grounding yourself is very important, because the slightest spark, including one so small that you can't even see or feel it, may damage the circuitry of your Mac or your card. Follow Apple's grounding instructions very closely. Pay close attention to the weather, too: dry, windy days often cause static electricity charges to build up in both you and your Mac, increasing the risk of damage.

Although I earlier said that installing a card can often be child's play, if your inner child has the least qualms about playing around inside of an often expensive and beloved piece of consumer electronics like a Macintosh, I strongly suggest you have an authorized Apple service representative install the Airport card. The peace of mind is worth the service charge.

Adding AirPort Extreme to Recent Macs

If AirPort Extreme were a movie, it might have the title, *AirPort 2: The Sequel.* While maintaining compatibility with Apple's original AirPort technology, AirPort Extreme provides faster network connections and stronger security than first-generation AirPort. Mac models introduced after the beginning of 2003 use AirPort Extreme cards.

Macs capable of using Airport Extreme include:

- iMac G5s
- Later flat-panel iMac G4s
- Later eMacs

- Power Macintosh G5s
- Power Macintosh G4s with FireWire 800
- Mac Minis
- iBook G4s
- Aluminum PowerBook G4s

Apple's AirPort is actually a version of the original *802.11b* standard for wireless networking, and AirPort Extreme is Apple's version of that standard's successor, the *802.11g* wireless standard. An 802.11b wireless connection can run at a peak speed of 11 million bits per second (Mbps), while an 802.11g connection can handle up to 54 Mbps. Although 802.11b uses a data-encryption standard that deters casual eavesdropping, the encryption that 802.11g uses is more sophisticated and much less easily circumvented, making it a popular choice for more sensitive business applications.

Gaining access to the AirPort Extreme card slot in various model Macs can range from trivially easy to more trouble than it's worth. In the following sections, I describe how to get the Mac open, so you can decide for yourself whether you want to undertake the task. If you do, I've provided links to Apple's detailed instructions, which you should follow.

iMac G5

The flat-panel iMac G5, the computer disguised as a screen, is one of the easier Macs to open, which is a blessing because you have to get completely inside the machine to install the AirPort Extreme card. You may not even have to, though: if your iMac G5 was one of the models that Apple released after May 2005, it comes with AirPort Extreme built-in as a non-replaceable component.

To open the iMac G5 you need the following:

- A towel or a blanket
- A sturdy flat surface
- A Phillips #2 screwdriver

You can find the complete instructions for opening an iMac G5 and installing an AirPort Extreme card at the following URL:

`http://docs.info.apple.com/article.html?artnum=300205`

Here's how you open an iMac G5:

1. **Disconnect all cables and power.**

2. **Spread the towel or blanket on the work surface you've chosen and lay the iMac face down on it.**

3. **Loosen the three screws at the bottom of the iMac.**

   You may need to raise the iMac's foot to get to the middle screw. The screws are captive screws, meaning that they are physically attached to the computer's case, so you won't risk losing them. The screws loosen when you turn the screwdriver counterclockwise.

   When the screws are completely loose, they'll stop turning. As soon as you feel resistance, stop turning the screwdriver so you don't damage the screws.

4. **Gently lift the cover off the iMac and set it aside.**

5. **Touch a metal surface inside the iMac to ground yourself.**

   It's okay to use your fingers for this.

   The bay that holds the Airport Extreme card is located near the left center of the main internal component that Apple calls the *mid-plane assembly*, and just to the right of the iMac's various port connectors. The AirPort Extreme card slides into the bay face down, so you can see the bar code label on the back of the card.

Later flat-panel iMac G4

The distinctive-looking flat-panel iMac G4 was introduced in January 2002; however, the first one that could use AirPort Extreme was the 17-inch 1GHz iMac G4 model that Apple released in February 2003. Any iMac G4 that runs at 1GHz or faster uses AirPort Extreme. To get at the iMac G4's AirPort card slot, you need to remove the cover plate at the bottom of the machine's base.

To open the iMac G4 you need:

- A towel or a blanket
- A sturdy flat surface
- A Phillips #2 screwdriver

You can find the complete instructions for opening an iMac G4 and installing an AirPort Extreme card at the following URL:

```
http://docs.info.apple.com/article.html?artnum=26264
```

You open an iMac G4 like this:

1. **Disconnect all cables except for the power cable.**

   Even though you are leaving the power cable connected, make sure that the iMac is shut down.

2. **Spread the towel or blanket on the work surface you've chosen and gently lay the iMac face down on it.**

   You should support the iMac by both the neck and the base as you lay it down. Make sure the display is face down to keep the iMac base from rolling.

3. **Loosen the four screws at the bottom of the iMac to remove the access cover plate.**

   The cover plate's screws are captive screws, meaning that they are physically attached to the computer's case so you won't risk losing them.

4. **Place the access cover plate aside.**
5. **Touch a metal surface inside the iMac with a finger to ground yourself.**
6. **Remove the power cable.**

The AirPort slot is located directly beneath the cover plate. The AirPort Extreme card slides into the slot face up so you can see the card's label.

Later eMac

Any of the white, bulbous eMac models introduced after May 2003 can use AirPort Extreme. Such eMacs also employ ATI graphics processors instead of the nVidia graphics processors used in earlier eMacs, which is another way to tell if the eMac can use AirPort Extreme. The AirPort card slot is located behind the optical drive door above the drive.

To access the eMac's AirPort slot, you need a #2 Phillips screwdriver.

You can find the complete instructions for opening an eMac and installing an AirPort Extreme card at the following URL:

```
www.apple.com/support/emac/doityourself/
```

Click the eMac (ATI Graphics) link to download the PDF that contains the installation instructions. This page also contains a link to a QuickTime movie that shows the process.

You access the eMac's AirPort card slot like this:

1. **Shut down your eMac.**
2. **With one finger, press in the optical drive door on one side as you pull the door open on the other side.**

   An access plate behind the optical drive door covers the AirPort card slot.
3. **Loosen the two screws that attach the access cover plate and remove the plate.**

   The access cover plate's screws are captive screws and remain attached to the cover plate.

The AirPort Extreme card slides face down — bar code label up — into the slot above the optical drive.

Power Macintosh G5

All of Apple's aluminum-clad Power Macintosh G5 towers use AirPort Extreme. Because Apple designed these machines to be opened, you don't need any special tools to install an AirPort Extreme card, although Apple's instructions recommend you have two soft clean cloths handy.

The Power Macintosh G5 comes with an AirPort antenna that you need to attach to the antenna socket on the tower's back. Make sure you can dig out the antenna from wherever you stashed it before you spend money on the AirPort Extreme card; without that antenna, the Mac will have difficulty joining a wireless network.

You can find Apple's instructions for opening a Power Macintosh G5 and installing an AirPort Extreme card at this URL:

```
http://docs.info.apple.com/article.html?artnum=26279
```

Follow these steps to gain access to the inside of a Power Macintosh G5:

1. **Shut down the Macintosh.**
2. **If the machine has been running, wait ten minutes.**

   The inside of a running G5 can become quite hot, and you run the risk of burning yourself if you don't let the machine cool down before you poke around inside of it.
3. **Disconnect every cable except the power cable from the Mac.**
4. **Ground yourself by touching one of the PCI access panels on the Mac's back.**
5. **Disconnect the power cable.**

6. **With one hand, hold the Mac's side access panel while lifting the panel release latch on the Mac's back.**

   The side access panel is on the Mac's right side when you view the machine from the front.

   Apple's instructions note that the edges of the side access panel can be rather sharp. As someone who has left blood in more machines than I can remember, I second their warning: Exercise care when removing and handling the side access panel.

7. **Carefully remove the side access panel and lay it aside on a flat surface.**

   Apple recommends you lay it down on a soft cloth.

8. **Remove the clear plastic air deflector and lay it aside.**

   Apple recommends you also place this component on a soft cloth.

   Don't forget that you need to replace the air deflector when you reassemble your Power Mac G5; the deflector helps direct the proper cooling airflow inside the Mac, and sensors inside the Mac can detect when it is missing.

The AirPort Extreme card slot resides near the front of the Mac below the drive bays. The card slides into the slot faceup.

Power Macintosh G4 with FireWire 800

The only Power Macintosh G4 tower that supports AirPort Extreme is the FireWire 800 model. You don't need any special tools to open this Mac — curiously, although Apple's PDF instructions mention a Phillips screwdriver, the procedure itself actually makes no use of it.

You can download Apple's instructions for installing an AirPort Extreme card in the Power Macintosh G4 with FireWire 800 from this URL:

`http://docs.info.apple.com/article.html?artnum=26262`

Follow these steps to gain access to the inside of this model Power Macintosh G4:

1. **Shut down your Mac and wait five minutes.**

   The wait allows the machine to cool and to dissipate any accumulated static electricity charges.

2. **Disconnect all cables, except the power cable, from the Mac.**

3. **On the Mac's rear, use a finger to touch one of the PCI card access covers.**

   This grounds you, further reducing the chance of any static electricity discharges that might harm the components inside the Mac.

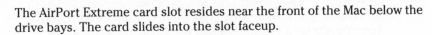

4. **Remove the power cord.**
5. **Lift the latch on the right side of the Mac and gently lower the Mac's side panel so it lays flat.**

   Many of the Mac's components are attached to this panel.

You can find the AirPort Extreme card slot near the top of the panel on the side that faces the front of the Mac when it's closed. The card fits into the slot face up — bar code label down.

Mac Mini

Although Apple does make AirPort Extreme an option on its itsy-bitsy Mac Mini, it does not consider the card a user-installable part.

Trust me, Apple is right about this: Opening the Mac Mini requires special tools and a delicate touch. Furthermore, after you get the Mini open, you have to attach the AirPort card to a daughter card inside the Mini, and the Mini does not have that daughter card inside unless AirPort or Bluetooth have already been installed. If that doesn't deter you and you want to buy the daughter card and forge blithely ahead, you hit another obstacle: Apple only sells the daughter card to Apple certified service providers.

In short, if you didn't buy your Mini with the AirPort Extreme card already installed, you should have an Apple service provider install it for you.

iBook G4

The first Mac to use AirPort was the original iBook, and the iBook is still one of the easiest to set up with a wireless card. You can find the iBook's AirPort Extreme card slot in a bay under the keyboard. Although the newest iBook G4s have AirPort Extreme included as a standard feature, slightly less recent models don't.

To access an iBook G4's AirPort Extreme card slot you need:

- A coin
- A clean flat surface
- A jeweler's screwdriver

You only need the last item if someone has locked the keyboard down — the iBook normally ships with the keyboard unlocked.

You can find the complete instructions for installing an AirPort Extreme card in an iBook G4 at the following URL:

`http://docs.info.apple.com/article.html?artnum=26277`

You access an iBook G4's AirPort Extreme card slot as follows:

1. **Shut down your iBook and wait 30 minutes.**

   The wait lets the iBook cool down and allows various electrical charges inside the iBook to dissipate, reducing the risk of static electrical discharges.

2. **Disconnect all cables, including the power cable.**
3. **On the bottom of the iBook, insert a coin into the battery latch and turn it a quarter turn counterclockwise.**
4. **Remove the battery and set it aside.**
5. **Turn the iBook over and raise the screen so you can access the keyboard easily.**
6. **With your fingers, slide the keyboard latches away from the screen and toward you.**

   One latch is to the right of the Esc key and the other is to the left of the Eject/F12 key. When you release the latches, you should be able to flip the keyboard up and toward you easily. If not, you need to release the keyboard locking screw, which you can find set into a plastic tab located to the left of the Num Lock/F6 key.

   The keyboard latches are spring loaded and snap back into position when you let them go; so if you need to perform the next step, you then need to come back and perform this step again.

7. **If the keyboard remains locked, insert the jeweler's screwdriver into the keyboard locking screw and turn it a half turn counterclockwise.**

   I've found that, in a pinch, the tweezers from a Swiss Army knife can double as a jeweler's screwdriver and turn the locking screw.

8. **Gently flip the keyboard up from the top toward you and lay it down across the palm rest and trackpad.**

The AirPort Extreme card slides face up — bar code label down — into the slot in the shallow bay toward the left of the iBook.

Aluminum PowerBook G4

If you have a 17-inch screen aluminum PowerBook G4, you already have AirPort Extreme because Apple includes it with that model. The card, however, *is* a user-installable option on the 12-inch and 15-inch screen models. On those aluminum PowerBooks for which AirPort is an option, you can find the AirPort Extreme card slot behind a flip-down door in the PowerBook's battery bay.

You need the following items to access the aluminum PowerBook G4's AirPort Extreme card slot:

- A coin
- A flat surface

Although Apple provides a different set of instructions for each of the two differently sized PowerBooks, the actual steps are practically the same for either model. You can find the complete instructions for installing an AirPort Extreme card in a 12-inch screen aluminum PowerBook G4 at this URL:

```
http://docs.info.apple.com/article.html?artnum=26261
```

You can find the complete instructions for installing an AirPort Extreme card in a 15-inch screen aluminum PowerBook G4 at this URL:

```
http://docs.info.apple.com/article.html?artnum=86518
```

Here's how you access the aluminum PowerBook G4's AirPort Extreme card slot:

1. **Place your PowerBook on a flat surface, shut it down, and wait five minutes.**

   The wait allows the PowerBook to cool.

2. **Disconnect all cables, including the power cable, attached to the PowerBook.**
3. **Close the lid and turn the PowerBook over.**
4. **Insert a coin into the battery latch slot and turn it a quarter turn counterclockwise.**
5. **Remove the battery and set it aside.**
6. **Touch a metal surface inside the battery bay to ground yourself.**
7. **Flip down the AirPort card slot door located on the right side of the battery bay's wall.**

The AirPort Extreme card slides into the slot face down — bar code side up.

Equipping Older Macs with AirPort Cards

Nothing lasts forever: In the case of the AirPort card, it took less than five years from its introduction before it marched into the history books. However, many older yet still serviceable Macs use them. I know — I personally have two such Macs.

The older Macs that have original AirPort card slots include:

- iMac G3 models with slot-loading optical drives, introduced after October 1999
- Flat-panel iMac G4 models running at 800 MHz or slower
- eMac models introduced before May 2003
- All but the last model of Power Mac G4
- Power Mac G4 Cube
- iBook G3 dual USB models
- iBook G3 original models
- Titanium PowerBook G4 models
- PowerBook G3 models with FireWire

Although Apple no longer sells the original AirPort card, you can shop around to find one. A number of vendors still carry them. You should be able to find one with a little online shopping.

Like the AirPort Extreme card slots in recent Macs, gaining access to the AirPort card slot in older-model Macs ranges from simple to nerve wracking. Use the following Mac-opening descriptions to help you decide if you feel up to the task of installing the card yourself. If so, use the links I provide to get Apple's detailed instructions, and follow those instructions.

iMac G3 with slot-loading optical drive

Here's the good news: The AirPort card slot on these iMacs lies directly behind an easily opened access door on the iMac's back. And here's the bad news: You need both the original AirPort card and the AirPort Card Adapter to install the card in an iMac G3. This means you must shop around to obtain two components that Apple no longer directly sells.

WARNING!

You cannot install an AirPort card into the low-cost Indigo 350 MHz iMac that Apple introduced in July 2000.

You need these items to open the iMac's Airport card access door:

- A coin
- A soft cloth

REMEMBER

Apple provides the complete instructions for installing an AirPort card and its required AirPort Card Adapter on the following Web page:

```
http://docs.info.apple.com/article.html?artnum=58537
```

You open the iMac's AirPort card access door by following these steps:

1. **Shut down the iMac.**
2. **Unplug every cable except the power cord.**
3. **Gently place the iMac screen side down on the soft cloth.**
4. **Use the coin to turn the latch on the iMac's access door, located on the back of the iMac, to unlock it.**
5. **Open the access door and touch the metal shield in the recessed area behind the door to ground yourself.**
6. **Disconnect the iMac's power cable.**

The Airport card slot is directly behind the access door.

iMac G4 running at 800 MHz or slower

The instructions for opening the earlier-model iMac G4s that use the original AirPort card are the same for opening the later-model iMac G4s given above in "Later flat-panel iMac G4." Refer to those instructions as you decide whether or not to install the card yourself.

Unlike the AirPort Extreme card used in later iMac G4 models, the AirPort card in early iMac G4 models fits into the card slot face down, with the bar code visible.

REMEMBER

Apple provides AirPort card installation instructions for the early iMac G4 models here:

```
http://docs.info.apple.com/article.html?artnum=75317
```

Earlier eMac

The first-generation eMac models use the original AirPort card. Like the later models, early eMacs locate the AirPort card slot behind the optical drive drawer above the drive.

To access the eMac's AirPort slot, you need a #2 Phillips screwdriver.

REMEMBER

You can find the complete instructions for opening an eMac and installing an AirPort Extreme card at the following URL:

```
www.apple.com/support/emac/doityourself/
```

Click the eMac link instead of the eMac (ATI Graphics) link to download the PDF that contains the installation instructions for the first-generation eMacs. This page also contains a link to a QuickTime movie that shows the process.

You can access the eMac's AirPort card slot like this:

1. **With your eMac running, press the Media Eject key to open the optical-drive door.**

   An access plate behind the optical-drive door covers the AirPort card slot.
2. **Shut down your eMac.**
3. **Loosen the two screws that attach the access cover plate and remove the plate.**

   The access cover plate's screws are captive screws and remain attached to the cover plate.

The AirPort card slides face down — bar code label up — into the slot above the optical drive.

Power Macintosh G4

Apple produced several different Power Macintosh G4 models that accept original AirPort cards. Apple refers to those models by the following names:

- the Mirror Drive Doors model
- the QuickSilver model
- the Digital Audio model
- the Gigabit Ethernet model

Although the features, internal component layout, and general appearance among these models vary, the process of accessing the AirPort card slot within each of them is the same. You don't need any tools to open the Power Macintosh G4 models.

Table 2-1 lists the URLs where you can obtain complete instructions for installing Airport cards in Power Macintosh G4 desktops.

Table 2-1 Finding Power Mac G4 AirPort Card Instructions

Model	*URL*
Mirror Drive Door	`http://docs.info.apple.com/article.html?artnum=26259`
QuickSilver	`http://docs.info.apple.com/article.html?artnum=75314`
Digital Audio	`http://docs.info.apple.com/article.html?artnum=75313`
Gigabit Ethernet	`http://docs.info.apple.com/article.html?artnum=75312`

To open the Power Macintosh G4 tower, follow these steps:

1. **Shut down the Mac and wait five minutes.**

   These Macs generate considerable heat and you need to give them a chance to cool down.

2. **Unplug every cable except the power cable from the Mac.**
3. **Touch one of the metal PCI card access covers on the back of the Mac with a finger to ground yourself.**
4. **Unplug the power cable.**
5. **Lift the cover latch on the right side of the Mac and gently lower the side access cover down.**

The Power Macintosh G4 towers have their AirPort card slots mounted on the side access panel toward the front of the Mac. On the Gigabit Ethernet model, the Digital Audio model, and the QuickSilver model, the card slot resides low on the panel, near the Mac's body when the panel is open. On the Mirror Drive Doors model, the slot is near the top of the access panel, further from the Mac's body when the panel is open. On all models, the card fits into the slot face down — bar code label exposed.

Power Macintosh G4 Cube

You have to get inside the compact and classy Cube to install an Airport card, but Apple made the Cube very easy to open, and you don't need any tools to do it — just a soft cloth to lay the Cube on as you work.

Apple provides both a PDF file and a QuickTime movie with instructions for installing the AirPort card in the Cube at this URL:

```
http://docs.info.apple.com/article.html?artnum=58688
```

Here's how you pop open the Cube:

1. **Shut the Cube down and wait five minutes for it to cool.**
2. **Disconnect every cable except the power cable.**
3. **Lay the Cube on its side and touch the exposed metal surface between the video ports on the Cube's bottom to ground yourself.**
4. **Disconnect the power cable.**
5. **Turn the Cube completely upside down.**
6. **Press the latch in to release it and let go, allowing the latch to extend.**

   The latch forms a handle when it extends completely.
7. **Using the latch as a handle, lift the Cube's core from its container.**

The AirPort card fits into a bay in the core's side, face up.

iBook G3 dual USB

Apple made a number of different models of the dual USB iBook G3, but the AirPort card installation procedure is the same for all of them, and, in fact, is the same as for the iBook G4 models, described earlier in this chapter. Take a look back at the section "iBook G4" to see how you get to the dual USB iBook G3's AirPort card slot.

Apple provides complete instructions for installing the AirPort card at:

```
http://docs.info.apple.com/article.html?artnum=50031
```

Original iBook G3

Apple introduced AirPort at the same time that it introduced the first of its brightly colored iBook G3s. The two were literally made for each other.

To get to the AirPort card slot of these curvy laptops, you need:

- A flat-blade screwdriver or a coin
- A clean flat surface
- A jeweler's screwdriver

REMEMBER

You can find the original iBook's Airport card installation instructions at:

`http://docs.info.apple.com/article.html?artnum=26244`

To get to the iBook's AirPort card slot, follow these steps:

1. **Shut down the iBook.**
2. **Disconnect all cables, including the power cable.**
3. **Turn the iBook upside down.**
4. **With a flat-blade screwdriver or a coin, turn the screws that hold the battery cover in place counterclockwise.**
5. **Remove the battery cover and the battery.**

   The battery has a plastic tab you can pull to lift it from the battery compartment.
6. **Turn the iBook right side up and open the lid.**
7. **With your fingers, slide the keyboard latches away from the screen and toward you.**

   One latch is to the right of the Esc key and the other is to the left of the Eject/F12 key. When you release the latches, you should be able to flip the keyboard up and toward you easily. If not, you need to release the keyboard locking screw that's set into a plastic tab located to the left of the Num Lock/F6 key.
8. **If the keyboard remains locked, insert the jeweler's screwdriver into the keyboard locking screw and turn it half a turn counterclockwise.**

   The keyboard latches are spring loaded and snap back into position when you let them go; so if you need to perform this step, you then need to perform Step 7 again.
9. **Gently flip the keyboard up from the top toward you, and lay it down across the palm rest and trackpad.**

The AirPort card slot is in a bay toward the left side of the iBook. You insert the card face down so that you can see the bar code.

Titanium PowerBook G4

As far as case design goes, Apple released three variations of the thin, light, sleek, and titanium-clad PowerBook G4:

- PowerBook G4
- PowerBook G4 (Gigabit Ethernet)
- PowerBook G4 (DVI) and (1 GHz/867 MHz)

These models share similar opening procedures, though the type and number of screws you must remove differs among them. They share another quality as well: Opening them takes a delicate touch if you don't want to damage the thin titanium case.

TIP: Though Apple considers the AirPort card a user-replaceable part for these PowerBooks, I recommend you have a certified Apple service provider install the card for you unless you have a light touch and confidence in your technical prowess.

You need the following items to open one of these PowerBooks:

- A soft cloth larger than the PowerBook
- A Phillips screwdriver (PowerBook G4)
- A Torx T8 screwdriver (Gigabit Ethernet, 1 GHz/867 MHz, and DVI)

REMEMBER: Table 2-2 lists the URLs where you can obtain complete instructions for installing Airport cards in titanium PowerBook G4 laptops.

Table 2-2 Finding PowerBook G4 AirPort Card Instructions

Model	*URL*
PowerBook G4	`http://docs.info.apple.com/article.html?artnum=95131`
PowerBook G4 (Gigabit Ethernet)	`http://docs.info.apple.com/article.html?artnum=26246`
PowerBook G4 (1 GHz/867 MHz) and (DVI)	`http://docs.info.apple.com/article.html?artnum=26245`

If you feel up to the challenge of opening a titanium PowerBook G4, review these steps to see what it involves:

1. **Shut down the PowerBook and wait 30 minutes to let the internal components cool.**
2. **Disconnect the power cable and all other cables.**
3. **Drape the soft cloth over a flat surface, letting some of the cloth hang over the surface's edge.**

   WARNING! Make sure enough of the cloth hangs over the edge to protect the PowerBook's display and that enough covers the flat surface when you follow the next step. Otherwise, you could damage the display or the keyboard.
4. **Open the PowerBook's display and lay the PowerBook — keyboard side down — on the cloth-covered surface, letting the display hang over the edge of the surface.**
5. **Slide the battery latch to the left, remove the battery, and slide the latch back to the right.**
6. **Remove the screws from the bottom of the case, using the appropriate screwdriver for the PowerBook model, in the order that Apple stipulates in their instructions.**

   The number of screws, their placement, their type, and the order in which you remove them differs among the three case variations. The PowerBook G4 and the PowerBook G4 (Gigabit Ethernet) have eight screws, and the PowerBook G4 (DVI and 1 GHz/867 MHz) models have seven.
7. **With the thumb of your right hand pressing inside the front edge of the battery compartment, and the thumb of your left hand pressing against the bottom left corner of the case, push gently away from you until the left side of the case releases.**

   WARNING! Do not press on the case's rubber feet, and push gently: too much pressure can warp or bend the thin titanium bottom case cover. Consult Apple's diagrams and instructions to see exactly how to do this.
8. **With the thumb of your left hand pressing inside the right front edge of the battery compartment, and the thumb of your right hand pressing slightly above and to the right of the display hinge, slide the PowerBook cover away from you and lift it up, pivoting it away from you.**

   WARNING! To avoid damaging the case, do not twist or bend the case bottom either left or right when you lift.
9. **Place the case bottom aside and touch the gray composite-metal inside framework of the PowerBook to ground yourself.**

You can find the PowerBook G4's AirPort card slot behind the battery on the left side of the PowerBook as you look at its bottom. The card slides into the slot label side down and bar code side up.

PowerBook G3 with FireWire

The only AirPort-ready PowerBook among the black-cased and smoothly curved PowerBook G3 models that Apple sold at the end of the last century is the FireWire-equipped model, which went by the code name *Pismo*. Like the iBooks introduced at roughly the same time, the Pismo's AirPort card fits into a slot under the keyboard.

You need the following tools to access the AirPort card slot:

- A jeweler's screwdriver
- A Phillips screwdriver

REMEMBER

You can download Apple's AirPort card installation instructions for the Pismo PowerBook, and view a QuickTime movie illustrating the procedure, from this URL:

```
http://docs.info.apple.com/article.html?artnum=50028
```

Accessing the Pismo's AirPort card slot requires that you follow these steps:

1. **Shut down the PowerBook and disconnect all cables, including the power cable.**
2. **Using the battery latch, eject the battery, and wait 30 minutes for the PowerBook to cool.**
3. **If necessary, use the jeweler's screwdriver to unlock the keyboard-locking screw by turning the screw counterclockwise.**

   The locking screw is set into a plastic tab located between the F4 key and F5 key.
4. **With your fingers, slide the keyboard latches toward you to release the keyboard and lift the top edge of the keyboard slightly up.**

   The keyboard latches are beside the F1 and F9 keys.
5. **Pull the keyboard slightly away from you until the small tabs that hold the keyboard's bottom in place clear the PowerBook's case.**
6. **Pivot the keyboard over so that it sits on the palm rests and trackpad.**
7. **Touch a metal surface inside the PowerBook to ground yourself.**
8. **Use the Phillips screwdriver to remove the two screws holding the PowerBook's internal heat shield in place.**
9. **Lift up the heat shield and set it aside.**

The AirPort card slot lies on the PowerBook's left side. The card slides into the slot face down so that you can see its bar code.

Considering AirPort Card Alternatives

Macs last a long time, and even though nearly all Macs made in the past few years support AirPort, more than a few older but otherwise usable Macs do not, such as the original iMac and the first several G3 PowerBook models. But that does not mean that pre-AirPort Macs need remain deaf to the wireless world.

In fact, the majority of this venerable Mac contingent can still participate in wireless networking. After all, AirPort employs standards-based wireless technology, and a variety of wireless network devices designed for non-Apple laptop and desktop computers also work with Macs. Two common types of devices that often work with Macs are

- PC card (also known as PCMCIA) wireless network adapters
- USB Wireless network adapters

Picking a PC card wireless network adapter

PC cards employ the PCMCIA (Personal Computer Memory Card International Association) standard and plug into the PC card slot that many laptop computers possess. Over the years, three types and sizes of PC card slot have seen use on laptops:

- **Type 1:** 3.3 millimeters thick and often used for memory expansion or storage
- **Type 2:** 5 millimeters thick and often used for network adapters
- **Type 3:** 10 millimeters thick and often used for small hard disks

With one exception, all models of non-AirPort-ready PowerBook G3s possess two Type 2 PC card slots that also double as one Type 3 slot. The more recent 400MHz PowerBook G3, which sports a bronze keyboard, only possesses one Type 2 PC card slot.

To use an AirPort compatible wireless PC card, follow these steps:

1. **Plug the card into an available PC card slot.**

   Most wireless PC cards fit into Type 2 slots and protrude about half an inch or more from the computer when inserted. The portion that protrudes contains the card's antenna.

   On computers with two stacked Type 2 PC card slots, you must take care to align the card in the slot properly to avoid damaging the card and the slot.

2. **Use either Apple's AirPort software or third-party driver software to activate the card and to join a wireless network.**

Although some wireless PC cards work directly with Apple's AirPort software — which is not surprising given that Apple obtains the chips inside its AirPort and AirPort Extreme cards from some of the same manufacturers that wireless PC card vendors do — other cards require special driver software. Before you purchase a wireless card, therefore, make sure either that the card is compatible with Apple's AirPort software, or that you can obtain a driver compatible with your Mac laptop and the version of the Mac OS that it runs. Table 2-3 lists some vendors of some popular PC cards that work with Macs and tells you where to get drivers for those cards.

Table 2-3 Some Mac-Compatible PC Card Vendors and Drivers

Vendor	*Model*	*Wireless Standard Supported*	*Driver*	*URL*
Agere	OrinocoGold and OrinocoSilver	802.11b	IOXperts	`www.ioxperts.com`
Asante	FriendlyNet AeroLAN	802.11b	Asante 1.2.5	`www.asante.com/support/incDownloads/level2/AL1011Revb.asp`
Belkin	F5D7011	802.11g	AirPort 3.4	`www.belkin.com`
Cisco	Aironet	802.11b	Cisco	`www.cisco.com/pcgi-bin/tablebuild.pl/aironet-utils-mac`
Dell	TruMobil1150	802.11b	IOXperts	`www.ioxperts.com`
Lucent	WaveLAN Gold or Silver	802.11b	AeroCard Universal Mac Driver	`www.macsense.com/product/broadband/aerouni_b.html`

Opting for a USB wireless network adapter

What's that? Your Mac has neither an AirPort card slot nor a PC card slot? You still may be in luck if your Mac has a spare USB port. A few vendors provide USB-based wireless network adapters that you can plug directly into your Mac's USB port — perfect, say, for the aging Bondi Blue iMac that you can't bear to discard.

It often makes sense to attach wireless USB network adapters to a powered USB hub connected to your Mac because the adapter may draw more current than the Mac's built-in USB ports provide. However, if you use a powered USB hub, make sure that *it* provides enough power for your wireless adapters. Some hubs don't.

USB-based wireless network adapters tend to be much more common in the Windows world than in the Mac world, but a few vendors do provide Mac OS X–compatible drivers for their devices. Table 2-4 lists a selection of these vendors.

Table 2-4 Some USB Wireless Network Adapters

Vendor	*Model*	*URL*	*Notes*
Belkin	F5D6050	`http://catalog.belkin.com/IWCatProductPage.process?Product_Id=122761`	
D-Link	DWL-122	`www.dlink.com.sg/products/?pid=175`	Not supported for Mac OS X versions later than 10.3.2
Macsense	AeroPad Plus	`www.macsense.com/product/broadband/WUA700_b.html`	
MacWireless	11b USB Stick	`www.macwireless.com/html/products/11g_11b_cards/11bUSB.php`	Works only with Mac OS 9.x

Of course, if none of these alternatives works for you, you may just have to bite the bullet and get a new, AirPort-ready Mac. That wouldn't be so bad, now, would it?

Chapter 3

Choosing Base Stationary

In This Chapter

- The AirPort Out-of-the-Box Experience
- Resetting Base Stations
- Working with Apple's Current Models
- Working with Older Apple Base Stations
- Choosing Non-Apple Access Points

Apple has made five AirPort base station models since the first flying saucer–shaped station touched down at the MacWorld Expo in San Francisco in 1999, and all but one have retained the distinctive, elegant look of the original. What lies beneath AirPort's smooth carapace, however, has seen changes aplenty, and those changes have made it faster, more secure, and more flexible.

But not harder to use. The original AirPort Base Station, along with its administration software, amazed grizzled networking gurus when they saw how easy it was to establish a complex wireless high-speed network right out of the box. And the latest version of that software is just as easy, if not easier, to use.

What's more, Apple's latest spin on the technology — the tiny yet mighty AirPort Express — puts a base station in the carry-on luggage of any traveling road warrior and a wireless jukebox in the dorm room of any college student.

Only a company that thinks differently could have looked at a complex component of a digital networking infrastructure and imagined a cool piece of consumer electronics — and then built it.

It's the friendliest flying saucer that's ever invaded Earth.

Stepping through a First-Time Setup

Remember Apple's TV commercial for the original iMac? In it, a Famous Voice explains that setting up an iMac takes three steps: First you it plug in, next you turn it on — and then the Voice chuckles in surprise as he discovers that there's no step three. Although setting up an AirPort Base Station takes more than three steps, that process also remains simple enough to leave you chuckling, "That's all there is to it?"

This simplicity comes courtesy of Apple's AirPort Setup Assistant program, which guides you through the task of setting up a new base station while shielding you from the ugly and complex details that underlie modern digital network configurations. You simply plug in a couple of cables, answer a few questions and you're on the air.

Of course, you *do* need to have the answers to those few questions. In particular, you need to know:

- What you want to call the AirPort Base Station.
- The name of the AirPort network you intend to create.
- The password you'll use to protect the station.
- The password you'll use to protect the network from unwelcome or unauthorized users.
- How, and if, you plan to connect the base station to the Internet.

  REMEMBER

  A base station can, and often does, share an Internet connection with the users of the wireless network it creates. Base stations can connect to the Internet through a modem connection or through an Ethernet connection to a cable modem, a DSL modem, or a local area network that has an existing Internet connection. DSL modems and cable modems are described more fully in Chapter 6.
- The setup information your Internet Service Provider gave you when you subscribed to your Internet service — this information usually comprises an account name, an account password, and, for dial-up Internet service, the service's phone number.

TIP

Even if you only plan to set up a casual home network, you should think carefully about what you want to name the station and the network, and what passwords you want to use. Unless you decide to create a closed network — one that doesn't advertise its existence — nearby wireless users can detect both your station's name and the network's name, and you don't want to choose names that can offend or confuse. The passwords you select should not be easy for others to guess but should let you easily remember and type them: A password serves little purpose if it's easily discovered, and is not much good if it is too cumbersome to use.

The majority of wireless home networks use their base stations to share an Internet connection among wireless network users. The following example shows how to set up a new AirPort Extreme Base Station to share an Internet connection provided by a DSL modem.

1. **Find a good location for the base station.**

   The location should be near both an electrical outlet and your Internet connection — such as a phone jack or the high-speed cable modem or DSL modem provided by your ISP — because you need to connect the base station to both. Chapter 5 provides more information about other issues affecting where you should place a base station.

2. **Connect your high-speed modem to the base station's Ethernet port, or the base station's modem port to the phone jack, and plug in the AirPort.**

   TECHNICAL STUFF

   Some AirPort Base stations have several Ethernet ports, and some lack a modem jack. Later on, this chapter describes the various connections available on different model AirPort Base Stations. Check the description of your AirPort Base Station in this chapter to see what cables plug in where on the base station you own.

3. **Connect the base station to the power outlet.**

   Because it works *much* better when it's plugged in.

4. **Open the AirPort Setup Assistant program, which you can find in the Utilities folder inside your Mac's Applications folder.**

   Figure 3-1 shows you where you can find the AirPort Setup Assistant. In Figure 3-2 you can see that the AirPort Setup Assistant tells you what steps you need to take before running it, just in case you didn't bother to read the first three steps in this section.

Figure 3-1: The AirPort Setup Assistant lives in the Utilities folder with Mac OS X's other utility programs.

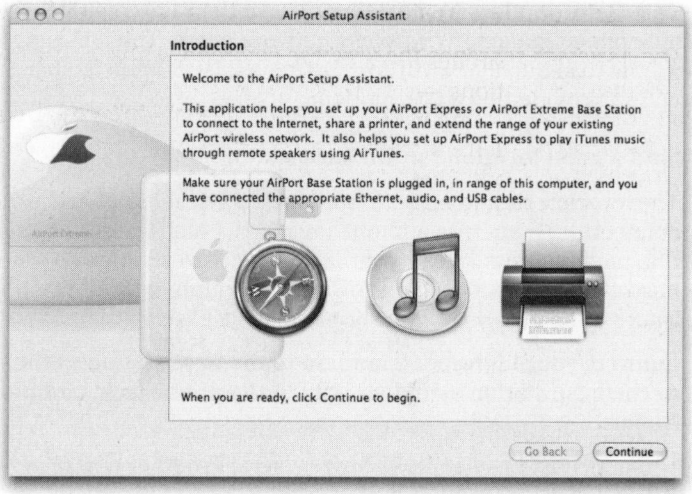

Figure 3-2: The Assistant tells you what you need to do before you begin the setup.

5. **Click Continue.**

   As shown in Figure 3-3, the Assistant asks you whether you want to set up a new base station or modify the setup of an existing station.

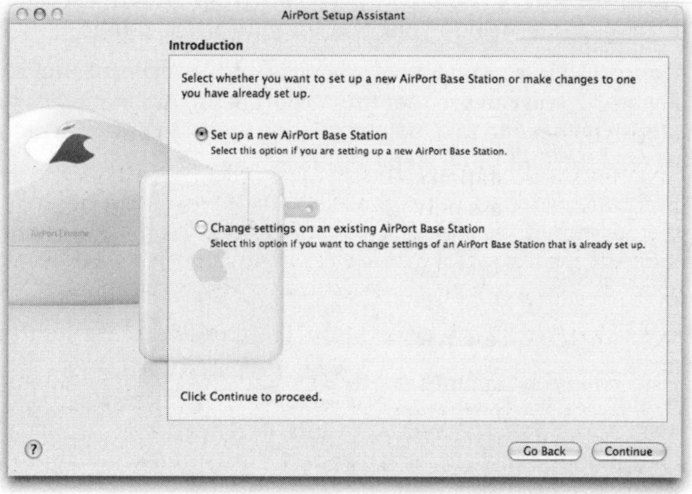

Figure 3-3: Tell the Assistant what you intend to do.

If you have turned off your AirPort card, at this point the Assistant sensibly displays a sheet that asks you if the Assistant should turn the card on and automatically make the necessary changes to your network settings. Unless you have created specialized AirPort card settings, you can safely click the sheet's OK button and proceed with the next step.

6. **Click Set Up a New AirPort Base Station and then click Continue.**

   The Assistant searches the wireless environment for a new AirPort Base Station. New stations — or base stations that have been reset to their factory settings — are named "AirPort Network" followed by a sequence of numbers and letters. In Figure 3-4, the Assistant has found a new base station, figured out what kind of base station it is, and asks you if it's the one you expected to see.

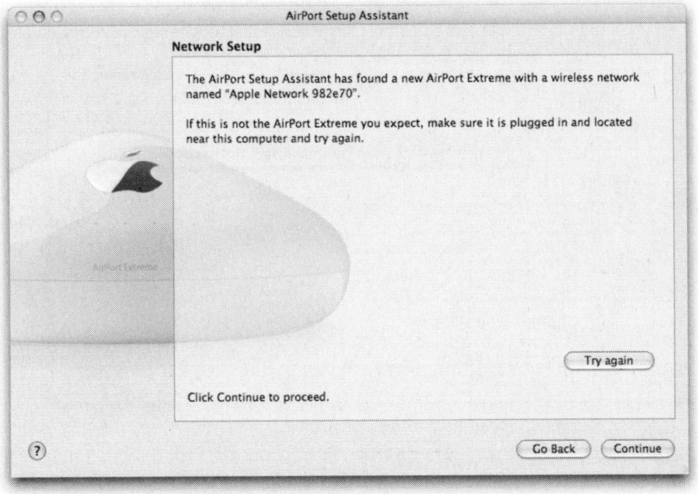

Figure 3-4: The Assistant finds a new base station.

7. **Click Continue.**

   Because a base station can either create its own wireless network or join an existing wireless network in order to increase that network's range, the Assistant asks you, as shown in Figure 3-5, what you intend to do with the base station that it found. In this example, you want to create a new network. Chapter 5 describes various ways to extend a wireless network.

8. **Click Create a New Wireless Network and then click Continue.**

   The Assistant examines the ports on the base station and presents some choices based on what it finds. In Figure 3-6, the Assistant sees that the base station used for this example has something plugged into its Ethernet port and asks how you want to handle that connection.

9. **Click the choice that's appropriate for your network and then click Continue.**

   You can see the choice I made in Figure 3-6.

The Assistant next gives you an opportunity to give your wireless network a password and to select its level of security. Figure 3-7 shows the choices that the Assistant provides. You can find out more about these security settings, and learn about some additional security choices available to AirPort wireless networks, in Chapter 4 and in Chapter 11.

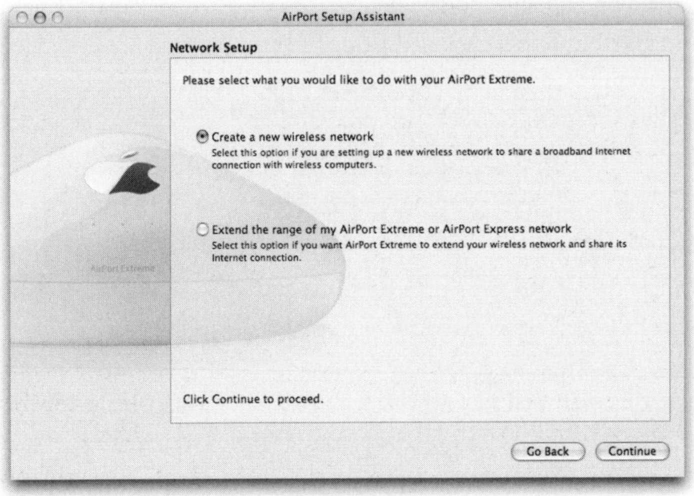

Figure 3-5: Tell the Assistant what you mean to do with the base station.

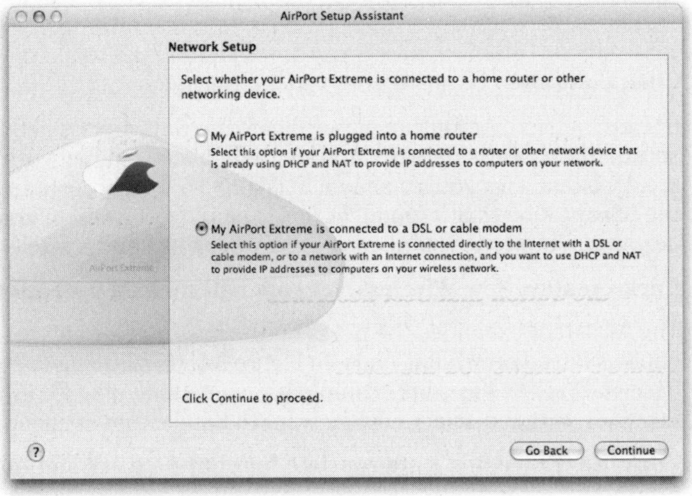

Figure 3-6: Specify how you want to handle the network connection the Assistant found.

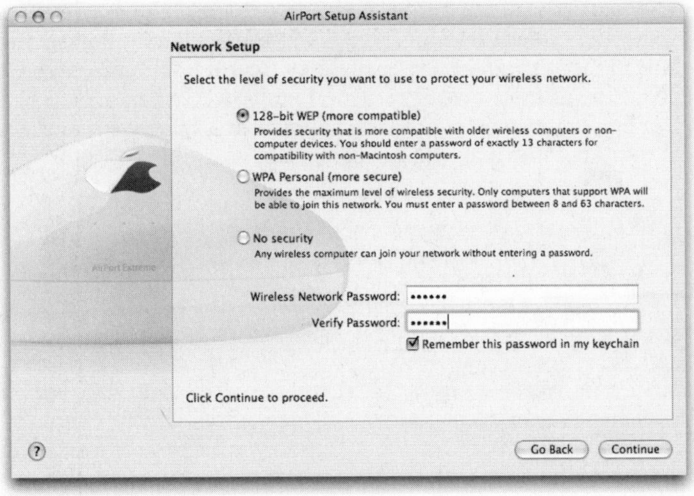

Figure 3-7: Pick your security settings and specify a password.

10. **Click a security setting, enter the password you want to give your wireless network, enter the same password again to confirm it, and then click Continue.**

    Although you can create a wireless network with no security — and far too many people do — you shouldn't. There's a word that describes those who use unsecured networks: *victims*. An unsecured network lets anybody use both it and the Internet connection that it shares, which could cost you money depending on your Internet service agreement. More frighteningly, an unsecured network does not encrypt any data traveling over it, which means that an unscrupulous person using easily obtainable software and equipment could eavesdrop on your network and steal the passwords or credit card numbers that you use online.

    Next, you need to give the Assistant some details about how you want to connect to the Internet. Figure 3-8 shows you the choices you have, including an option to defer the choice until later.

11. **Click the button that best describes how you plan to connect to the Internet, then click Continue.**

    Figure 3-8 shows you the choice I made for this example — if you make a different choice the rest of the process will differ slightly from what you see here. In particular, if you choose not to connect to the Internet, the Assistant skips to Step 13. Otherwise, the Assistant shows you a version of the Internet Setup pane, shown in Figure 3-9, that matches the type of connection you told it you want to create.

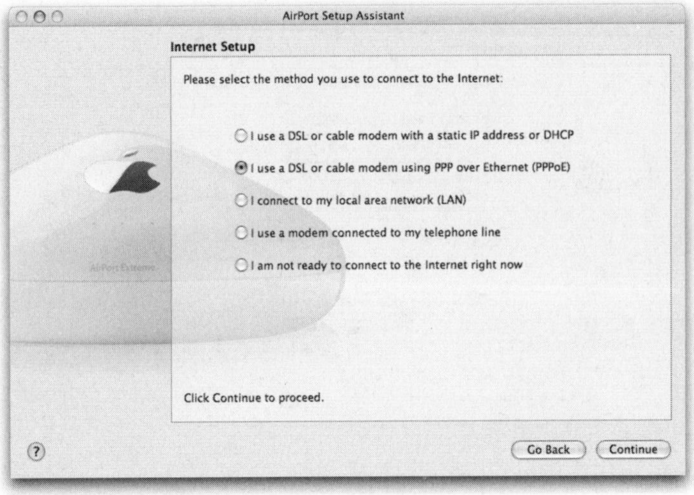

Figure 3-8: Choose how you connect to the Internet.

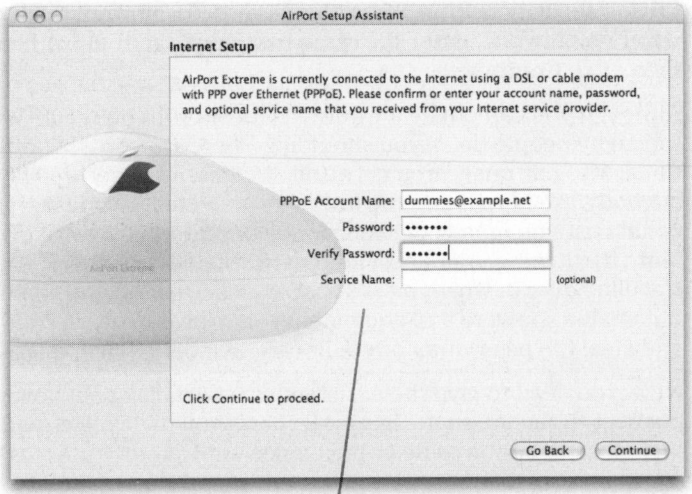

Figure 3-9: Provide your Internet account information so the base station can use it.

12. **Enter your Internet setup information and then click Continue.**

    The Assistant wants you to assign a password to the base station's settings, as illustrated in Figure 3-10. You use this password when you want to change the base station's configuration.

    As the Assistant tells you, you should not use the same password that your wireless network users will employ when they join the network. If you ignore this advice, you make it very easy for any one of your network's users to change your base station's settings, and permitting such shenanigans is officially Not A Wise Move.

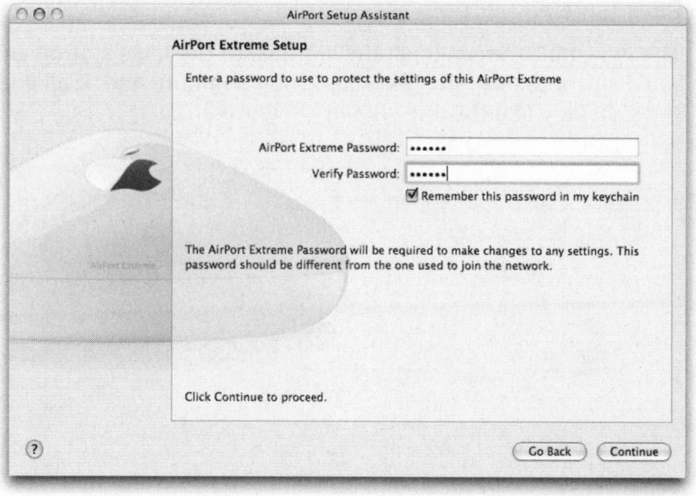

Figure 3-10: Lock your settings down to prevent tampering.

13. **Enter a password, enter it again to confirm it, and then press continue.**

    Now you're in the home stretch. The Assistant summarizes the settings you've made, laid out as in Figure 3-11. If any of them seem wrong, you can use the Go Back button, which appears on every pane the Assistant displays, to go back to the appropriate setting pane and make changes. If you need to do this, however, you won't have to reenter everything, because the Assistant remembers all the settings you don't change. Simply use the Continue button to get back to this pane after you've changed what you need to change.

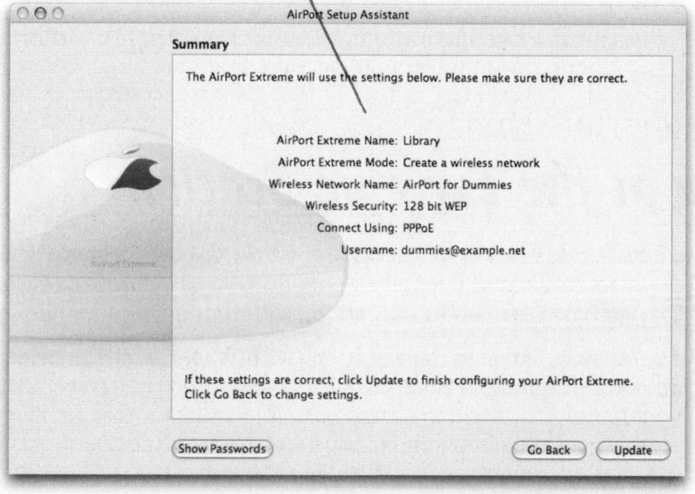

Figure 3-11: Review your settings one final time.

14. **Click Update.**

The Assistant transmits all the settings to the base station and restarts it. This process usually takes less than a minute and, if all goes well, you see a display like the one shown in Figure 3-12.

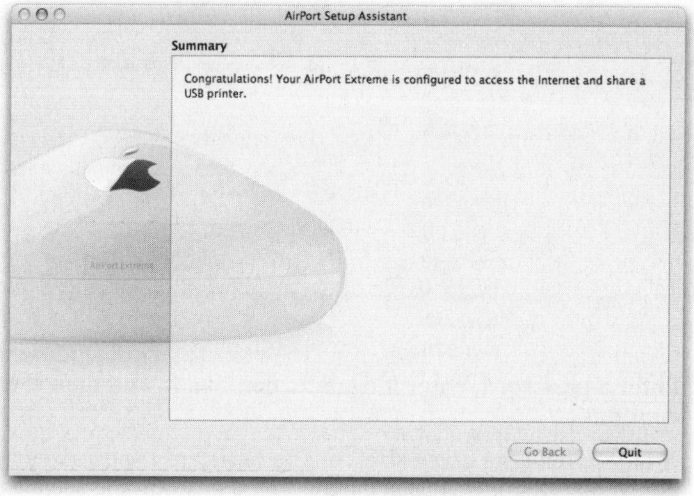

Figure 3-12: Behold the screen of success.

15. **Click Quit.**

The AirPort Base Station is now on the air. Yes, the setup process requires more than three steps, but the questions the Assistant asks aren't very hard to answer, and the entire procedure actually takes less time to perform than it does to read about it here.

As wireless network configuration procedures go, Apple's AirPort setup really does rank among the simplest you can find.

Arriving at the Extreme Station

Apple makes two models of its top-of-the-line AirPort Extreme Base Station:

- The AirPort Extreme Base Station (with modem and antenna port)
- The AirPort Extreme Base Station (with Power over Ethernet support and conformance to UL 2043)

On the air in the air ducts

Apple tailors the wireless base station model that it calls AirPort Extreme Base Station (with Power over Ethernet support and conformance to UL 2043) to meet one specific need of its educational and institutional customers: The need to place wireless base stations in air ducts.

Why in air ducts? Consider that campus networks often cover irregularly laid-out spaces — often comprising labs, classrooms, and offices set up in whatever space may be left over — and usually serve areas populated by large numbers of irregularly supervised individuals — that is, students. As a consequence, campus networking personnel have often strung their network cables through air ducts: after all, air ducts tend to reach nearly every place where network users might be found, and placing network cables in the air ducts keeps them relatively secure from the tampering hands of curious students and yet makes them relatively accessible for maintenance and upgrades.

So, when wireless networking technology came along, it made sense to put the wireless networking equipment in the same places as the wires, especially because wireless base stations often serve as bridges between the wired and wireless portions of the network.

Unfortunately, base stations require electrical power, and you don't usually find electrical outlets in air ducts. Furthermore, building regulations very sensibly require that air ducts not house any equipment that can produce smoke or other noxious gases in any harmful quantity, even when on fire.

Apple's educational model AirPort Extreme Base Station gets around both those problems. First, Ethernet cables can deliver electrical power when attached to network switches that follow the IEEE 802.3af standard, and Apple's educational base station can receive power over its WAN Ethernet port when connected to such a cable. Second, Apple designed the base station to meet the UL 2043 standard, which specifies the maximum amount of smoke and gas that a burning electrical device housed in an air duct can produce and still fall within the provisions of the National Electric Code, NFPA 70.

However, the educational AirPort Extreme Base Station does lack two features that its more mainstream sibling possesses: It has no modem port, and it doesn't support USB printing when it receives power over Ethernet. Luckily, there are few phone jacks in air ducts, and fewer USB printers in them, so most purchasers of this model base station probably won't mind losing these two features.

Chances are you'll only encounter the first of these models because Apple currently sells the second one only to educational customers — although you can find them available for sale from third-party vendors if you hunt around a little. Apple also made a stripped-down version of the AirPort Extreme base station that lacked the antenna port and modem port, but discontinued the model when it introduced the AirPort Express base station, which is described later in this chapter.

Out of the box, the AirPort Extreme Base Station, shown in Figure 3-13, looks opaquely simple: a rounded white cone, just over three inches high and just under seven inches in diameter, which sports an Apple logo and three unobtrusive status lights. All of the base station's ports and connectors reside on an inset panel behind it — we'll get to those in just a bit, in the section called "Perusing the ports."

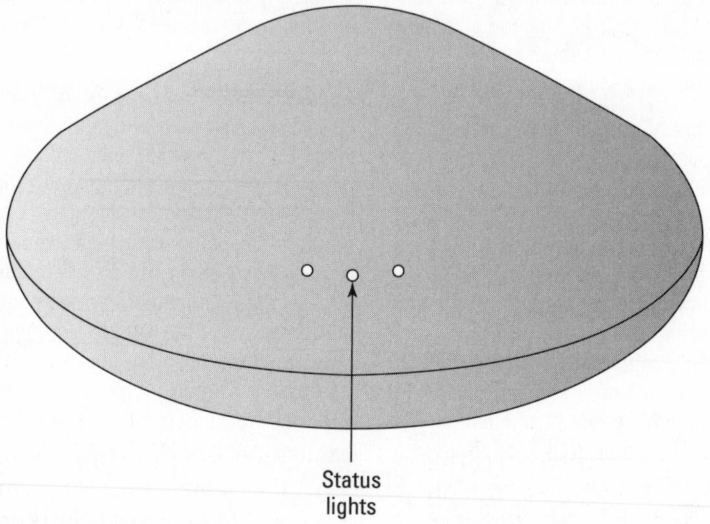

Figure 3-13: Full frontal view of the AirPort Extreme Base Station.

The station comes with a clear plastic mounting bracket and two screws that fasten the bracket — and, hence, the base station — to a wall. Prongs on the mounting bracket hook into slots on the base station's bottom, and cutouts let you route the cables that connect to the base station through the bracket. The cutouts can accommodate as many as six cables, which, coincidentally, just happen to match the number of ports on the rear of the base station.

If you don't want to fasten the base station to a wall, you can set it on any flat surface: It also has three soft plastic feet that shouldn't scratch the furniture.

Both models of AirPort Extreme Base Station suit small-business and home-office users as well as large businesses — the station can support as many as 50 simultaneous users. Its two Ethernet ports, described in the next section, allow it to serve both wired and wireless network clients; its mounting wall bracket makes it easy to place it in a location that allows its signal to reach as far as possible; and the station's external antenna connector, also described below, lets you extend the signal even farther. This is the base station to get if you're serious about your wireless networking needs.

Perusing the ports

The back of the AirPort Extreme Base Station is where it's all happening, as shown in Figure 3-14. You can find six connectors and a tiny recessed reset button, arranged in a nice, tidy row on the lower half of the base station's rear.

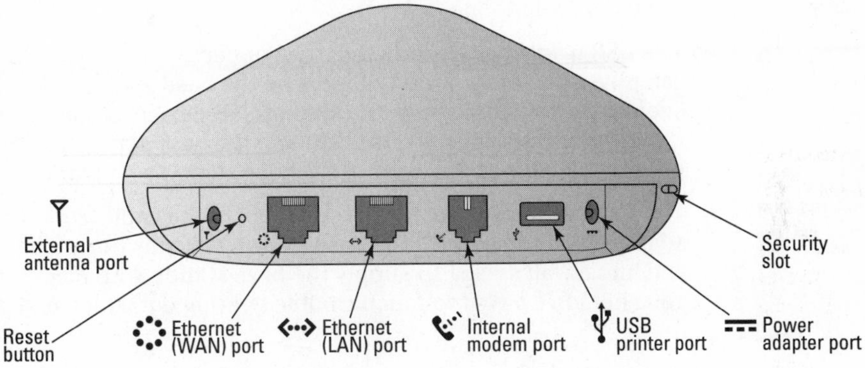

Figure 3-14: The AirPort Extreme Base Station's ports hide in the back.

- **The antenna port:** This lets you extend the base station's transmission range — which ordinarily extends no more than 150 feet from the station — by attaching a third-party antenna. Apple itself does not make external antennas for the base station, but it does sell compatible antennas made by other companies through the Apple Store.

  You should only use an Apple-approved antenna with the base station, and you must disconnect the base station's power before connecting or disconnecting an external antenna.

- **The Ethernet (WAN) port:** No, WAN isn't the sound of a base station crying: WAN stands for *Wide Area Network*. The Wide Area Network it refers to is the Internet, which, you must admit, covers a pretty wide area. You usually connect the WAN Ethernet port to an Ethernet cable that leads to the Internet in some way: either via a cable modem, a DSL modem, or an Ethernet network that provides Internet access.

- **The Ethernet LAN (Local Area Network) port:** This connects to a local network that has no Internet access. For example, a typical small office's local network might consist of a couple of desktop computers and a printer connected with Ethernet cables to a small Ethernet hub. Connect this network to the LAN Ethernet port on the base station, and the devices on this network can then share the base station's Internet connection — as well as communicate with the base station's wireless network users.

- **The Internal modem port:** This lets the base station connect to the Internet through a dial-up connection, using what telephone technicians refer to as POTS: Plain Old Telephone System. It's not a fast connection, but it's a widely available one. Note that the educational model AirPort Extreme Base Station with Power over Ethernet does not have a modem port; see the sidebar "On the air in the air ducts."

  You can damage the base station if you connect the modem port to a PBX or other digital telephone system. Only connect the modem port to a standard telephone jack.

- **The USB printer port:** Although the connectors on most USB devices can physically fit into the USB port on the base station, the base station knows how to handle only one kind of USB device: a USB printer. Plug a USB printer into this port, though, and the base station makes the printer available to any user of the base station's network.

- **The Power adapter port:** Insert one end of the cable from the power adapter that comes with the base station into this port. Plug the other end into a wall socket to supply the base station with electrical power. You shouldn't have too much trouble figuring out which end goes where.

Resetting the station

Occasionally you may discover you need to reset an AirPort Extreme Base Station. You can either reset it partially — to remove its passwords — or completely — to reset it to its factory settings. To accomplish both kinds of reset, you use the base station's reset button in conjunction with the AirPort Admin Utility or the AirPort Setup Assistant.

You can find the base station's reset button between the external antenna connector and the WAN Ethernet port, as shown in Figure 3-14. The button is small and set flush with the station's housing, so you'll need a slim pointed object, such as the end of an unbent paper clip, to press it.

The AirPort Admin Utility, described in detail in Chapter 4, lets you adjust all of the base station's settings, but you can't adjust those settings if you don't know the password that gives you access to them. Resetting the station's passwords with the reset button gives you a five-minute grace period during which you can access the settings, using the AirPort Admin Utility without needing a password.

Here's how you temporarily reset the base station's passwords:

1. **Make sure the base station is on and operating.**
2. **Unbend a paper clip and use it to press the reset button and hold it for one complete second.**

   The station's middle status light begins slowly blinking.

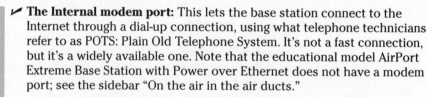

3. **Release the button.**

   You now have five minutes to fire up the AirPort Admin Utility and to change any settings you like, including giving the station a new set of passwords. The station retains any settings that you don't change.

If you ever have need to reset the station to its factory settings — for example, you plan to give it to someone else and don't want your settings to go along with the station — follow this procedure:

1. **Make sure the base station is on and operating.**
2. **Unbend a paper clip and use it to press the reset button and hold it for five complete seconds.**

   The station's middle status light blinks three times.
3. **Release the button.**

   The station loses all its settings, including its name, which becomes *Apple Network ######* (where ###### is a sequence of letters and numbers). The base station's password changes to *public.*

You can now use AirPort Setup Assistant to configure the base station from scratch, as described earlier in this chapter.

Reading the lights

Ordinarily you can ignore the status lights on the AirPort Extreme Base Station because the AirPort software on your Mac can tell you more about the base station than they can — that is, unless the station stops working, or if you are resetting it. Then the lights can tell a story, as detailed in Table 3-1.

Table 3-1 The Tale of the Blinking Lights

Left Light	*Center Light*	*Right Light*	*What It Means*
off or blinking	on	off or blinking	Everything is working fine. The blinking left light indicates wireless network traffic and blinking right light indicates Ethernet network traffic.

(continued)

Table 3-1 *(continued)*

Left Light	***Center Light***	***Right Light***	***What It Means***
First this blinks	then this one blinks	then this one blinks, then back to the first	You see the three lights blinking in sequence from left to right when the base station starts up.
on	on	on	The base station is running an internal check.
off	slow blink	off	You see this when you reset a base station's passwords.
off	three blinks	off	You see this when you reset a base station to its factory settings.
off	off	off	This means that the base station is either unpowered or broken. At least one light is on when the base station is working.
slow blink	slow blink	slow blink	The base station hasn't passed its power-on test, and you should have it checked by a service technician.

Riding the Musical Express

In order to write this section of the chapter, I had to sacrifice a major portion of the signing payment I received for this book in order to buy an AirPort Express Base Station. Do you feel sorry for me? Well, you shouldn't: Although this one is the fourth AirPort Base Station in our home, it has quickly become our most favorite.

As shown in Figure 3-15, the Express has no separate power adapter — the AirPort Express *is* its own power adapter: just unfold the plug on the back of its housing and stick it into a power outlet. In fact, to a casual observer this compact device looks very much like the power adapter for an iBook or iPod.

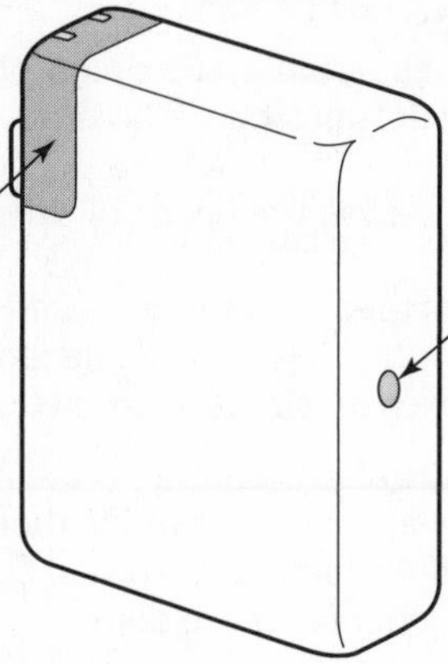

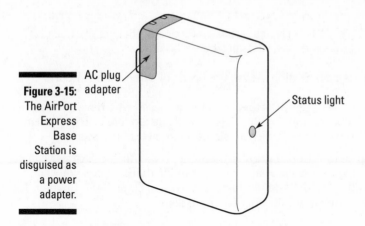

Figure 3-15: The AirPort Express Base Station is disguised as a power adapter.

Weighing in at well under half a pound and small enough to fit into an overcoat pocket, the tiny AirPort Express packs quite a wallop:

- It can share an Internet connection with as many as ten simultaneous users.
- It can broadcast over the same range — about 150 feet — as its pro sibling, the AirPort Extreme Base Station.
- It can extend the range of an existing AirPort network.
- Finally, and the most fun of all, the AirPort Express can connect to a hi-fi system or a set of powered speakers so you can play any of the songs in your iTunes library through it, using the iTunes AirTunes feature shown in Figure 3-16. Chapter 7 provides more information about using AirTunes with the AirPort Express.

Figure 3-16: iTunes can transmit music to an AirPort Express with its AirTunes menu.

TIP

After you have your AirPort Express set up, you can easily turn on AirTunes by opening up iTunes and, in iTunes Audio preferences, clicking the Look for remote speakers connected with AirTunes check box.

The AirPort Express Base Station serves the needs of three classes of user:

- Home users possessing AirPort-ready computers and who have a broadband Internet connection supplied either by cable modem or DSL modem or who wish to play their iTunes music through their hi-fi systems.
- Business travelers with AirPort-ready laptops — the AirPort Express can store the settings for as many as five different locations, which you can manage with the AirPort Admin Utility, as shown in Figure 3-17. Chapter 4 describes the AirPort Admin Utility in more detail.
- Home or business users who already have a wireless network and wish to extend its range, a task you can easily accomplish with the AirPort Setup Assistant, as Figure 3-18 illustrates.

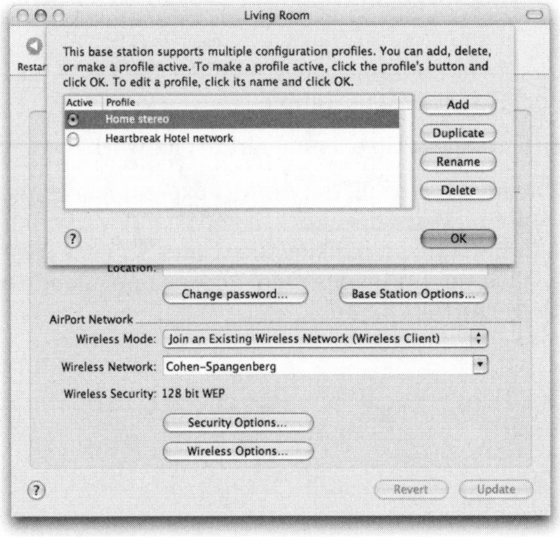

Figure 3-17: AirPort Admin Utility lets you manage as many as five different location profiles.

Perusing the ports

You can find Express's three ports on its bottom — that is, the side that is the bottom when you plug the Express into a typical wall socket so that the Apple logo on the side appears right side up. Figure 3-19 shows the Express's port arrangement.

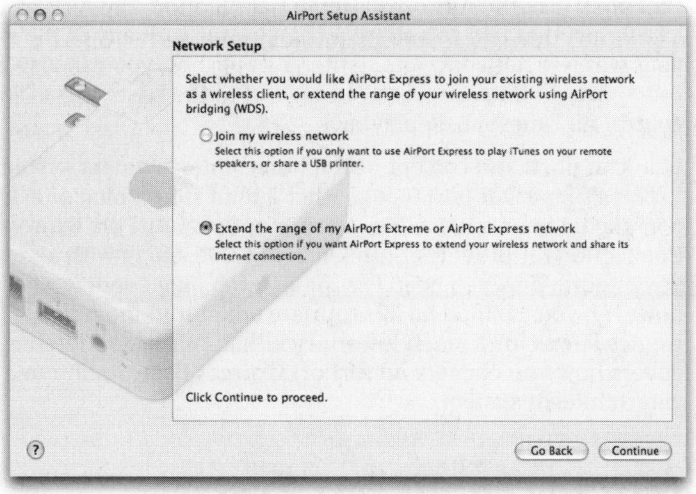

Figure 3-18: Use an AirPort Express to extend the range of an existing AirPort network.

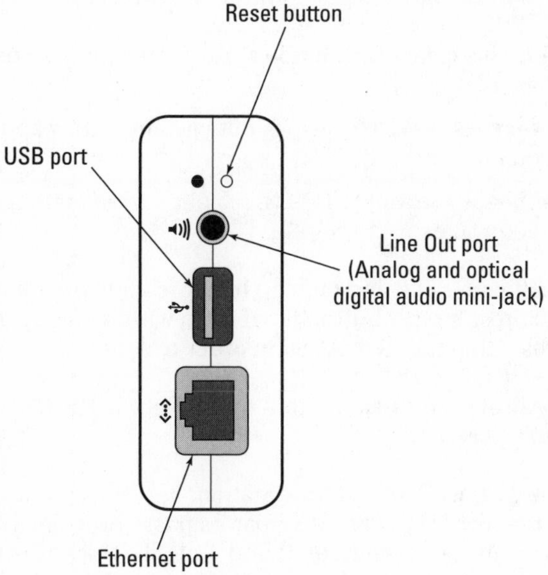

Figure 3-19: The three ports, and the reset button, found on the bottom of the AirPort Express.

- **Ethernet port:** The Airport Express possesses a single Ethernet port, which you usually connect to an Ethernet cable that leads to the Internet, either by way of a cable modem, a DSL modem, or an Ethernet network that provides Internet access. Unlike the AirPort Extreme Base Station, the Express has neither a separate LAN Ethernet port — making it less suitable for serving as a bridge between a wired and wireless network — nor a modem port — so you can't use it to share a dial-up Internet connection.

- **USB port:** Like the AirPort Extreme Base Station, the AirPort Express has a USB port that lets you share a USB printer with any of the users on your wireless network. As Chapter 7 describes, you can also attach a third-party USB infrared remote control to the base station's USB port to control AirTunes music playback.
- **Line Out port:** You can connect a home entertainment system to the Express's Line Out port using either a mini-stereo plug or a Toslink-to-mini-digital fiber-optic cable. Apple's optional AirPort Express Stereo Connection Kit provides both kinds of cable, along with an optional power cord. If all you need is a mini-stereo plug connection to your hi-fi, however, you can find an appropriate adapter at most music and electronics stores for a much lower price than Apple's kit fetches. Chapter 7 covers how you connect an AirPort Express Base Station to a home entertainment system.

Resetting the AirPort Express

The AirPort Express Base Station's reset button provides three levels of reset:

- **A simple reset:** This clears the station's passwords but retains all other current settings.
- **A reset to factory settings:** This wipes out current settings but retains any profiles you have created.
- **A full reset to factory settings:** This wipes out current settings and eliminates all stored profiles.

TIP

Find yourself a small paper clip and unbend it before you attempt any of these resets: the Express's reset button is set flush with the body of the base station and requires a thin pointed object in order to press it.

You can find the base station's reset button to the right of the Line Out port, as shown in Figure 3-19 earlier.

REMEMBER

Unlike resetting the AirPort Extreme Base Station, described earlier in this chapter, resetting the AirPort Express to clear its passwords does so indefinitely, rather than for just a five-minute period. But, like the AirPort Extreme station, you can enter new passwords and adjust any additional settings, using the AirPort Admin Utility, which you can find out more about in Chapter 4.

Here's how you clear the AirPort Express Base Station's passwords:

1. **Make sure the base station is on and operating.**
2. **Press the reset button with the end of a paper clip and hold it for one complete second.**

   The station's status light blinks.

3. **Release the button.**

   After you've performed this reset, you can use the AirPort Admin Utility, described in Chapter 4, to give the base station shiny new passwords.

You can perform a much more extensive reset like this:

1. **Make sure the base station is on and operating.**
2. **Press the reset button with the end of a paper clip and hold it for five complete seconds.**

   The station's status light turns yellow and blinks.
3. **Release the button.**

   Though the base station retains any profiles you've stored, it loses all its current settings and its name becomes *Apple Network* ###### (where ###### is a sequence of letters and numbers). The base station's password changes to *public*.

   You can use AirPort Setup Assistant to configure the base station from scratch, as described earlier in this chapter, or you can use the AirPort Admin Utility, which is described in Chapter 4.

You perform the fullest reset like this:

1. **Unplug the AirPort Express.**
2. **Press the reset button with the end of a paper clip and hold it in while you plug in the base station.**

   The station's status light turns yellow and blinks.
3. **Release the button.**

   Again, the station's name becomes *Apple Network* ###### (where ###### is a sequence of letters and numbers), and the base station's password changes to *public*. All profiles vanish. You can now use the AirPort Setup Assistant to set up the base station as though it were brand spanking new.

Deciphering the status light

The AirPort Express Base Station's single status light usually just glows greenly and steadily. Occasionally, however, you may see something else. Table 3-2 describes how the status light may sometimes appear, and what it means.

Table 3-2 The Meanings of the AirPort Express's Status Light

Appearance	*What It Means*
off	The AirPort Express is either unplugged or not receiving power.
blinking green	You see this when the AirPort Express starts up.
steady green	The AirPort Express is on and working.
blinking yellow	The AirPort Express can't find an Internet connection or, if the station is part of a larger AirPort network, it is not within range of that network.

Tuning into the Oldie Stations

Right now — which, when you read this, is at least a few months ago — the only two Apple AirPort base stations you can buy from Apple are the AirPort Extreme Base Station and the AirPort Express Base Station. That doesn't mean that all the previous AirPort base station models have suddenly twinkled brightly and dissolved in the haze of a transporter beam: You can still find many older AirPort base stations out in the wild.

AirPort base stations went through two previous generations before the AirPort Extreme Base Station landed on the planet:

- **AirPort Base Station (Graphite):** The original model, introduced in mid-1999.
- **AirPort Base Station (Dual Ethernet):** Also known as the Snow base station, the second-generation model was introduced in late 2001.

TECHNICAL STUFF

Both of these models adhere to the slower 802.11b wireless networking standard rather than the faster 802.11g standard used by the AirPort Extreme and AirPort Express base stations. However, both can communicate quite handily with newer Macs that have AirPort Extreme cards because the 802.11g standard supports compatibility with 802.11b. See Chapter 2 for more information about the 802.11b and 802.11g standards.

Gazing at Graphite

Shaped much like its successors, the original AirPort Base Station (Graphite) has a shiny gray plastic housing from which it derives its name.

Not surprisingly, this model seems primitive by today's standards. It possesses only two ports — an Ethernet port and a modem port — and it can support only ten simultaneous users. Of course, when this model first came out, the chances of finding ten wireless network users within a 150-foot range of any base station were rather low.

You can think of the Graphite base station as being similar to a slower, heavier, non-musical version of the AirPort Express, although it actually exceeds the Express in one regard by providing the ability to share a dial-up Internet connection. The Graphite base station can suit a small home network or even a small office network quite adequately. In fact, my wife and I used one for our home network until quite recently, and we still keep it around just in case.

Setting up a Graphite base station

The latest version of the AirPort Setup Assistant that works so well for the newest base stations does not work with the Graphite base station. At one time, of course, Apple did produce an AirPort Setup Assistant that worked with the Graphite station, but that earlier Assistant — which, when you install the latest version of the Assistant, is automatically renamed AirPort Setup Assistant for Graphite or Snow — does not work with Mac OS X 10.4.

As a consequence, if you have a Mac running Mac OS X 10.4 and a Graphite base station, you have two setup choices:

- Find a Mac running Mac OS X 10.3 or earlier and a copy of the AirPort Setup Assistant for Graphite or Snow, and use them to set up the Graphite station.
- Use the AirPort Admin Utility to set up the Graphite base station.

Although the second option may sound a little daunting, the limited number of networking options supported by the Graphite stations has the advantage of reducing the number of settings that AirPort Admin Utility makes available when you use it with a Graphite station.

REMEMBER: Chapter 4 describes the AirPort Admin Utility.

Resetting a Graphite base station

Apple made the process of resetting the original base station long and complex — not to mention that it requires a good deal of manual dexterity and strong fingers.

Apple's instructions specify that you have to press the Graphite's reset button for at least 30 seconds as you reconnect the power (the developers must have thought that resetting a base station would not be a common occurrence). What's more, Apple placed the reset button about half an inch deep inside a pinhole on the station's bottom, so you have to hold the base station steady with one hand, carefully insert the paper clip deep into the pinhole with the other, and use a third hand to plug in the power. The instructions also require that you connect the base station to your Mac with an Ethernet cable, and that you make a number of tedious changes to your Mac's network settings.

WARNING!

Apple's instructions also say that you should reload the AirPort base station software with the AirPort Setup Assistant. However, as noted above, the Graphite-compatible version of that program does not work with Mac OS X 10.4. In fact, the program could either unexpectedly quit or possibly even upload erroneous settings to the base station. If you run Mac OS X 10.4 on your Mac, you should disregard Apple's online instructions regarding this part of the reset process and use the AirPort Admin Utility to reload the base-station software.

REMEMBER

If you ever find yourself with a Graphite base station and decide that you need to reset it, you can obtain the complete Graphite reset instructions here:

```
http://docs.info.apple.com/article.html?artnum=58613
```

Seeing Snow

Both of Apple's names for its second-generation AirPort Base Station — Dual Ethernet and Snow — fail to differentiate it adequately from its successor, the AirPort Extreme Base Station. Both models have two Ethernet ports. Both models have the distinctive shiny white plastic cone-shaped housing. Of course, the Extreme station is faster, more secure, and more flexible than the Snow, just as the Snow station improved upon the features and performance of the original Graphite station.

Here's how the Snow base station improves upon the Graphite station:

- It supports 50 simultaneous users instead of the Graphite's ten.
- It has a separate LAN Ethernet port, allowing the station to bridge a wired and a wireless network without requiring you to buy a separate Ethernet hub.
- It supports 128-bit WEP (*Wired Equivalent Privacy*) encryption instead of the 40-bit encryption offered by the Graphite station, making it more secure from technically savvy electronic eavesdroppers.
- It incorporates software that makes its modem port more compatible with America Online's proprietary dial-up networking protocol.

Like the Graphite base station, the Snow station serves the needs of home- and small-business users, and does it better than the original model.

Setting up a Snow base station

Both the Graphite and the Snow base station use the same, older version of the AirPort Setup Utility — described in the section "Setting up a Graphite base station" earlier in the chapter — which means that the same limitations and caveats apply to setting up the Snow base station.

Resetting a Snow base station

Although still not as easy to reset as the current Apple base stations, the Snow base station improves upon the Graphite version by having a more accessible reset button, and one that doesn't need to be pressed for nearly as long to reset the station. Also, the Snow base station allows both a *soft* reset — which simply clears out passwords — and a *hard* reset — which returns it to factory settings.

But, like the Graphite base station, a hard reset requires you to connect your Snow base station to your Mac with an Ethernet cable, adjust various network settings, and — for Mac OS X 10.4 users — reload the station's software and configure it with the AirPort Admin Utility. On the other hand, a soft reset allows you to connect to the Snow base station over a wireless connection, an option not possible with the original Graphite model.

REMEMBER

You can find Apple's instructions for resetting a Snow base station here:

```
http://docs.info.apple.com/article.html?artnum=106602
```

Picking Third-Party Stations

Apple is not the only fruit in the wireless base-station orchard, and some folk who are both adventurous and light of pocketbook find themselves picking the products of other vendors. As it has with so many other digital consumer products, Apple blazed the wireless network trail with AirPort, and, after AirPort proved successful, other companies rushed to create lower-cost, more feature-laden competing products. Today, you can find base stations — or *wireless access points* as they are more commonly known outside of the Macintosh marketplace — that cost less than half of Apple's price for an AirPort Extreme Base Station, and that deliver equal, or even better, performance.

Non-AirPort wireless access points tend to share certain characteristics:

- **Multiple Ethernet ports:** Commonly four, in addition to the port corresponding to the AirPort Extreme's WAN port.
- **Built-in Firewall:** Software that blocks unauthorized access from outside the wireless network to services and data provided by the wireless network's users.
- **Web-based configuration:** The access point presents its setup controls on a Web page that it serves over the local wireless network, which allows you to set up the access point, using a standard Web browser, like Firefox or Safari.
- **Windows favoritism:** Although the wireless standards that these products use are Mac-compatible, the documentation and software provided with them assumes that you use Windows, and the companies' tech-support staffs, should you call them, frequently assume the same.

This last item does not mean that Mac owners can't avail themselves of these lower-cost alternatives. In many cases, you can access all the wireless access point's features, and perform any software upgrades they require from your Mac. You just shouldn't assume that you can do all that you need to do with it from your Mac until you read the product's data sheets and consult a review or two.

Table 3-3 lists some of the manufacturers of wireless access points and their Web sites:

Table 3-3 A Few Wireless Access Point Manufacturers

Company	*URL*
Asante Technologies, Inc.	`www.asante.com`
D-Link Corp.	`www.dlink.com`
Linksys, a division of Cisco Systems, Inc.	`www.linksys.com`
U.S. Robotics	`www.usr.com/home.asp`

Part II
Knitting a Network

"Do you think that this will match our router?"

In this part . . .

Once you have your wireless pieces, you have to put them all together. That's what this part is all about.

The three chapters that follow show you how to assemble a home network, how to deal with the invisible terrors of radio interference, and how to make an AirPort network on the cheap.

Chapter 4

Building the Basic Home Network

In This Chapter

- Tabbing through AirPort Admin Utility
- Connecting to the Internet three or four ways
- Keeping uninvited visitors off your network
- Probing the anatomy of a SOHO network

Whenever I think about networks, I think about plumbing. And I don't believe I'm the only one who does that, either: Technology writers, network administrators, and equipment vendors all talk about things like "pipelines," "information flow," "data pumps," and, of course, the final resting place for information, "the bit bucket."

Like plumbing, the job of a network is to get stuff flowing from one place to another, and to do it unobtrusively and without leaks. Like plumbing, sometimes getting one part of a network hooked up to another part can be a little tricky. Like plumbing, sometimes the details of how the parts of the network all work together can be less than pretty. And, like plumbing, after you get everything on the network in place and connected, you just want it to work: You don't ever want to think about it again.

Of course, you don't have to crawl under the house with a flashlight, a pipe threader, and a monkey wrench to get your AirPort network working. Instead, you'll use Apple's AirPort Admin Utility and maybe a few other software tools. Best of all, you can do most of this work from a comfortable chair — except for maybe having to crawl behind your desk to untangle and disconnect the network cable you no longer need.

Going Deep into AirPort Admin Utility

In Chapter 3, you get a good look at the AirPort Setup Assistant, which companionably takes you by the hand and walks you through setting up an AirPort Base Station right out of the box. More often than not, the Assistant provides all the help you need to get your base station up and running on the

Internet, and, after you use it, you don't have to think about your base station again: it just works. Networks, however, have a way of growing and changing, and when they do, you require a tool that lets you adjust your base station's settings to accommodate your new networking needs. The AirPort Admin Utility is that tool.

Opening the Admin Utility

Like most of the utility programs that Apple ships with Mac OS X, you can find the AirPort Admin Utility among its utilitarian brethren in the Utilities folder that Apple tucks away into an odd corner of the Applications folder.

TIP

In the Finder you can always get to the Utilities folder with a simple keyboard shortcut: Shift+⌘+U. You can also open it directly from the Finder's Go menu.

When you launch the AirPort Admin Utility, you first see the Select Base Station window shown in Figure 4-1. The AirPort Admin Utility displays all the nearby AirPort base stations it can detect in the panel on the left. On the right, the AirPort Admin Utility displays details about the base station selected in the left panel, including a picture of the type of base station selected. In the toolbar along the window's top you can find the AirPort Admin Utility's not-so-vast array of tools.

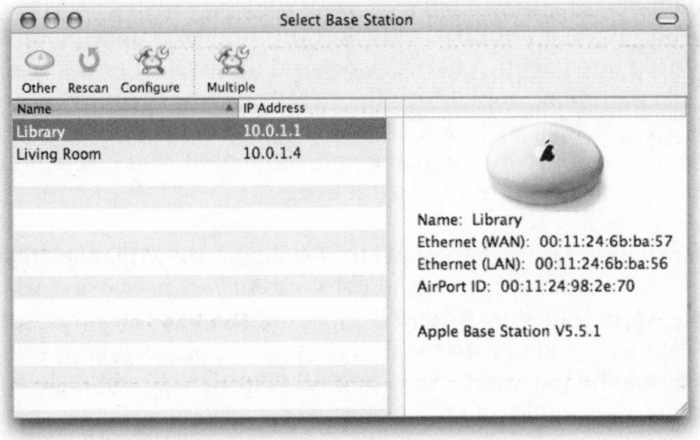

Figure 4-1: The AirPort Admin Utility wants you to select a base station.

TECHNICAL STUFF

The long strings of numbers shown in Figure 4-1's right-hand panel are called MAC addresses. In this case, MAC doesn't mean *Macintosh* but *Media Access Control*. A MAC address identifies individual network nodes and no two of the addresses can be the same. The base station selected in the figure has three MAC addresses because the AirPort Extreme Base station incorporates three network connections: one for the LAN port, one for the WAN port, and one for the base station's wireless connection. You can find out more about the AirPort Extreme's ports in Chapter 3.

The tool buttons on the toolbar go from left to right:

- **Other:** Click this to specify the IP address of a base station if it doesn't show up in the list of detected base stations.
- **Rescan:** Click this to have the AirPort Admin Utility scan the airwaves for base stations again. You might want to use this if you plugged in your base station after you launched the AirPort Admin Utility.
- **Configure:** This button should wear a scuff-guard because it gets used so much: It produces a base station configuration window that contains on its various tabs all the check boxes, fields, pop-up menus, and option buttons that you use to change the settings of the base station that's selected in the Select Base Station window's list.
- **Multiple:** Click this button when you want to upload new software to a bunch of selected base stations all at once, or if you need to apply identical settings to more than one base station, using a previously saved configuration file. Network administrators in particular find this tool useful.

TIP

If the labels below each of the tool buttons on the toolbar don't quite help you remember what the tool does, just hover your mouse over a tool button to see a yellow help tag appear. It'll have a slightly fuller description of the tool's function.

You can customize the toolbar by choosing View➪Customize Toolbar, but most people may not find much point in doing so: The buttons shown by default in the AirPort Admin Utility toolbar comprise all the tools the program has. On the other hand, I suppose it's nice that you can hide the Multiple button if you only have one base station and don't want to be reminded that some people have more.

Conquering the configuration controls

At the heart of the AirPort Admin Utility lies the base station configuration window, which, as described a couple of paragraphs back, provides the controls you employ to set up a base station just the way you want it.

Select a base station in the Select Base Station window's list, click Configure, and a configuration window named for the base station you selected appears, as shown in Figure 4-2, containing several tabs along the top, just below the window's toolbar. The number of tabs in the configuration window varies depending on the type of base station you chose to configure. Figure 4-2 shows the base station configuration window for an AirPort Extreme base station.

Base Station Configuration Window

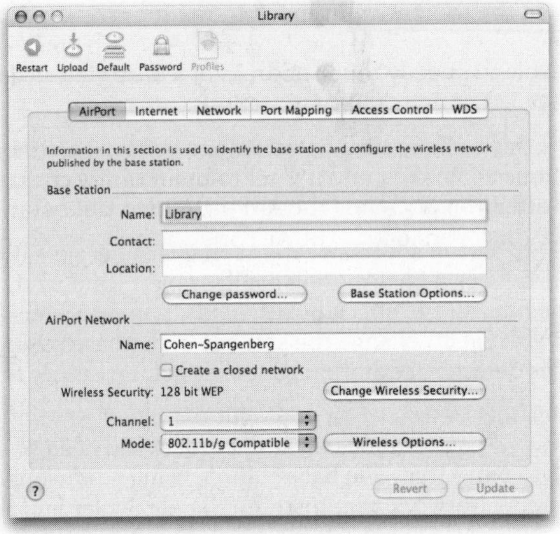

Figure 4-2: The configuration window for a base station named *Library*.

REMEMBER

Any configuration changes made in this window do not take effect until you click Update. Click Revert if you want to discard your changes but continue working in the configuration window for the base station. You can click the configuration window's close button to discard your changes and to close the window.

TIP

You can tell if you have made any changes to the base station's configuration by looking at the base station configuration window's red close button. If the button contains a dark dot, it means that changes have been made in the window. Many programs on the Mac, by the way, use this same indication to tell you that the window contains unsaved changes.

Tackling the toolbar

The toolbar along the top of the base station configuration window contains the same tool buttons no matter which tab you click in the window. You use these tools not to alter the base station's configuration but to control, or to get information about, the base station's operation.

The tool buttons in Figure 4-2 are shown from left to right:

- **Restart:** As you might expect, clicking this button restarts the base station. You may wish to click this button if the base station doesn't seem to be communicating properly with its clients. Of course, if the base station has become *very* confused, it may not respond to the Restart button, but it's a good place to start.

- **Upload:** You click this tool button when you need to install software, such as a patch from Apple to fix a bug in the base station's software. The tool produces a standard file sheet that you use to select the software file to upload.
- **Default:** Click this tool button to upload the current default base station software to the base station. The AirPort Admin Utility contains the default software for every model of Apple's AirPort base station. When Apple releases new versions of the AirPort Admin Utility, these new versions often contain revised default software for one or more base station models.
- **Password:** Click this tool button to produce a sheet that displays the base station's network password in the user-hostile format that some non-AirPort equipped Windows or Linux wireless clients may have to use to gain access to the AirPort network. The password appears as a long string of letters and numbers, such as *8FB9438D4FB99A3D0E15397291*. If you don't share your AirPort network with non-Mac users, you can click this button to remind yourself of why you chose to use a Mac.
- **Profiles:** Some base stations, such as the portable AirPort Express base station, can store as many as five different configurations, called *profiles*. Click this tool button to see a sheet like the one shown in Figure 4-3 that you can use to create, remove, rename, or choose different configuration profiles. When you choose a profile, any configuration changes you subsequently make with the AirPort Admin Utility are stored with the chosen profile. When, for example, you go on a business trip, you can click this button to create, modify, and save a profile for your lodging's Internet service. Later, when you return home, you can click the Profiles button to switch the AirPort Express base station back to your home configuration. The Profiles button remains visible but inactive, as in Figure 4-2, for base stations that don't support multiple profiles.

Exploring the AirPort tab

In case you can't quite make out the small print in Figure 4-2, Apple describes the AirPort tab in the configuration window like this: "Information in this section is used to identify the base station and configure the wireless network published by the base station." What Apple doesn't say is that this tab provides more settings per square inch than any other tab in the configuration window. Yet, believe it or not, they're all arranged in a way that makes sense when you work with them.

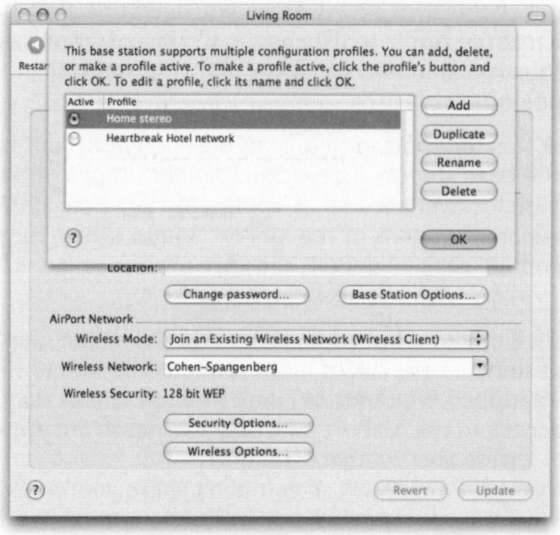

Figure 4-3: Some base stations can store multiple configuration profiles.

The base station configuration window's AirPort tab contains two sections:

- **Base Station:** This section provides three text fields that you use to specify the name of the base station, who to contact to administer it, and where the base station is. For most home networks, only the Name field matters — this is the name that you see in the AirPort Admin Utility's Select Base Station window. You use the Change Password button in this section to set or change the password needed to administer the base station. Other users on your home network don't need to know this password — only you, the administrator, need to know it. The Base Station Options button produces a richly featured sheet of setting options, described a little later in this section.
- **AirPort Network:** This section provides a text field in which you enter the public name of the base station: Wireless network users see this name when they use the AirPort status menu or the Internet Connect application. The AirPort Network section of this tab contains a number of other options that you use to set up your network's security, its broadcast channel, and some other options. These options are described later in this section as well.

Cf. 348 password required to administer the base station (cf. 92)

Base Station Options

The Base Station Options button in the Base Station section of the AirPort tab produces a sheet containing various option controls, and, like the window from which it was spawned, the sheet also has its own set of tabs.

Open AirPort Admin Utility → Select a Base Station → Click Configure → Click AirPort tab → Click Base Station Options

The first tab on the sheet provides the WAN Ethernet port options, shown in Figure 4-4. As the text near the top of this sheet tab points out, in somewhat more technical language, the options on this tab limit how much information about your base station and wireless network Internet users elsewhere can obtain. What it *doesn't* say is that some of these options also limit what users elsewhere on the Internet can *do* with your base station.

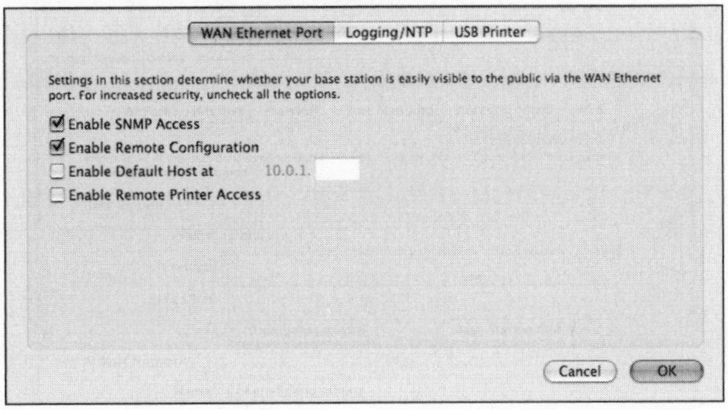

Figure 4-4: You can limit your station's visibility on the Internet with these options.

Here's what the options on the WAN Ethernet Port tab mean:

- **Enable SNMP Access:** *SNMP* stands for *Simple Network Management Protocol*, and it's a method that network administrators use to collect information about a network device and to control it in various ways. By default, Apple enables this option, but home users should disable it because it otherwise could allow someone from anywhere on the Internet to mess around with your base station settings.
- **Enable Remote Configuration:** Similar to the SNMP access option, this option lets users elsewhere on the Internet configure your base station. By default, this option is on. Unless you want allow other Internet users to configure your base station, turn this option off.
- **Enable Default Host at:** An AirPort Base station usually acts as a firewall between its client computers and the Internet. You use the Enable Default Host at option to establish a permanent network address for one of the base station's clients so it can act as what network professionals sometimes call a *DMZ*. You often use default hosts for Internet games and certain kinds of videoconferencing. When you enable a default host and give it an address, you also must manually set to the same number the IP address on the computer that you want to act as the default host. The most common IP address used in AirPort networks for a default host is 10.0.1.253, probably because the AirPort Admin Utility offers it as the default. Typically, you also use the configuration window's Port Mapping tab, described later in this chapter, to route specific types of Internet traffic to the default host computer. If you are a gamer or a network administrator, you likely know what I'm talking about. If not, just leave the option disabled.

- **Enable Remote Printer Access:** Enable this option to allow others on the Internet to see and use a USB printer connected to your base station. Unless you feel like donating your printer supplies to any one of the millions of people who might wish to use them, disable this option.

The Logging/NTP tab, shown in Figure 4-5, provides options that you can use to send a base station status log to another computer, and to set how much detail the status messages contain. This tab also provides an option to set your base station's internal clock from a time server on the Internet.

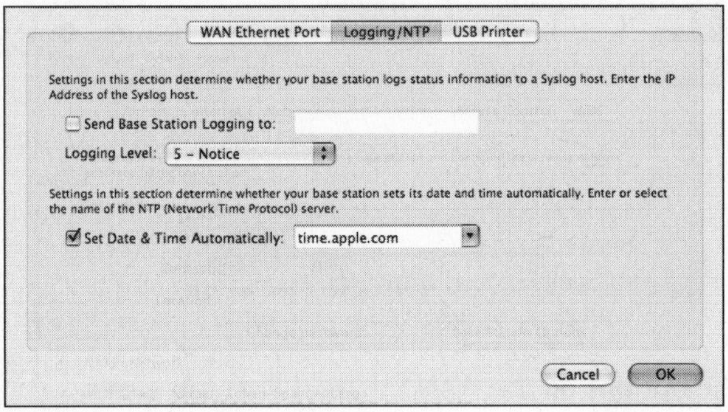

Figure 4-5: You can store status logs and set the base station's clock automatically.

If you put your Mac's wireless network IP address in the Send Base Station Logging to field and enable the option, the base station sends its status messages to your Mac's system log, which you can read with the Console application in the Utilities folder. You should have the base station set its clock from a time server if you wish to use the logging option.

Figure 4-6 shows the final tab in the Base Station Options sheet, the USB Printer tab. As the text in the tab points out, you double-click the printer's name to assign a new name.

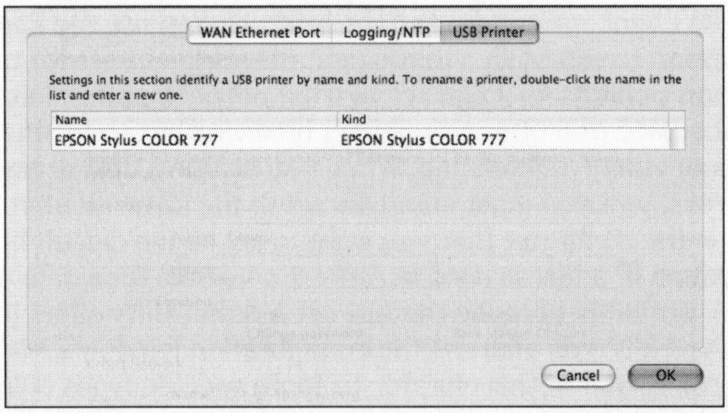

Figure 4-6: Rename the USB printer connected to the base station.

Two things you should note:

✔ You should turn your printer on before you open the AirPort Admin Utility or you won't see your printer listed on this sheet.

✔ You can have only one USB printer connected to a base station's USB port. Perhaps that explains why the tab's label says *USB Printer* instead of *USB Printers*.

Base station security options

Open AirPort Admin Utility → select a Base Station → Click Configure → Click AirPort tab

The AirPort Network section of the configuration window's AirPort tab contains several security options. These options let you control how wireless users can detect and join your base station's wireless network.

The simplest security option is the Create a Closed Network option, visible in the window shown in Figure 4-2 and in Figure 4-7. Enabling this option conceals the name of your base station's wireless network: The network's name appears neither in the AirPort status menu nor in the Internet Connect application of wireless Macs. Windows wireless users can't see the network name using their network software either. However, wireless users can still join the base station's wireless network by typing the network name into the connection dialog box or window that their network software provides. Think of this option as being similar to giving your network an unlisted phone number.

p.114, 17
cf. 348

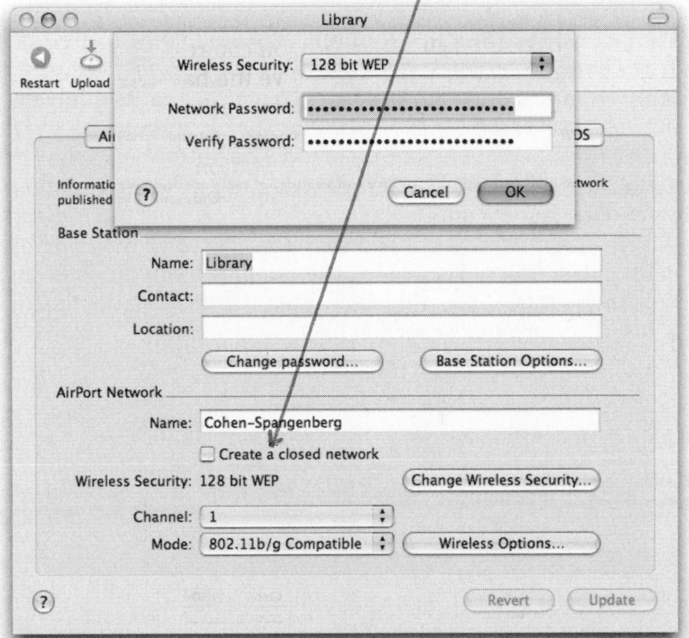

Figure 4-7:
One face of the Wireless Security sheet — it has several.

Open AirPort Admin Utility → Select a Base Station → Click Configure →
Click AirPort tab → Click Change Wireless Security button

92 Part II: Knitting a Network

cf. 348

The Change Wireless Security button provides a much more comprehensive set of security options. Click it to see a sheet, like the one shown in Figure 4-7, from which you can change wireless security settings. This sheet changes its appearance and options depending on what you select from the sheet's Wireless Security pop-up menu.

Here are the choices that the Wireless Security pop-up menu offers:

WARNING!

✔ **Not Enabled:** As you might expect, this choice makes your network accessible to anyone. Users wanting to join your network won't need to use a password or other authentication technique. Use this choice if you live in farmhouse in the middle of a very large barren tract of land, or on an ice floe somewhere out in the ocean. Otherwise, you should choose one of the other options on the menu.

✔ **40 bit WEP/128 bit WEP:** These two options employ the original security methods available when Apple first introduced AirPort, and they are the only two that the original AirPort card can use. WEP (which stands for *Wired Equivalent Privacy*) provides a relatively simple form of data encryption that hackers monitoring your network for a few minutes can crack. 128-bit WEP, shown in Figure 4-7, provides slightly stronger protection than 40-bit WEP, but it, too, can be cracked by a hacker with the right software tools. Both methods use a simple password. Use WEP security if you have no other choice: Like a chain lock on your front door, it's better than nothing, but a determined intruder can quickly break it.

✔ **WPA Personal:** Designed for home and small office wireless networks, WPA (*Wi-Fi Protected Access*) provides more robust security than WEP, but users must have an AirPort Extreme card or its equivalent to use a network protected by WPA. You can specify either a password or a Pre-Shared Key, consisting of from 8 to 63 hexadecimal digits (that is, the digits from 0-9 and the letters from A-F, used by programmers to express base-16 numbers), that users must use to access the network. The Wireless Security sheet's Show Options button reveals the items shown in the lower half of Figure 4-8. The number that you enter into the Group Key Timeout field specifies how often the base station changes its encryption key, to further foil hacking attempts.

password required to
join the network
(cf. 89)

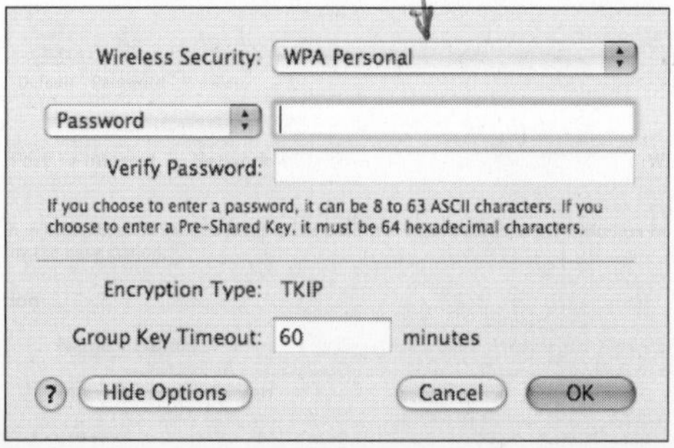

Figure 4-8:
WPA
Personal
security
provides
modern
security for
home users.

Wireless Security:	WPA Personal ⬦
Password ⬦	
Verify Password:	

If you choose to enter a password, it can be 8 to 63 ASCII characters. If you choose to enter a Pre-Shared Key, it must be 64 hexadecimal characters.

Encryption Type: TKIP

Group Key Timeout: 60 minutes

(?) (Hide Options)   (Cancel) (OK)

How does this differ from
the security options discussed
on pp. 172-173?

If you plan to share your WPA Personal-protected network with Windows users, select the Password option and use a password that consists of exactly 13 hexadecimal digits. Although this sounds inconvenient, it's not as inconvenient as having intruders breaking into your network.

- **WPA Enterprise:** You won't use this option at home. WPA Enterprise requires a separate RADIUS (*Remote Authentication Dial-in User Service*) server, which is usually set up by an experienced network administration professional. You can see the options the sheet provides for WPA Enterprise security in Figure 4-9, if you are curious.

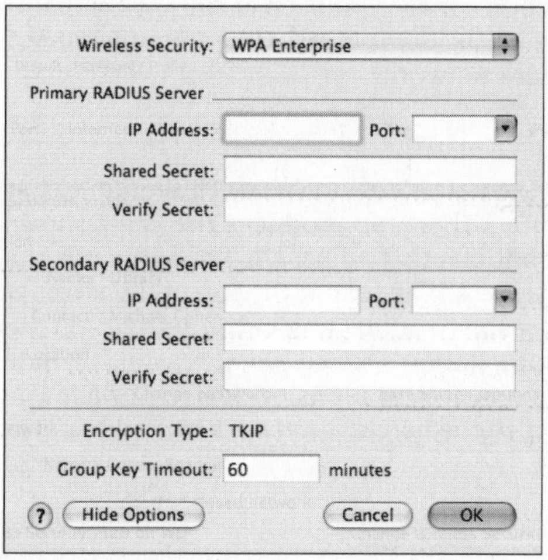

Figure 4-9: You usually use WPA Enterprise security in large business settings with IT staff.

Channel and mode options

The Channel pop-up menu, near the bottom of the AirPort Network section of the base station configuration window's AirPort tab, lets you select a broadcast channel for your network. The AirPort network can use one of 11 different channels. If possible, pick a channel that is two or three channels away from the ones used by any nearby wireless base stations. You can also choose Automatic from this menu and let the base station decide what channel to use. For more information about choosing a channel for your base station, refer to Chapter 5. pp. 124 – 127

A free program called iStumbler, available for download from `www.istumbler.net`, can show you the names, types, and channel numbers of nearby base stations. It can also show you the signal strength of each station that it detects. Cf. 123

Below the Channel pop-up menu is the Mode pop-up menu. Use this menu to specify how your base station communicates with wireless clients:

- **802.11b Only:** This menu item makes your base station completely compatible with the oldest AirPort cards, and restricts the station's maximum data transfer rate to 11 megabits per second, even for computers with AirPort Extreme cards that use the faster 802.11g protocol.
- **802.11b/g Compatible:** Picking this menu makes your base station compatible with both the old AirPort cards and the new AirPort Extreme cards. The station transfers data at the appropriate rate for each card.
- **802.11g Only:** This choice makes your base station compatible only with AirPort Extreme cards. Computers using the original AirPort cards cannot join the wireless network.

Wireless options

The Wireless Options button in the AirPort Network section of the base station configuration window's AirPort tab produces the sheet shown in Figure 4-10. This sheet provides options that adjust the base station's transmission and reception characteristics. As the text at the top of the sheet points out, you should have a good reason to adjust any of the settings on this sheet because they affect the base station's range. Good reasons include wanting to overcome interference from nearby stations or devices, or allowing distant wireless clients to connect to the wireless network at slower data rates than normal.

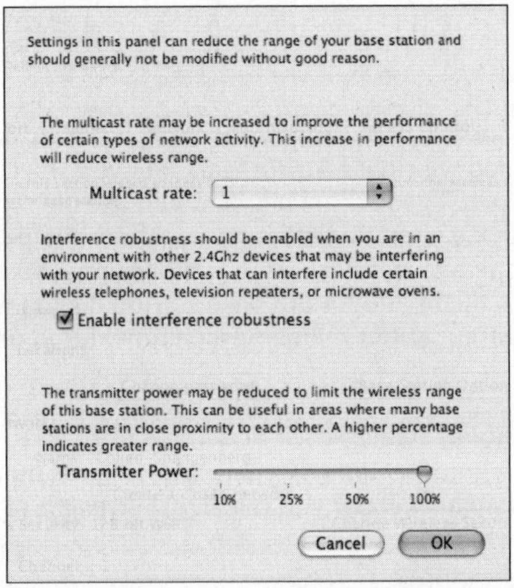

Figure 4-10: You control wireless transmission and reception abilities on this sheet.

These are the options that this sheet provides:

- **Multicast rate:** This option specifies rate at which the base station can send the same message to multiple clients on the network. The lower the number, the slower the rate, and the further away from the station a wireless client can be placed. Conversely, the higher the setting, the smaller the base station's effective range, but the greater the station's maximum performance.
- **Enable Interference Robustness:** This option helps the AirPort base station overcome data dropouts caused by interference from wireless phones that use the same frequency range as the base station, or from devices like microwave ovens that emit signals in the AirPort broadcast frequency range. Wireless clients using AirPort Extreme cards also need to enable interference robustness on their computers for this option to help.

  The Use Interference Robustness item on the AirPort status menu enables this option on AirPort Extreme–equipped Macs.
- **Transmitter Power:** Use the slider to increase or decrease the transmitter power of an AirPort Extreme or AirPort Express base station. Lowering the transmitter power does decrease the base station's range but can, paradoxically, make the connection more reliable for clients in an environment that contains a number of other base stations. You may also want to reduce transmitter power to make your station less accessible to distant users, such as neighbors in an apartment building or college dormitory.

Wireless Mode options

Although a fully capable network base station in its own right, the AirPort Express can also act as digital music receiver for iTunes. With an AirPort Express, you can play songs transmitted wirelessly from your Mac's iTunes music library through a home entertainment system. Apple calls this capability *AirTunes*.

To accommodate AirTunes, the AirPort Express's configuration window provides a pop-up Wireless Mode menu in the AirPort Network section of the AirPort tab. The pop-up menu provides items that you can choose to specify how the AirPort Express handles wireless network activity.

The AirPort Express's Wireless Mode pop-up menu offers three items:

- **Create a Wireless Network (Home Router):** Choose this menu item when you want to use the AirPort Express as a typical base station that can create a wireless network and allow wireless clients to share its Internet connection. This choice does not prevent you from using the AirPort Express as a digital music receiver for iTunes at the same time.

- **Join an Existing Wireless Network (Wireless Client):** Choose this menu item if you already have an AirPort network established and you want to use the AirPort Express as a digital music receiver connected to powered speakers or a home entertainment system. As you can see in Figure 4-11, selecting this menu item reduces the number of tabs in the configuration window as well as the number of options provided in the AirPort tab of the window.
- **Wireless Disabled:** Choose this menu item when you have connected the AirPort Express to a wired Mac with an Ethernet cable, and you merely want the Express to play the Mac's iTunes music through a set of powered speakers or a home entertainment system.

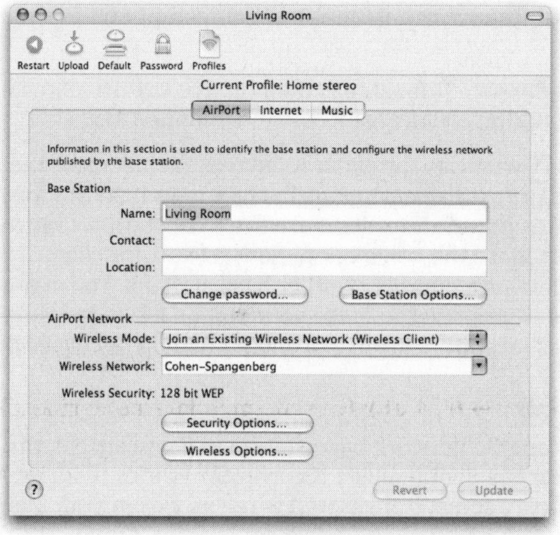

Figure 4-11: You see fewer tabs when an AirPort Express is a wireless client.

Going online with the Internet tab

Just as a Mac can connect to the Internet with a modem, or a *local area network (LAN)* connection, or a cable modem, or a *digital subscriber line (DSL)* modem, so can an AirPort base station. And, just as a Mac can "remember" how to connect to the Internet when it is turned off and then turned back on, so can an AirPort base station. You use the Internet tab in the base station configuration window to specify how the AirPort base station creates and maintains its connection to the Internet.

REMEMBER

The number of Internet connection methods offered depends on the type of AirPort base station you have. For example, an AirPort Express does not have a modem port, so the base station configuration window for an AirPort Express doesn't show you any modem connection methods under its Internet tab. Chapter 3 describes the connection features available with different AirPort base stations.

You use the Connect Using pop-up menu, visible in Figure 4-12 just below the configuration window's tabs, to see a settings pane for an Internet connection method.

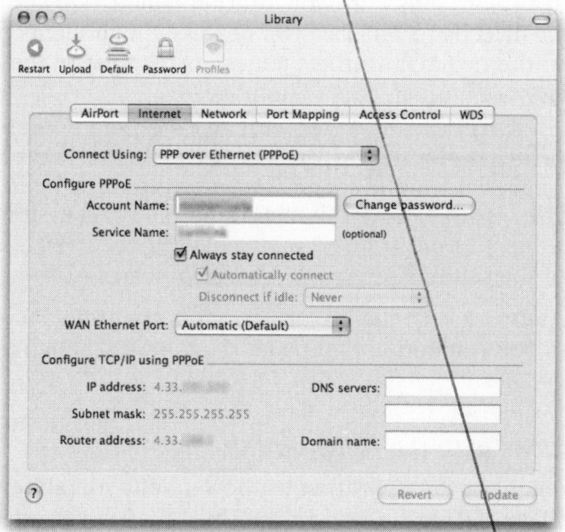

Figure 4-12: Use the Internet tab to specify how the base station connects to the Internet.

Here's a quick look at the Internet connection options you may see:

- **PPP over Ethernet (PPPoE):** One common Internet connection method is PPPoE, the settings pane for which is shown in Figure 4-12. DSL Internet connections, for example, frequently use PPPoE connections.

  See "Connecting with a DSL modem" later in this chapter to see how to use PPPoE with a DSL modem to establish an Internet connection.

- **Ethernet:** Your AirPort can use an Ethernet connection to the Internet in one of two situations: Either the AirPort connects to a local network that has an Internet connection, or it connects to another network device that has a direct Internet connection. Unlike a PPPoE connection, you don't need to supply an account name or password with an Ethernet connection to the Internet.

  See "Connecting to the Internet over a LAN" and "Channeling the Internet with a cable modem" in this chapter for more about using the Ethernet Internet connection method.

- **Modem (V.34)/Modem (V.90):** Depending on the model of AirPort base station you have, the Connect Using pop-up menu may offer you two modem connection setting panes, one for each of the base station's available modem connection speeds. You won't see any modem choices at all, however, if you have an AirPort Express which doesn't have a modem.

The section "Dialing in with the built-in modem" in this chapter describes the available modem settings for this Internet connection method.

- **America Online (AOLnet V.34)/America Online (AOLnet V.90):** AirPorts with modem ports may also provide Internet connection setting panes for connecting directly to America Online. Like the modem connection setting panes, the two AOL settings panes correspond to the AirPort station's two available modem connection speeds.

  See this chapter's "Accessing America Online" section to find out how to use these AOL Internet connection methods.

- **AirPort (WDS):** Use this connection method to have the AirPort base station obtain its Internet connection from another AirPort base station that is part of the same *Wireless Distribution System (WDS).*

  pp. 103-104 "Widening a network with the WDS tab" in this chapter summarizes what a WDS is, and you can find out much more about setting up a WDS in Chapter 5.

Sharing connections with the Network tab

The network to which the configuration window's Network tab refers is the network that the base station creates. Under this tab, you can specify how the base station identifies its network clients so it can route information both among those clients and to the rest of the Internet. Or, as the text near the top of the Network settings pane shown in Figure 4-13 puts it, "Settings in this section determine how this base station's Internet connection is shared with computers connected to its AirPort network."

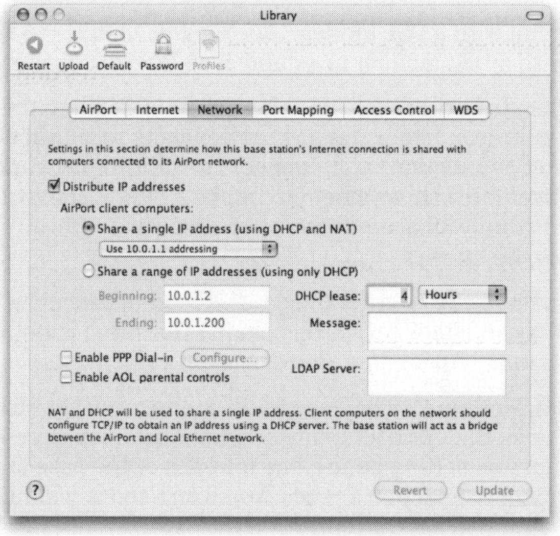

Figure 4-13: The Network tab. You see this tab again later in the chapter.

Different base stations may have somewhat different setting controls available in this pane. For example, an AirPort Express base station lacks the Enable PPP Dial-in check box and its associated Configure button, because that base station doesn't have a modem, which PPP dial-in requires.

The PPP dial-in feature can come in very handy because it enables a modem-equipped AirPort base station to share its Internet connection, as well as other resources on its network, with clients that dial in to the phone line to which the AirPort base station's modem is connected. For example, if you take a laptop computer on a business trip, you can phone into your home network and get any files you may have left behind, or print messages on your home network printer. You use the Configure button to set a username and password to join the network when you dial-in, as well as to set a few other phone-related options. Naturally, this feature is not available if your base station uses its modem connection to get its Internet connection: The station's phone connection will be busy every time you dial-in.

Most of the Network tab's settings have to do with whether the base station assigns network addresses to the network's clients, and how it assigns those addresses. When the base station assigns network addresses to AirPort network clients, it uses its built-in *DHCP server.*

A DHCP (*Dynamic Host Control Protocol*) server is a common technology used by network devices called *routers* to supply network addresses to the devices on a LAN. Your base station can do one of two things:

- Generate addresses and assign them to network clients.
- Assign addresses from a manually entered address range, which you may obtain from your ISP. Because a network may have more possible clients than addresses it can share, the DHCP server *leases* addresses to connected computers for a specified time: when the lease expires, the client computer requests an address again, and receives one if an address is available.

"Putting It All Together," later in this chapter, shows how a base station on a typical home network uses these settings.

Directing data with the Port Mapping tab

See that Technical Stuff icon to the left? It more or less pertains to this entire section about the base station configuration window's Port Mapping tab. Unless you plan to use a computer on your AirPort network as a server that can be accessed by the outside world — that is, any computer elsewhere on the Internet — feel free to skip right on by.

The *ports* to which the Port Mapping tab refers are not the physical, plug-a-cable-into-them kind of port like those on the back of your computer or on a base station, but the logical — that is, imaginary — numbered network ports that act more like channels on a TV. For example, Web pages usually arrive through port 80, port 25 often handles mail, port 3689 is the channel where other computers find shared iTunes music, and so on.

An AirPort base station that shares its Internet connection with the client computers on its wireless network can maintain an internal list of which computers on the network are supposed to receive what data from the Internet. It does this by keeping track of the requests that those computers make. For example, if a computer on the AirPort network requests a Web page, the base station remembers the computer that made the request so it can pass the Web page back to it from the Internet. This technique is referred to as *Network Address Translation (NAT)*, and the base station uses NAT when it shares a single IP address, using DHCP, which you set under the configuration window's Network tab as described in the previous section, "Sharing connections with the Network tab."

However, when a computer on the AirPort network acts as a server — for example, it has been set up as a mail server or a Web server — computers elsewhere on the Internet won't be able to find that server: Those computers can only see the AirPort base station itself, which shields the computers on its network from the rest of the world. And, unfortunately, the internal list that the base station creates for requests coming *from* the AirPort network does no good when requests come *into* the AirPort network.

You can use the Port Mapping tab to assign requests coming into the AirPort network through a particular port — such as requests for Web pages through port 80 — to go always to a specific computer on the AirPort network. In turn, this means you must give the specific computer a manually assigned IP number, which you set with System Preferences on that computer.

Figure 4-14 shows the settings pane under the Port Mapping tab. In the figure, the pane doesn't yet have any entries added to the list.

The action is all on the right, hidden in those six buttons:

- **Add:** Click this button to add an entry to the Port Mapping list.
- **Edit:** Select an entry in the list and click this button to edit it.
- **Delete:** Select an entry in the list and click this button to delete the entry.
- **Export:** Click this button to save the entries in the Port Mapping list in a file. If you need to set up several AirPort networks the same way, you can save the settings, and then apply them to a different AirPort network with the next button.

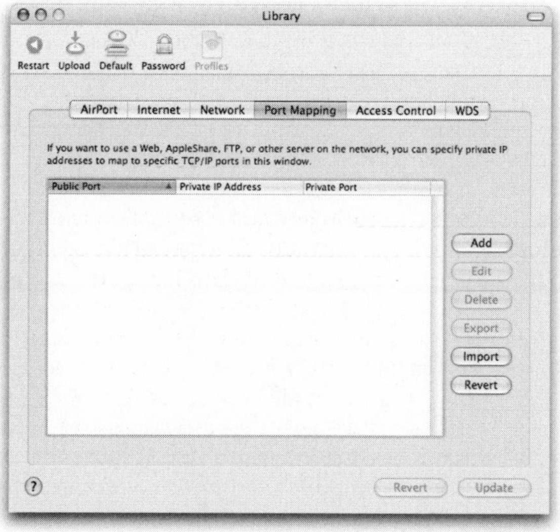

Figure 4-14: Use the Port Mapping settings to direct digital traffic.

- **Import:** Click this to import a Port Mappings file that someone, possibly you, previously saved by clicking the Export button.
- **Revert:** Click this button to return the Port Mapping list to the way it was when you first clicked the Port Mapping tab.

Figure 4-15 shows the sheet that appeared in the Port Mapping tab right after someone — okay, that someone was me — clicked the Add button on the right-hand side of the settings pane. The Public Port number on the sheet is a port number used in requests that arrive at the base station from computers on the Internet. The Private Address is the address of a Mac connected to the base station's network that will handle those requests arriving at the base station that use the specified public port number. The Private Port number is the port number on that Mac which actually receives the forwarded request. Think of it as the network equivalent of call forwarding.

For example, suppose you have a Mac on the AirPort network serving Web pages through port 8080. Further suppose that you want the AirPort base station to handle standard Web page requests from computers on the Internet, who expect Web pages to arrive through port 80. You enter 80 in the Public Port field, you enter 8080 in the Private Port field, and you enter the last part of the Web-serving Mac's IP address in the Private Address field. When the AirPort base station receives a Web page request from the Internet through port 80, it sends that request to the Web server's port 8080.

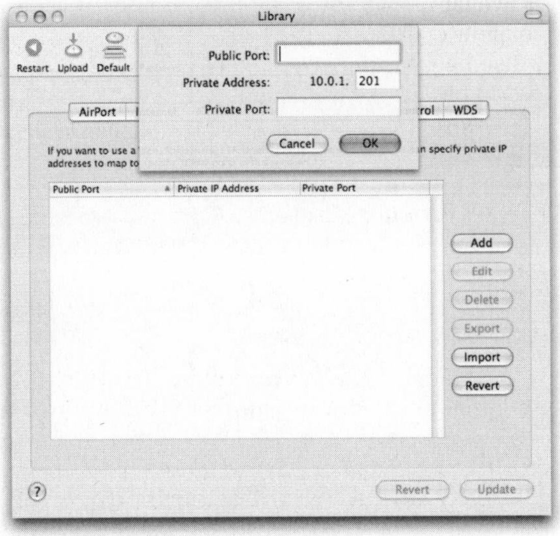

Figure 4-15: Mapping a public port to a private place on the AirPort network.

Gatekeeping with the Access Control tab

Although you can set up an AirPort base station with a password and even hide it from nearby computers, sometimes you may want to limit the computers that can join the base station's network to a specific list. That's where the Access Control tab, shown in Figure 4-16, comes into play.

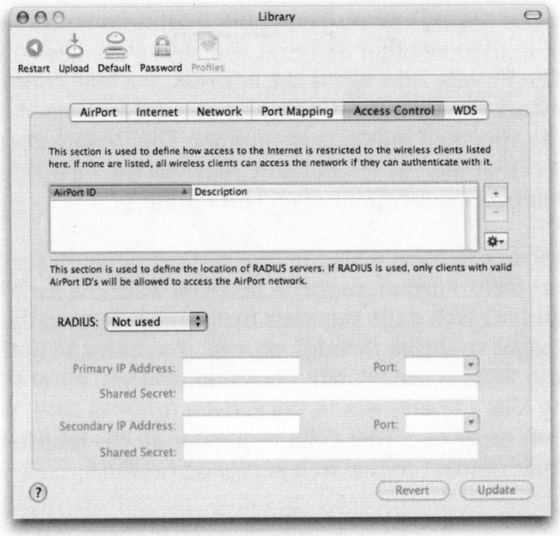

Figure 4-16: Add computers to this list to limit network access to a special, favored few.

You use the settings available under this tab to create and edit a list of the computers you intend to allow on the network and, optionally, to require that each of these computers also be approved by a *RADIUS* (*Remote Authentication Dial In User Service*) server, which provides additional security. Most home users won't have a RADIUS server: You tend to find such beasts in large businesses and university settings.

You can read more about how to secure an AirPort network, about RADIUS servers, and about how to use the base station configuration window's Access Control tab in Chapter 10.

Widening a network with the WDS tab

Apple advertises the range of an AirPort base station as being around 150 feet. Although a base station that doesn't have to transmit through walls, ceilings, floors, house wiring, bookcases, metal file cabinets, and other obstacles can usually attain such a broadcast range, not many base stations broadcast in an obstacle-free environment. Often, base station ranges fall far short of 150 feet, and even when they don't, some wireless networks may need to extend beyond 150 feet — how else can you log on from the pool house behind your Malibu mansion?

When you need a larger network than a single base station can provide, you can use more than one AirPort base station to extend the range of your wireless network by creating a *Wireless Distribution System (WDS)*. Figure 4-17 shows the settings available under the WDS tab. As you can see, the settings pane explains the different roles a base station can play in a WDS.

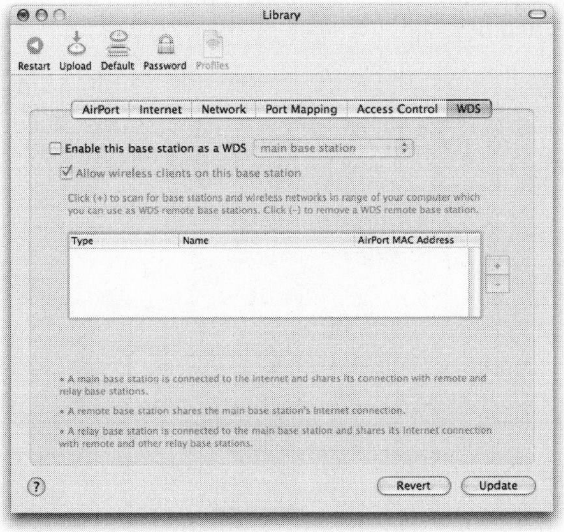

Figure 4-17: Build a larger wireless network by creating a Wireless Distribution System.

If you can't read the small type, here are the three roles a base station can play:

- **Main:** Provides the Internet connection for all the other base stations and computers on the wireless network.
- **Remote:** Shares the Internet connection from the main base station with all the wireless clients in the remote base station's broadcast area.
- **Relay:** Shares the Internet connection from the main base station both with the wireless clients and with any remote base stations in its broadcast area.

Chapter 5 delves deeply into the specifics of setting up a WDS and explains how to use the settings under the base station configuration window's WDS tab.

Rocking out with the Music tab

The AirPort Express Base Station includes a collection of technologies that Apple brands as AirTunes. As Chapter 3 points out, this base station comes with some music-savvy circuitry and an audio output port so you can play music from your iTunes library through a home entertainment system or a set of powered speakers. For example, I can play the music collection I keep on my iMac in our office wirelessly through our living room stereo with an AirPort Express. When you bring up the base station configuration window for an AirPort Express, it has a Music tab, which you use to set up wireless music sharing.

Figure 4-18 shows the settings available under the Music tab. The pane makes a good attempt at explaining itself, but you don't have to rely just on it: Turn to Chapter 7 for a walk through the settings on the tab that transforms iTunes into your wireless jukebox.

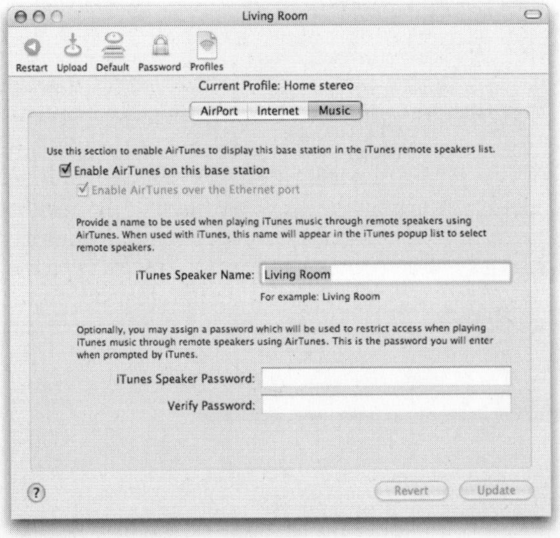

Figure 4-18: You can find a Music tab in the AirPort Express configuration window.

Connecting to the Internet

As "Going online with the Internet tab" in this chapter explains, the Connect Using pop-up menu under the base station configuration window's Internet tab provides a number of Internet connection options, which can vary depending on the base station being configured. This section describes the connection options provided by the AirPort Admin Utility in more detail.

Dialing in with the built-in modem

Of the AirPort base station models that Apple has released to date, all but the AirPort Express and the Power over Ethernet (PoE) AirPort Extreme models have modems that you can use to establish an Internet connection — see Chapter 3 for more information about AirPort base station models. The base station modems use the PPP (*Point-to-Point Protocol*) connection method to establish the connection with your *Internet Service Provider (ISP)*.

PPP is a connection method that directly connects two points, also known as *nodes*. In this case, one point is your AirPort Base Station and the other is a modem at the ISP. The protocol provides authentication methods, which usually require the node requesting the connection to supply a username and a password.

For those base stations that have modems, the two possible modem setting panes provided by the AirPort Admin Utility look identical: Figure 4-19 shows the pane for the Modem (V.90) modem connection standard. The difference between the two modem connection standards is in their speed and reliability: Modem (V.34) standard works with older and somewhat slower modem connections, but can handle noisier phone lines than the somewhat faster Modem (V.90) standard.

Here are the kinds of connection speeds you can get with a modem connection: A V.34 modem connection can download the amount of text contained in one-and-a-half pages of this book in about a second, and a V.90 connection can download about two pages of such text in the same time. When an AirPort shares a modem connection to the Internet with its wireless clients, each client shares a connection that only provides enough speed for some of the more simple Internet uses, such as sending and receiving e-mail, or accessing Web pages that don't contain a lot of pictures or rich media.

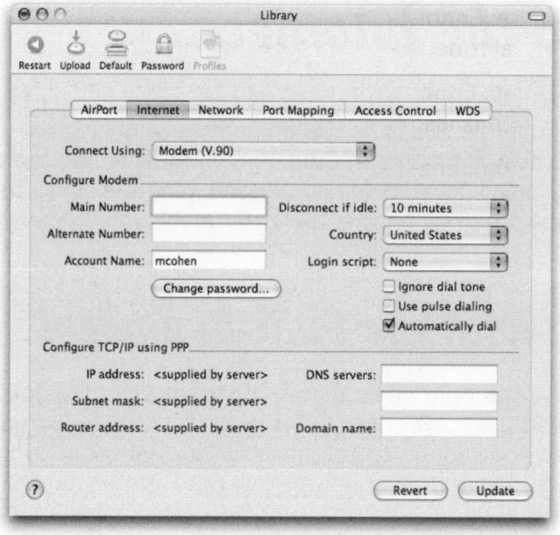

Figure 4-19: Sharing a modem connection to the Internet is a leisurely experience.

The modem pane provides these settings:

- **Main Number/Alternate Number:** Enter your ISP's connection phone number in the Main Number field and an optional second number in the Alternate Number field should the first number produce a busy signal.
- **Account Name:** Enter the account name your ISP gave you in this field.
- **Change Password:** Click to enter your Internet account's password in a sheet.
- **Disconnect if Idle:** Use this pop-up menu to choose how long the AirPort's modem will wait before hanging up the phone after it detects no Internet use by any of the base station's wireless clients.
- **Country:** Select the country you are in from this pop-up menu so the modem can use the appropriate dialing and connection commands.
- **Login Script:** Use this menu to select a script, supplied by your ISP, that may be required to facilitate the login process. This menu almost always says None. What's more, the AirPort Admin Utility has no feature to add a script, either. You need to use the AirPort Setup Assistant and have it copy your modem connection settings to the base station: If your Mac's modem settings use a script, the Assistant can copy it to the AirPort Base Station. Such scripts, by the way, can be found, if they exist at all, in the Terminal Scripts folder inside your Mac's Library folder. Chapter 3 describes the AirPort Setup Assistant.

- **Ignore Dial Tone:** Enable this if for some reason your phone line does not provide a dial tone, which is pretty rare, really.
- **Use Pulse Dialing:** Enable this if your phone system does not support touch-tone, which is also pretty rare.
- **Automatically Dial:** Enable this when you want the base station to call your ISP whenever the base station detects that one of its clients is attempting to access the Internet.

The bottom of the modem pane, which you use to configure the network connection after the base station has created it, provides three fields. You normally leave these fields blank, unless your ISP tells you otherwise:

- **DNS Servers:** Use these two fields to specify *Domain Name System (DNS)* servers. DNS servers translate user-friendly Internet addresses like `www.example.com` into the numeric addresses they actually represent. Usually, the ISP provides its own DNS server.
- **Domain Name:** Use this field to enter the domain name, such as `example.com`, that the base station adds to any Internet address that lacks one. This option is used quite rarely.

Accessing America Online

One of the most popular ISPs in the United States is America Online, which serves tens of millions of dial-up customers daily. Unlike most ISPs who serve dial-up customers, America Online uses a special method for establishing and maintaining dial-up connections — a connection method incorporated into the America Online software installed on customers' computers. However, the connection method used in the America Online software is proprietary, causing problems for devices like the original AirPort Base Station, which uses the PPP standard for dial-up connections. The Snow and Extreme AirPort base stations, though, contain software licensed from America Online to allow these base stations to dial and connect to America Online directly.

Figure 4-20 shows the AOLnet V.90 version of the America Online connection method settings, which are identical to the AOLnet V.34 settings. Like the regular modem connection methods described in the previous section, the only difference between the two America Online choices is the speed at which the base station's modem transfers data between itself and America Online.

Here are the options you can set for an America Online Internet connection:

- **Main Number/Alternate Number:** Enter the local America Online dial-up phone number for your chosen America Online connection method in the Main Number field: America Online sometimes provides different numbers to dial for V.34 and V.90 connections. Enter an optional second number in Alternate Number field should the first number produce a busy signal.

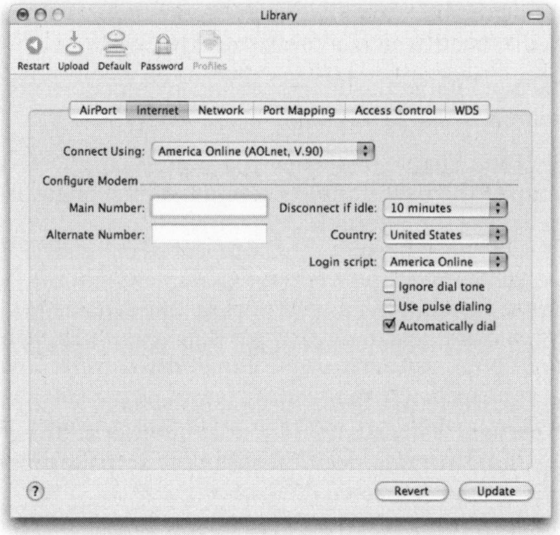

Figure 4-20: Have your base station dial up AOL with this connection method.

- **Disconnect if Idle:** Use this pop-up menu to choose how long the AirPort's modem will wait before hanging up the phone after it detects no Internet use by any of the base station's wireless clients.
- **Country:** Select the country you are in from this pop-up menu so the modem can use the appropriate dialing and connection commands.
- **Login Script:** This always lists the America Online script and you should never need to touch it.
- **Ignore Dial Tone:** Enable this if for some reason your phone line does not provide a dial tone.
- **Use Pulse Dialing:** Enable this if your phone system does not support touch-tone dialing.
- **Automatically Dial:** Enable this when you want the base station to call America Online whenever the base station detects that one of its clients is attempting to access the Internet.

REMEMBER

You may be wondering, "What about the America Online member name and password? How does the base station know what they are? And how can it log in to America Online without them?" Well, the answer is, the base station can't. When you use an America Online connection method, the only thing the base station does is dial up the America Online number and get things ready for you to log in using your America Online software on your Mac, just as you used to do when you connected to America Online directly from your Mac. This also means that if you select one of the America Online connection methods for your AirPort base station, all users on the AirPort network who want to share that connection must have their own America Online accounts and use the America Online software to log in.

Connecting with a DSL modem

One of two popular high-speed Internet connection technologies available to home users, DSL (*Digital Subscriber Line*) provides a fast Internet connection over regular telephone lines. DSL service from an ISP usually includes a DSL modem, which you connect to a telephone jack. DSL modems also include an Ethernet port, which you connect to your base station's WAN port.

TECHNICAL STUFF

PPPoE extends the PPP connection method so that a high-speed network connection over an Ethernet port, such as a base station's connection to a DSL modem, can use the same connection method as a low-speed modem — only, of course, faster. As with PPP, the two points involved in a PPPoE connection are your ISP and your base station. And, as with PPP, PPPoE connections require that the connecting point, such as an AirPort base station, identify itself with an account name and password before the ISP establishes the Internet access.

Figure 4-21 shows the settings pane for a PPPoE connection — if it looks familiar to you, that may be because you saw it earlier in this chapter as Figure 4-12. When you obtain a DSL account from your ISP, you normally receive information that helps you make the right settings in this pane.

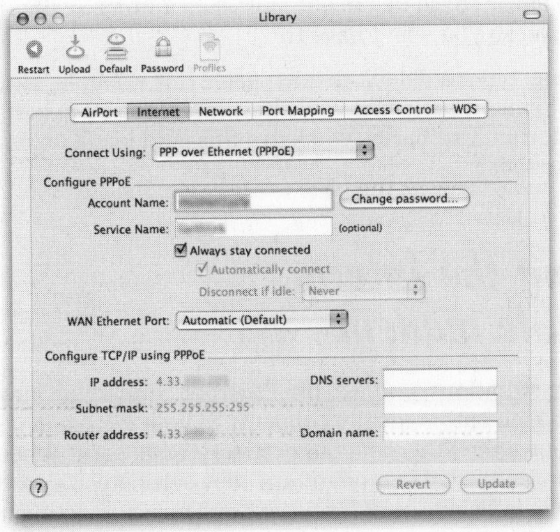

Figure 4-21: You've seen this before — it's the PPPoE connection settings from Figure 4-12.

Here are the PPPoE connection settings that you can modify:

- **Account Name:** Enter the account name you received from your ISP when you subscribed to your Internet service.
- **Change Password:** Click this button to bring down a sheet in which you enter the account's password.
- **Service Name:** You can usually leave this blank unless your ISP tells you otherwise. If so, the ISP's instructions tell you what to put here.
- **Always Stay Connected:** Unlike dial-up modem accounts, you can usually leave your DSL modem continually connected to your ISP, so you can leave this setting enabled unless your ISP recommends otherwise.
- **Automatically Connect:** If your DSL modem doesn't always stay connected, enable this setting to have the base station reestablish the connection whenever it detects that a client on the AirPort network requires an Internet connection.
- **Disconnect if Idle:** If you want your DSL modem to disconnect when the base station detects no Internet activity on the AirPort network, use this pop-up menu to choose how long the base station waits before telling the DSL modem to disconnect.
- **WAN Ethernet Port:** Use this pop-up menu to set the speed at which the base station's WAN port sends and receives data from the DSL modem. Usually, you need to set the WAN port speed only when connecting to a very old and finicky DSL modem that can't automatically detect a connection's speed.
- **DNS Servers:** You can enter the Internet addresses of two DNS servers you want to use in these fields, but most ISPs automatically supply DNS service for you so you won't have to.
- **Domain Name:** Enter a default domain name (for example, `example.com`) in this field to add it to partial Web site addresses you may type while using the Internet. Ordinarily, you leave this field blank unless your ISP suggests otherwise.

Channeling the Internet with a cable modem

In nearly all home situations, you use the Ethernet connection method with a cable modem, which automatically connects to your ISP when you turn it on. Figure 4-22 shows the Ethernet connection settings pane. When connecting your base station to a cable modem, the only setting you usually need to choose in this pane is Using DHCP from the Configure pop-up menu: The cable modem supplies the necessary settings for the pane's various fields.

For a cable modem Internet connection, the cable modem is either a router that includes its own DHCP server, or is directly connected to a router and a DHCP server at the cable company. Your AirPort Base Station gets its Internet address from that DHCP server. See "Sharing connections with the Network tab" earlier in this chapter for more about DHCP servers.

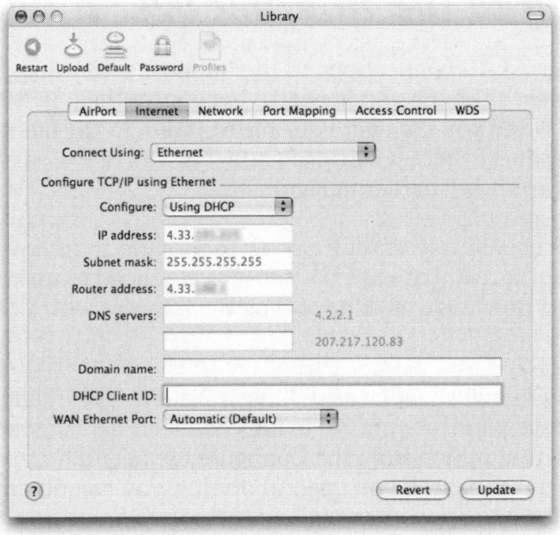

Figure 4-22: Cable modems usually use the Ethernet connection method.

Here are some additional settings, other than selecting DHCP from the Configure pop-up menu, that you may need to make for a cable modem connection:

- **Domain Name:** You can enter the domain name, such as `example.com`, to have the base station add it to any partial Internet address that lacks a domain name. Most people don't use this option unless their ISP tells them otherwise.
- **DHCP Client ID:** You seldom need to put anything into this field, but if your ISP tells you to, you enter the MAC (*Media Access Control*) address of base station's WAN port here. You can find the WAN port's MAC address in AirPort Admin Utility's Select Base Station window, as explained in "Opening the Admin Utility" earlier in this chapter.

  You can also find the AirPort base station's MAC address quickly by choosing View⇨Summary. Choose View⇨Details to return to the base station configuration window.

- **WAN Ethernet Port:** This pop-up menu offers settings for the speed at which the base station's WAN port sends and receives data from the cable modem. Your ISP or the cable modem's instruction manual should tell you if you need to choose a particular setting.

Connecting to the Internet over a LAN

Few residences have LAN connections to the Internet, other than, say, some college dorm rooms, although you may also encounter them in hotel rooms when you travel. When you connect your base station to the Internet over a LAN, you use the same Ethernet settings pane that you use for a cable modem connection, shown earlier in Figure 4-22.

Such connections usually use a DHCP server to provide an Internet address to the base station, and in that case the settings you need to make are the same as described previously in "Channeling the Internet with a cable modem."

However, you may not be so lucky as to have a DHCP server filling in the fields of this settings pane for you: Some LAN connections may require that you choose a different option from the Configure menu and that you fill out the pane's settings manually. If you need to do this, you can obtain the necessary settings for both the pop-up menu and the pane's fields from the network's administrator. (Of course, if the LAN is part of a business network, and you *are* that LAN's administrator, you already know what settings to use — but if you don't, you'd better hope that the boss doesn't find out.)

Putting It All Together

Though not as complex as large business networks, small office and home office networks — called *SOHO networks* by the computer industry, which really doesn't know what it would do without its acronym generators — don't lack for complexities of their own. SOHO networks often mix and match equipment of various vintages and capabilities, and can include both wireless and wired network devices. You can use the various AirPort Admin Utility settings described previously in this chapter to hook together rather elaborate SOHO networks. All you need to know is which settings to apply in which situations.

To make it easier to wrap your mind around some of the issues you may encounter in setting up a SOHO network of your own, in this section I have invented the fictional Hackett family and given them a SOHO network that demonstrates many of the techniques and concepts you may use when you set up your own SOHO network. Your home network may not match theirs in all particulars, but you probably can see some similarities.

The Hacketts have the following equipment on their SOHO network:

- **iMac G5:** This is Saffron Hackett's new computer. She runs her home business on it. It has a built-in AirPort Extreme card.
- **Sage iMac G3:** Saffron Hackett used to run her home business on this computer until she bought the iMac G5. Although the Sage iMac can take a wireless card, Saffron never got one for it. Last year, she finally upgraded this computer to Mac OS X. It now acts as a file server and sits in a corner of the home office.
- **Titanium PowerBook G4:** Saffron's husband, Tad Hackett, uses this computer at home, at the company he works for, and on business trips. It has an AirPort card. Tad often uses it at his desk in the Hackett's home office.
- **iBook G3:** The son, 14-year-old Merbert Hackett, got this computer on a two-year loan as part of his school district's Digital Advancement Initiative program. He does his homework on it. It has an AirPort card.
- **iMac G4:** 17-year-old daughter Citronella Hackett bought this from money earned working in the food service court at the local mall last summer. It has an AirPort Extreme card. It is located in her bedroom.
- **LaserWriter 8500:** Saffron bought this printer years ago for her business and she continues to use it to this day. It has an Ethernet connection and uses AppleTalk over Ethernet.
- **Cable modem:** The Hackett home obtains Internet service from CoyoteCableCo, the local cable company serving their community. The cable modem resides in their home office.
- **AirPort Extreme Base Station:** Tad bought this so that he wouldn't have to get separate phone lines and Internet accounts for each of the children. It connects to the Internet through their cable modem.
- **Ethernet hub:** An inexpensive six-port hub that formerly connected Saffron's iMac G3 and the printer. Tad also formerly connected to it with his PowerBook until he bought the AirPort base station. Currently, the hub connects the iMac G3 file server, the printer, and the base station, which connects to the hub via the station's LAN port.

- **AirPort Express Base Station:** Merbert won this by being the seventh caller to a recent Summer Musicfest Hit Rodeo broadcast. His sister is jealous.

Doesn't look very simple, does it? Yet all these devices can be made to communicate with each other quite easily and happily on the Hacketts' SOHO network.

Configuring the base stations

The first step in putting together the Hacketts' network is to set the AirPort Extreme base station's AirPort configuration settings. "Exploring the AirPort tab" in this chapter describes all the settings that this tab offers.

Figure 4-23 shows the settings that the Hacketts have made under this tab. Although some of the settings, such as the Contact and Location fields, are optional, the Hacketts have provided this information anyway.

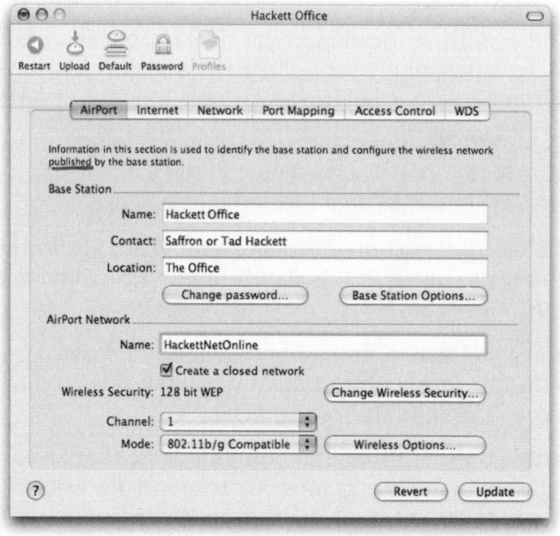

Figure 4-23: Configuring the SOHO base station.

Here are some of the other settings they've made and why they've made them:

- **Create a Closed Network:** The Hacketts have enabled this setting to provide additional network security, partly to compensate for the Wireless Security setting they need. (See the next item.) Setting up a closed network means that the network name won't appear in the AirPort status menu or in Internet Connect. The Hacketts have set their other computers'

Network preferences to automatically look for and join this network, as shown in Figure 4-24.

- **Wireless Security:** Although WPA provides stronger protection, the Hacketts have chosen 128-bit WEP in order to be more compatible with older wireless network devices. Note that this setting does not affect the Sage iMac or the printer, which are on the wired segment of the Hacketts' SOHO network and therefore neither require nor use a wireless security setting.
- **Mode:** The Hacketts have selected 802.11b/g Compatible, again so that the iBook and the PowerBook (which have AirPort cards that use only the 802.11b protocol) can connect to the wireless network.

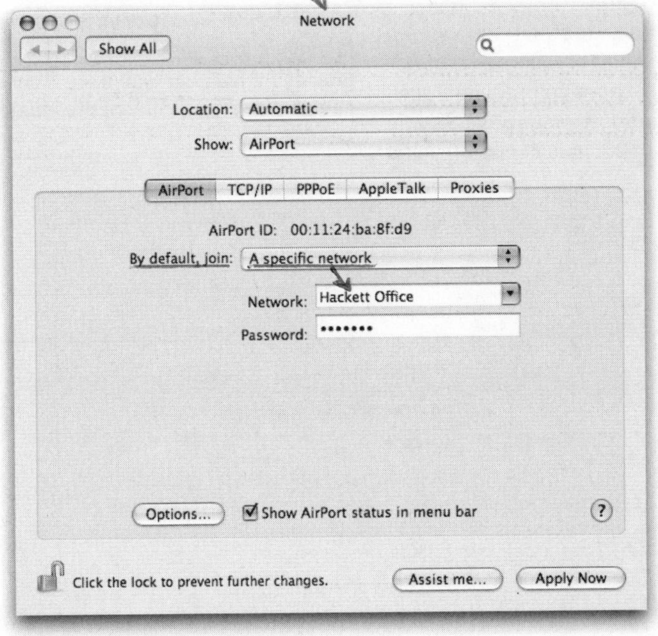

Figure 4-24: Specifying the default network to join.

Aside from the AirPort Extreme base station, the Hacketts' network also includes the AirPort Express base station that Merbert won. They have placed this base station in their living room and connected the AirPort Express's audio port to their entertainment system. Using AirPort Admin Utility, they have set that base station's Wireless Mode to Join an Existing Wireless Network (Wireless Client), as described in "Exploring the AirPort tab" earlier in this chapter. They did this because they simply want to use the AirPort Express to play music from any of their iTunes libraries: Their other base station has a wide enough range to cover their entire home, so they don't need to use the AirPort Express to extend their network or to establish its own network.

System Preferences → Network preference window
Apple menu → Location submenu

Location, location, location!

When you move a computer from one place to another, you usually have to make a visit to System Preference's Network preference window and adjust one or more settings to connect the computer to the network. If you regularly move the computer — for example, an iBook that you take from home to work or school and back almost every day — the visit to System Preferences can become quite tiresome.

You can avoid this regular preference-setting trek by creating a *location* in your network preferences for each network you regularly join. The Network preference window in System Preferences provides a Location pop-up menu for choosing between, or to create, specific network locations.

When you create a new location, your Mac's current network settings are saved under the location's name. Then, the next time you want to apply or to change those settings, choose the location from the Network preference window's Location pop-up menu (see the following figure).

You don't even have to open System Preferences to apply a location's settings: The Apple menu provides a Location submenu from which you can quickly choose any of the locations you've created.

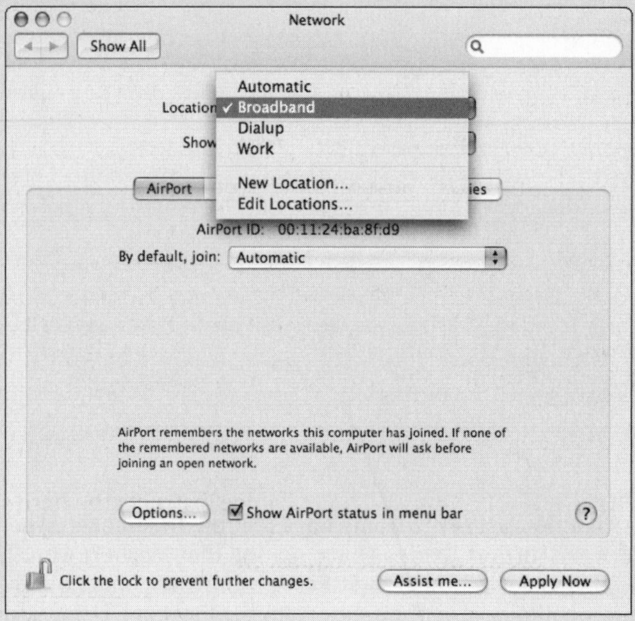

Under the Music tab in the AirPort Express base station's configuration window, the family has selected Enable AirTunes on this base station so that it can play music, and they have entered the name "Living Room" in the iTunes Speaker Name field. This is the name they select from iTunes' remote speaker pop-up list when they want to play music through the AirPort Express's connection to their entertainment system. Merbert wanted to add a password to the Express, but the rest of the family voted him down.

Connecting the network

The Hacketts have placed their main base station, the AirPort Extreme, in their home office. They have connected an Ethernet cable to its LAN port and have connected the other end of the cable to the Ethernet hub. This allows the wired devices on the network — the printer and the Sage iMac — to join the network created by the base station. They have connected the base station's WAN port to their CoyoteCableCo cable modem with a second Ethernet cable to give the AirPort Extreme base station its Internet connection. They have also connected the base station's modem port to their office's fax phone line. Now they have to modify the AirPort Extreme base station's Internet and Network settings with AirPort Admin Utility.

They have chosen Ethernet from the Connect Using pop-up menu under the Internet tab, and have chosen Using DHCP from the Configure pop-up menu. These are the most common settings for connecting a base station to a cable modem, as "Channeling the Internet with a cable modem" in this chapter explains, and works just fine with the Hacketts' CoyoteCableCo service. p. 111

With the Internet settings squared away, the Hacketts' next stop is the base station configuration window's Network tab, where they specify how the base station assigns network addresses to its network clients, as outlined in "Sharing connections with the Network tab" earlier in this chapter. Figure 4-25 shows the settings that the Hacketts have made under the Network tab. pp 98-99

Here are the settings the Hacketts have made and the reasons why:

- **Distribute IP Addresses:** This choice instructs the base station to act as a router and a DHCP server, assigning network addresses to devices on the AirPort network that request them. These addresses apply only on the AirPort network itself: no devices beyond the AirPort network can access them directly, which provides an additional measure of security to the Hacketts' network.

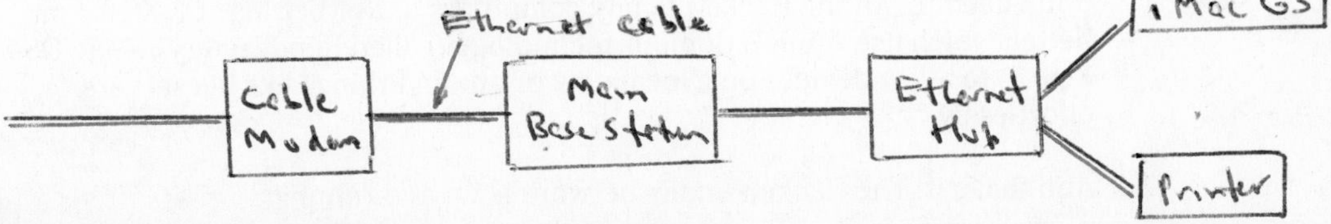

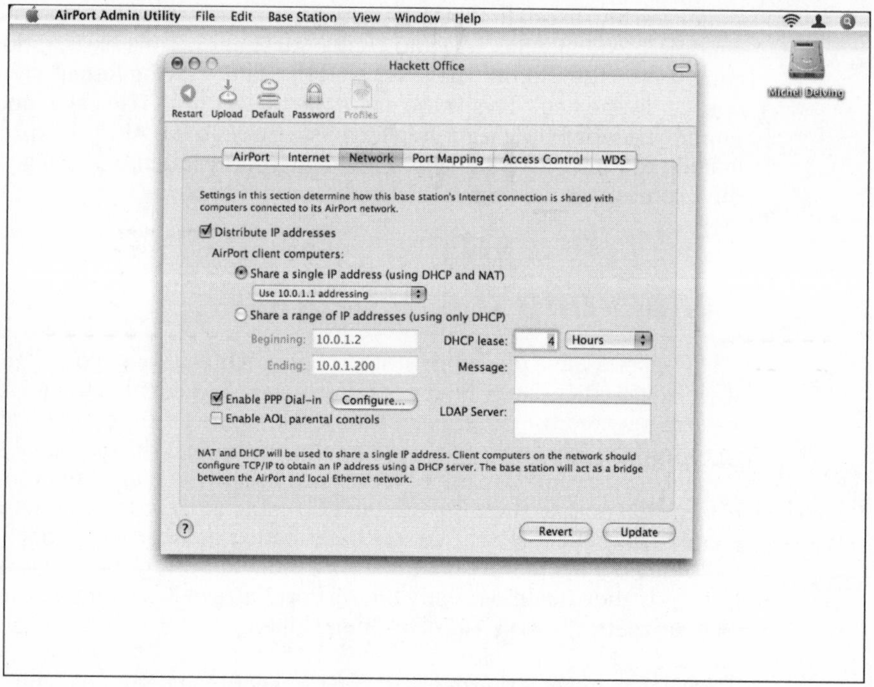

Figure 4-25: Setting an AirPort Express to act as an entertainment system component.

- **Share a Single IP Address (Using DHCP and NAT):** The single IP address being shared is the one that CoyoteCableCo assigns with *its* DHCP server to the AirPort Extreme base station. In turn, the base station uses its own DHCP server to assign addresses to devices on its AirPort network. The base station coordinates communications between devices on the AirPort network and the Internet using NAT (Network Address Translation), which is described earlier in "Sharing connections with the Network tab."
- **Enable PPP Dial-in:** The Hacketts have turned this option on and have set a username and password with the Configure button to the option's right in Figure 4-25. Tad uses this sometimes to call the home office's fax line to connect to the SOHO network with his PowerBook's modem when he's traveling and needs to get some files from the SOHO network's file server, the Sage iMac.

REMEMBER

And speaking of file servers: Although the Hacketts could set up port mapping as previously described in "Directing data with the Port Mapping tab" to enable the Sage iMac's file-sharing capabilities on the home network, they don't need to. All the Hackett family computers use either Mac OS X 10.3 or better, which use Apple's Bonjour technology (called Rendezvous in Mac OS X 10.3) to allow devices on a local network to find things like file servers effortlessly.

And that's it. The Hackett family network is up and running.

Chapter 5

Picking a Location . . . or Two . . . or More

In This Chapter

- Changing channels
- Envisioning the Base Station as light bulb
- Raising the antenna
- Roaming on the range
- Spreading out your network

A very old joke tells of a traveler in Maine who stops and asks a local resident how to get to a particular town. After going on for while with a long and detailed recital of circuitous directions, rich with local place names and obscure landmarks, the local pauses, scratches his head for a moment, and then says, "Yeh can't get theah from heah!" Sometimes wireless computer networking can seem a lot like that.

However, just like traveling through the real Maine and not the one of regional tradition, when it comes to networking you usually *can* get there from here. What you need are good directions and a willingness to explore alternate routes.

In this chapter, you find various ways of getting your data from there to here on your wireless network, no matter the number of hills it has to climb, the streams it has to cross, and roads it has to travel. And you won't have to pack a lunch.

Choosing a Channel

Your base station most likely is not perched high on a hill overlooking your town, or crowning a skyscraper above your city, yet it is a broadcast transmitter nonetheless, and a close relative to the radio and TV towers that have dotted the landscape for most of the last century. And like them, your base station broadcasts only over a limited frequency range, portioned into channels and assigned by government decree.

Colors that you can't see

Pull up a chair for a quick and painless combination physics and history lesson to help you understand base station channels. It goes something like this: The signals that your AirPort Base Station broadcasts are nothing more or less than light — just light in a range of colors that you can't see. Each color that your eyes detect has a specific *frequency*, just like each radio station has *its* specific frequency. If you look at a rainbow, nature shows you the colors your eyes can detect neatly arranged by frequency: violet has the highest frequency, red has the lowest frequency, and green is right in the middle. Radio signals simply have much, much lower frequencies than the rainbow's deepest red, the lowest frequency of light that your eyes can see. A *frequency range* refers to a range of colors, visible or not, that are close together, like the range of colors, say, from orange-yellow to yellow in a rainbow.

The reason scientists use the word *frequency* when talking about light — or, to be technical, *electromagnetic energy* — is because all light is made up of waves. The frequency of any particular color refers to how many of those light waves pass by each second. The higher the frequency, the more waves; the lower the frequency, the fewer waves. The deepest red light you can see sends about 428 trillion waves past your eye each second. An AM radio station midway on your dial sends 1 million waves each second past your radio's antenna — its "eye," if you will.

Soon after radio broadcasting was invented, national governments began taking control of what is often called the *broadcast spectrum*, that range of colors much, much deeper than the deepest red you can see, which are used for AM radio, FM radio, television, and all the other wireless broadcast uses that we've come to know and love — or, at least, tolerate — here in the first decade of the 21st century. To avoid confusion among devices and manufacturers, governments have dictated which frequency ranges can be used for which purposes. They started by assigning the frequency ranges allowed for AM radio, and as other broadcast technologies developed over time, governments allotted those technologies their portions of the broadcast spectrum, too.

The frequency range allotted to wireless base stations like your AirPort Base Station falls between 2,400 – 2,483 MHz, and that range is further divided up into 11 channels that run from 2,412 MHz to 2,462 MHz, spaced 5 MHz apart. By comparison, FM radio's frequency range runs from 88 MHz to 108 MHz, with individual FM station frequencies spaced 0.2 MHz apart.

And, finally, what do those letter-number combinations that I just reeled off mean? *Hz* is short for *Hertz*, an abbreviation that pays homage to the physicist Heinrich Hertz, and refers to a single cycle or wave: a frequency of 1 Hz means that one wave passes by each second. *MHz* is the abbreviation for *megahertz*, where *mega* means one million: A frequency of 2400 MHz means that 2.4 billion waves pass by each second.

Browsing your local airwaves

Wireless network base stations broadcast over any one of 11 separate channels. When, as described in Chapter 3, you set up your base station with the AirPort Setup Assistant, you don't get to choose which channel your base station uses: The Assistant makes that choice for you. In most cases you don't need to worry about, or even know, the channel that the Assistant selects because the Assistant is designed to examine the airwaves for nearby wireless broadcasts and to choose an appropriate setting. But if you intend to set up your AirPort Base Station with the AirPort Admin Utility, as described in Chapter 4, and if you know or suspect that other wireless networks are located nearby, you need to find out what channels those other base stations use so you can avoid creating interference.

To be precise, the 802.11 standard (discussed in Chapter 2), which governs AirPort and AirPort Extreme networks, specifies 14 channels, but the number of base station channels you can use varies depending on the country in which you live. In the United States and Canada, you can use channels 1 through 11. In Japan, on the other hand, you can use only channel 14, and in Spain channel 13 is your only option.

You might think that it's not too hard to detect if nearby wireless networks exist. And, in the majority of cases, it isn't hard: The Internet Connect application can show you what's on the air in your neck of the woods on its Network pop-up menu, as shown in Figure 5-1.

Figure 5-1: This pop-up menu shows you nearby wireless networks and lets you join them.

Here's what you do to see this pop-up menu:

1. **Open your Application folder.**
2. **Open Internet Connect.**

3. **On the toolbar, click AirPort.**
4. **In Internet Connect's AirPort pane, click Turn AirPort On.**

   You need to do this step only if your AirPort card is turned off.
5. **Click the Network pop-up menu.**

TIP

Here's an even faster way to see nearby wireless networks: Put the AirPort status menu on your menu bar and use it any time you want to see the list of nearby wireless networks, as shown in Figure 5-2. The AirPort status menu provides some other useful functions, too.

- Turn your AirPort card on or off.
- Join one of the displayed networks.
- Enable Interference Robustness, if your base station and AirPort card support this feature.
- Create an *ad-hoc* network, discussed in Chapter 6.
- Open Internet Connect.

To get the AirPort status menu on your menu bar, in Internet Connect's AirPort pane click the Show AirPort status in menu bar check box.

Figure 5-2: You can also see and join nearby networks with the AirPort status menu.

These techniques work great for letting you see nearby networks — except for those networks that don't want to be seen. As you saw in Chapter 4, you can tell your AirPort Base Station to create a *closed network* with AirPort Admin Utility, and the configuration tools provided with non-AirPort Base Stations offer similar abilities. A closed network doesn't appear in Internet Connect's Network pop-up menu nor in the AirPort status menu's list of networks you can join.

Furthermore, neither Internet Connect nor the AirPort status menu shows you which channels are used by nearby networks, including your own. If you want to find out that information, you need to look somewhere else.

TIP

Alf Watt's iStumbler can show you detailed information for all the nearby wireless networks, whether closed or not, including the channels that they use. You can download the free iStumbler application from `www.istumbler.net`. The iStumbler window depicted in Figure 5-3 shows the information you get when you run the application in the presence of several wireless networks.

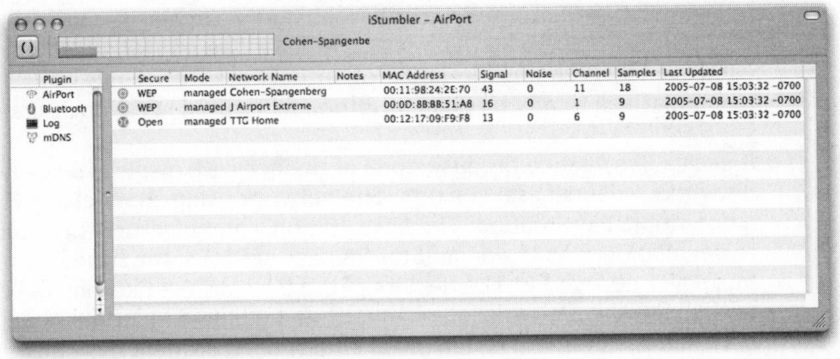

Figure 5-3: iStumbler gives you the lowdown on neighboring networks.

You see these other items in the iStumbler window:

- **Secure:** The type of security encryption, if any, used by the network.
- **Mode:** Whether the network is provided by a base station, indicated by *managed* in the figure, or whether it is an *ad-hoc* network created by an individual computer. See Chapter 6 for more about computer-to-computer networks.
- **Network Name:** The wireless network's name, which, for an AirPort Base Station, you assign with the AirPort Setup Assistant or AirPort Admin Utility. For a third-party base station, you use the setup utility provided by the manufacturer.
- **MAC Address:** *MAC* stands for *Media Access Control,* not *Macintosh*, and every network device has a unique MAC address. You can find out more about MAC addresses in Chapter 4.
- **Signal/Noise:** These two columns show the strength of the network's signal and the amount of *noise* — that is, distortion or static — detected in that signal. For a good wireless connection, you want much more signal than noise. This display updates periodically as iStumbler monitors each network's signal.

- **Channel:** The channel on which the wireless network broadcasts.
- **Samples:** How many times iStumbler has examined the network's signal.
- **Last Updated:** The last year, month, day, and time that iStumbler updated its report on the network. This timestamp also includes the local time zone shown as an offset from UTC (Coordinated Universal Time).

Keeping channels separate

In Figure 5-3 earlier, the three networks that iStumbler detected use channels 11, 1, and 6 respectively. As a result, at least five unused channels separate each network's channel from the other networks' channels. This arrangement has not arisen totally by chance, although it might seem so at first.

It seems like chance because none of the network owners have chosen their networks' channel assignments. The network using channel 6 in Figure 5-3 is a non-Apple base station: The base station's manufacturer originally decided to set channel 6 as the default for its base stations; the base station's owner has never changed it. The owners of the two AirPort Extreme base stations that create the networks on channel 1 and channel 11 have not explicitly assigned their networks' channels either: Each of them has chosen the Automatic channel assignment from AirPort Admin Utility's Channel pop-up menu, as shown in Figure 5-4. That choice is the reason why the channel assignments are separated as way they are.

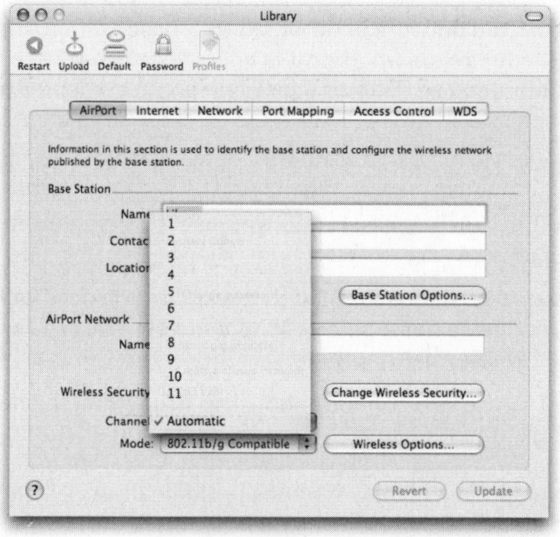

Figure 5-4: Setting an AirPort station's channel to Automatic can produce good results.

Five degrees of separation

The 802.11 standard assigns Wi-Fi network channels to frequencies that are spaced 5 MHz apart. However, the signal that a base station transmits does not limit itself to the exact frequency of the channel to which it has been assigned. Rather, its transmission *centers* on that channel's frequency, which means that the channel's assigned frequency is the frequency where that station's signal is the strongest.

However, some of that signal leaks over to nearby frequencies, too, in much the same manner that it does with AM radio, where you don't have to tune exactly to a radio station's frequency to pick up its broadcast. The closer you tune your radio to the radio station's assigned frequency, the stronger the signal becomes, and the farther away from that frequency you tune, the weaker the signal becomes.

In the case of a base station, you have to find a frequency 25 MHz away from the station's assigned broadcast frequency before you stop picking up any of its signal. Because channel frequencies have a 5 MHz separation, you have to select a channel number that is five numbers away from the base station's channel number to reach a frequency that is 25 MHz away from that base station's frequency.

Choosing Automatic from AirPort Admin Utility's Channel pop-up menu instructs an AirPort Base Station to look for all network channels already in use and to choose a non-conflicting channel. Because telecommunications specialists recommend a five-channel separation between wireless networks when their broadcast areas overlap, in order to avoid signal interference and improve each network's performance, the AirPort Base Station software tries to follow that recommendation whenever it can.

If you follow this recommendation stringently when you set up base stations, it turns out that the only channels you can use among the 11 available in the United States and Canada are channels 1, 6, and 11. Figure 5-5 illustrates how you can set up a bunch of neighboring base stations and not have adjacent base stations' channel assignments conflict.

The closer the channel numbers used by two or more neighboring base stations, the more the base stations' transmissions may interfere with each other. The more interference a wireless network experiences, the slower that network can transfer information among its clients. For the best network performance, you should assign your network to a channel that is five channels away from the channels used by neighboring networks.

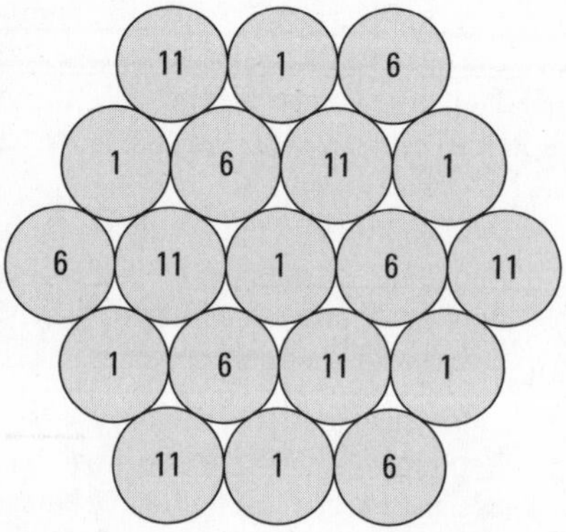

Figure 5-5: How to assign channels for a whole bunch of neighboring base stations.

Relaxing channel separation rules

Sometimes you can't maintain the recommended five-channel separation. If you live in an apartment building, for example, you have no control over the channel numbers your neighbors choose for their wireless networks. That does not mean that you are necessarily doomed to substandard network performance. Several factors affect how much base stations can interfere with one another.

The five-channel separation guideline applies most strongly when base stations are physically located close to one another: The farther apart they are, the less interference they can cause one another. Just as two radio stations can use the same frequency with no problem if they are out of range of each other, so, too, if two base stations are more than a few hundred feet apart it doesn't matter if they use the same channel: Their signals won't interfere because they don't reach far enough to cause a problem.

Nor do the base stations have to be completely isolated from one another. You can, for example, choose a channel number for your base station that's separated by only two or three channels from another station's channel — if the other base station sits far enough away so that its signal is significantly weaker than the signal coming from the base station that the computers on your network use. For example, if you place your computer no more than 25 feet away from your base station, and the other base station is 75 feet away from your computer, your computer can pick up that other base station's signal, but the signal won't be strong enough to interfere very much with the closer base station's signal.

Figure 5-6 depicts an iStumbler display that illustrates the most extreme case: two base stations, both using the same channel number within range of each other. But the situation appears more dire than it is. Although the computer running iStumbler does receive a signal from both base stations on channel 1, the signal strength from one base station exceeds the signal strength from the other by more than 300 percent. In such a case, it doesn't matter much to the computer that it can detect two base stations on the same channel: The base station creating the network that the computer uses is effectively "loud" enough at that computer's location to drown out the signal from the other base station without much resulting interference.

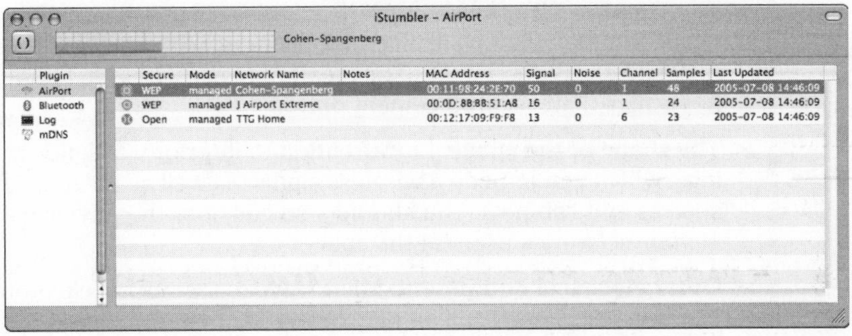

Figure 5-6: When channel numbers collide.

REMEMBER

In addition, modern base stations, like the AirPort Extreme and the AirPort Express, have more sophisticated signal-filtering abilities than earlier models, allowing base stations to be placed closer together without causing much interference. Some telecommunication specialists feel that a four-channel separation usually suffices to reduce interference from neighboring base stations to an acceptable level.

Whether or not your wireless network experiences interference from other wireless networks, therefore, depends upon the following:

- How physically close your network is to other wireless networks
- How numerically close the channel number used by your network is to the channel numbers used by other nearby wireless networks

Even if you can't adhere strictly to the five-channel guideline or even to the more relaxed four-channel guideline, you can usually find a channel for your network that provides acceptable, even if not theoretically perfect, performance. When in doubt, experiment.

Placing the Base Station

Where you place your AirPort Base Station has much to do with how well your AirPort network does its job. Obviously, the distance between the AirPort Base Station and the computers that communicate with it can affect performance, but distance is not the only factor you need to consider when you decide where to place your base station.

Among the other factors to consider when deciding on your AirPort Base Station's placement are these:

- **Nearby wireless networks:** "Choosing a Channel" in this chapter covers how to find nearby Wi-Fi networks and how to avoid of the performance issues these can create for your AirPort network.
- **The wired environment:** An AirPort network usually involves some wired equipment, such as network printers, wired work stations, and the AirPort Base Station itself. The location of the wires that connect these items may limit your placement options.
- **The shape of the space:** Pay attention to interior walls, floors, and ceilings, which can impede a wireless signal, as well as to things like metal shelves and cabinets. You should also consider the physical distribution of your network's users in the space.
- **Other electrical equipment:** Microwave ovens, electrical motors, and wireless telephones can all play havoc with your network.

Don't let this list of considerations make you think that an AirPort network is some sort of hothouse flower that needs just the right temperature, soil conditions, humidity, and piped-in background music to flourish and bloom. AirPort networks can thrive in most home and office environments. Paying some attention to the items in this list, however, can help make your network perform at its best, and possibly help you sidestep some easily avoidable problems.

Mapping the location

You may think you know the layout of where you live and where you work, but it never hurts to take a close look around anyway before you place your AirPort Base Station. A site diagram can help you better visualize how the AirPort network will fit into your environment and can help you keep track of all the environmental details you might want to consider.

The site diagram need not be a big production number, involving architectural blueprints, elevation projections, and the like. Just grab a pencil and a blank sheet of paper and sketch it out. Figure 5-7 depicts the kind of sketch I mean.

Although this sketch won't win any art prizes (unless it gets entered into an American-Primitive Bad Drafting competition), it contains the points your site diagram should include:

- Equipment locations
- Dimensions of the area to be networked
- Electrical outlets
- Phone, cable, and network jacks
- Physical obstructions, such as walls, filing cabinets, bookcases

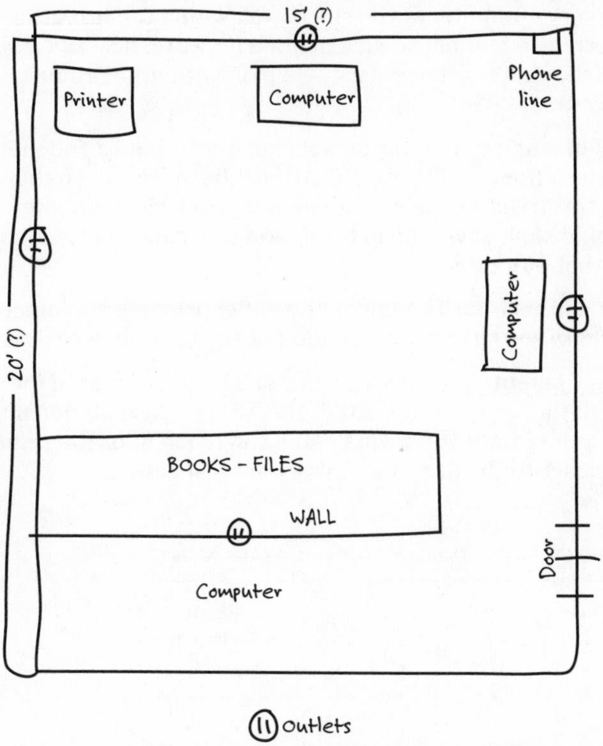

Figure 5-7: You don't need to be Frank Lloyd Wright to make a site diagram.

With this information in hand, you can more easily decide upon the best place to set your AirPort Base Station.

Figure 5-8 offers a version of the same site diagram shown in my hand-drawn Figure 5-7 sketch. It has been cleaned up so you won't have to decode my idiosyncratic printing style as you read the following example of how you might analyze a site.

In the figure, you can see a typical small-business site layout, consisting of a back-office area with two computers, a printer, some metal file cabinets, and a front office area with a third computer, separated from the back area by an interior wall.

Using this diagram, you can analyze the site and determine the conditions that could limit where the base station ends up:

- **Power:** Power outlets are placed at regular intervals along the wall, so the AirPort Base Station will probably be placed along one of the walls as well. A central ceiling mount would require extra wiring work.

- **Space:** Every computer at the site is well within the broadcast range of an AirPort Base Station, so an extended network, described later in this chapter, shouldn't be necessary — barring other insurmountable obstacles, of course.
- **Physical obstacles:** An interior wall, lined with books and metal file cabinets, divides the site. Placing the AirPort Base Station so that its signal must pass through the metal cabinets to reach the computer in the front office could slow down the network and cause intermittent connections for the front-office computer.
- **Internet connection:** The telephone outlet provides the office's Internet connection. The AirPort Base Station needs to be able to connect to it.
- **Other equipment:** A network printer sits in one corner of the office. To use it with the base station, either the AirPort Base Station must be placed so the printer's network cable can reach it, or the printer must be moved closer to the base station's eventual location.

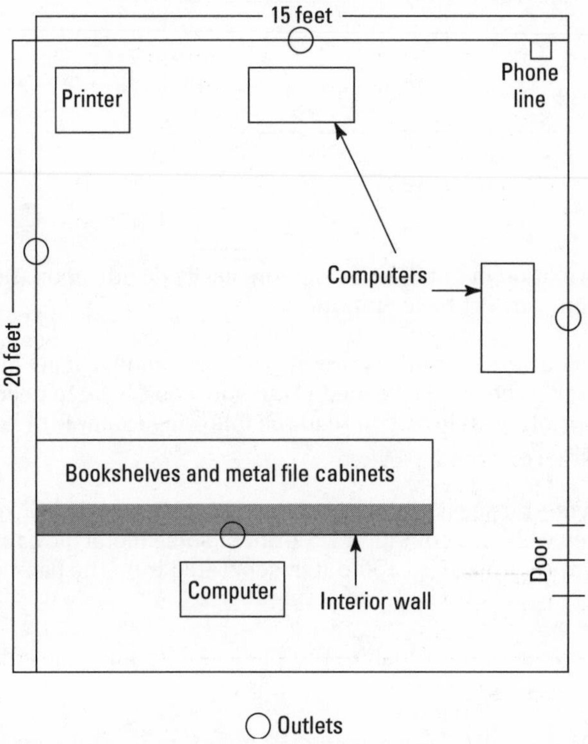

Figure 5-8: A neater version of the site diagram, provided as a courtesy to the reader.

The physical obstacles — the metal cabinets lining the interior wall — are the biggest problem the site diagram in Figure 5-8 reveals. Although AirPort signals can penetrate most interior walls and lose only a modest amount of signal strength, the metal file cabinets act almost like shields, reflecting rather than passing a wireless signal. Figure 5-9 offers a good placement solution.

As Figure 5-9 shows, placing the AirPort Base Station on the wall beside the doorway allows unobstructed line-of-sight connections between the base station and each computer.

To accommodate this placement, the office requires some slight rearrangement: In particular, the printer must be moved so its cable can reach the AirPort Base Station's *local area network (LAN)* port. Luckily, the cables connecting the AirPort Base Station to the phone line and the power outlet are long enough to reach; if not, longer cables can be obtained. In fact, a longer network cable can allow the printer to remain in its previous location, too.

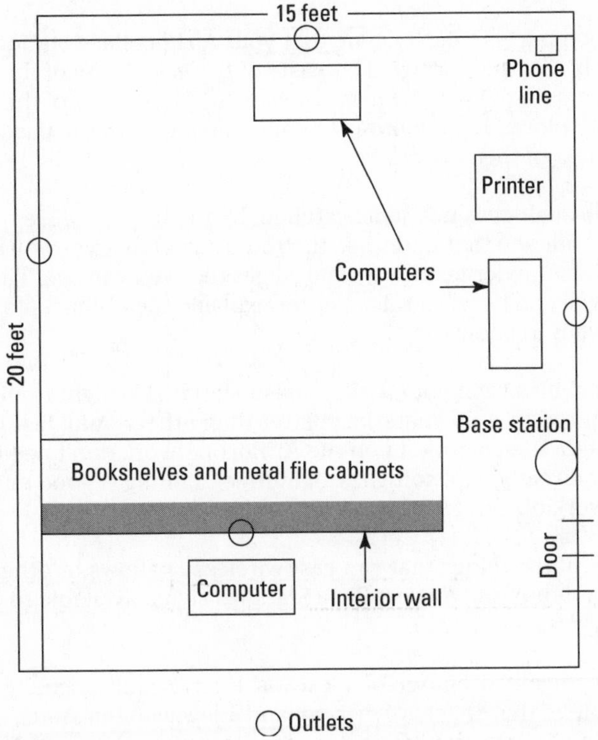

Figure 5-9: Proposed office layout based upon site analysis.

Don't forget about the third dimension — height — which the site diagram in this example does not indicate. If, in the example site, the metal cabinets that create a radio obstacle don't reach very high, you can place the base station along any wall in the depicted site as long as the AirPort Base Station is mounted higher than the cabinets reach. The interior wall by itself does not impede a normal Wi-Fi signal enough to cause the front office computer to experience reception problems.

Seeing shadows

When a computer on a wired network exhibits intermittent network connection problems, you can often trace the problem to a bad cable, a physical object that you can easily see and, usually, easily replace. In a wireless network, a weak wireless signal can cause the same problems as a bad cable. Weak signals are harder to see than a physical cable because to visualize them you have to use your imagination and, perhaps, a software tool.

The imagination part is this: Imagine that your AirPort Base Station is a colored light bulb, and that its signal consists of the brightness of that colored light. Computers on the AirPort network have to "see" some of the AirPort Base Station's colored light in order to communicate with the base station and with the rest of the network.

This light bulb analogy is not so far-fetched, by the way, because the same physical phenomenon that underlies the radio signal that your AirPort Base Station emits also underlies the ordinary light that you can see. The sidebar "Colors that you can't see" in this chapter explains these concepts in more detail if you want to know them.

The light that shines from your AirPort Base Station, like light from a light bulb, passes through some materials better than others. And, just like light from a light bulb, the computers on the AirPort network don't need to see the AirPort's light directly: The soft glow of indirect lighting is good enough to keep the network up and running.

Here are some of the things that can cast wireless shadows or otherwise reduce the amount of the AirPort Base Station's signal available to computers on the network:

- **Concrete, brick, or cinder-block walls:** Interior walls, usually consisting of a couple of thin sheets of drywall and a few wooden studs, are usually no more opaque to an AirPort Base Station's signal than is a cloth lampshade to a light. Exterior walls, on the other hand, tend to be thicker, and their materials more dramatically reduce the strength of the AirPort signal passing through them.

- **Lead-based paints:** These are not only unhealthy, for those of you who insist on eating paint-chips, but surprisingly opaque to radio waves. A few coats of lead-based paint can make even a thin interior wall cast a wireless network shadow as dark as the one cast by a brick wall. Latex-based paints, on the other hand, pose much less of problem for your AirPort Base Station's signal.
- **Metal plumbing and wiring conduits:** Your AirPort Base Station's signal has a harder time penetrating a wall filled with metal conduits than one that doesn't, which is something you should take into account when deciding where you want to put your base station.
- **Insulation:** Although radio signals pass more easily through many solid substances than visible light does, the more stuff a signal has to pass through, the weaker it becomes, so a wall filled with insulation (foil-wrapped insulation in particular) casts a darker shadow than one that doesn't. Furthermore, metal generally blocks radio signals more than wood or plaster, so foil-wrapped insulation acts almost like a sunshade as far as your wireless signal is concerned.
- **Glass:** Yes, glass. Although transparent or, at least, translucent to visible light, radio signals have a harder time passing through glass. Thick glass, or double-glass doors and windows, can sometimes significantly weaken a wireless signal.

REMEMBER

When you're trying to figure out why some part of your home or office seems to cause problems for wireless connections, keep the list of materials in mind. Sometimes moving the AirPort Base Station, or the computer, just a foot or two can improve a wireless connection dramatically as you move out of the wireless shadow that you can't see.

Performing a site survey

This approach described in the previous section works well enough for a small office or home setup, when just some visualization and a little fiddling about can solve your network's weak connection problems. However, when you have to provide AirPort network coverage over a larger area and to more than four or five users, you may want to take a more methodical approach to finding and resolving wireless *dead zones*.

To make a more methodical site survey, you need the following:

- **A Mac laptop with an AirPort card:** You use this to move around the area you intend to cover so you can measure the strength and quality of the wireless signal. Ideally, you should have a laptop with a well-charged battery and an AirPort Extreme card.

- **A detailed diagram of the area to be covered:** You don't need a highly refined architectural diagram, but you do need something reasonably accurate, drawn to scale, and large enough to contain the notes you'll make. The diagram should include the locations of the various computers in the area.

  TIP

  The simple methods are often the best: You can draw a properly scaled site diagram much more easily on a sheet of graph paper. If you have Excel, you don't even have to buy the paper because Microsoft makes several graph paper templates you can download and print. Go to Microsoft's Templates page at `http://www.microsoft.com/mac/resources/templates.aspx?pid=templates` and look for the Graph Paper link in the Students and School Activities category.

- **Apple's AirPort Client Monitor utility:** This program displays graphs of the strength, speed, and quality of the AirPort wireless signal over time. You can obtain the program from Apple's support site; go to `www.apple.com/support/airport/` and look for a link labeled AirPort Management Tools in the list of Additional Resources on the support page. The link goes directly to a disk image containing both the AirPort Client Monitor and the AirPort Management Utility.

  You may also want to obtain Apple's document *Managing AirPort Extreme Networks* which covers how to perform a site survey. The document is available in PDF form from the Apple Support page at `www.apple.com/support/airport/`.

Your main survey tool is the AirPort Client Monitor program running on the Mac laptop. Figure 5-10 shows the AirPort Client Monitor in action.

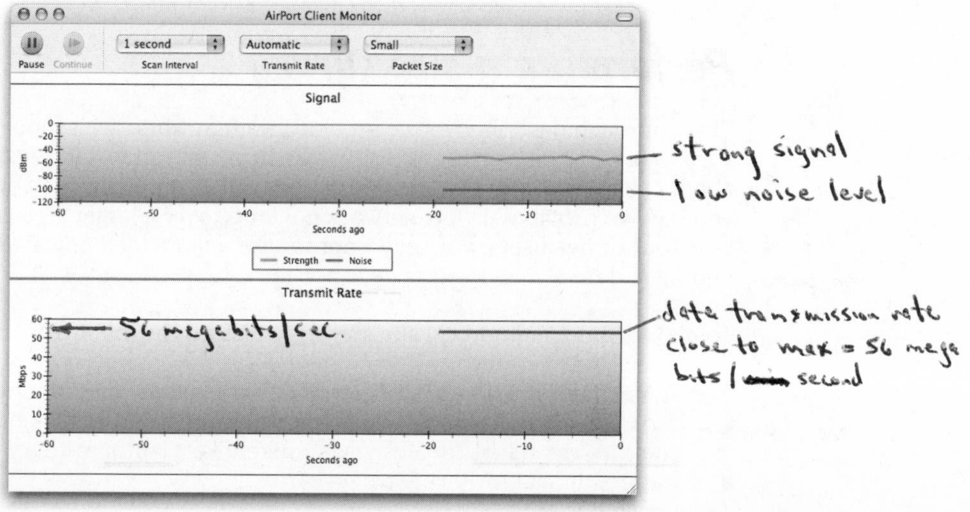

Figure 5-10: The AirPort Client Monitor measures wireless signal strength in real time.

The AirPort Client Monitor's window displays two graphs: The top graph shows both the strength of the wireless signal from the AirPort Base Station and the amount of noise in that signal, and the bottom graph shows the data transmission rate. The graphs in this particular figure, by the way, display an almost perfect set of readouts, showing a strong signal, little noise, and a data transmission rate very close to the AirPort Extreme's top speed of 56 mega bits per second — and a good thing, too, seeing as how that screen shot was taken on a Mac sitting a scant 20 inches away from the AirPort Extreme base station it was monitoring!

To perform a methodical site survey, do the following steps:

1. **Place the AirPort Base Station where you think it will work best.**

   I know. The whole point of this exercise is to find out where the base station *should* go, but you probably have a pretty good idea of the locations that are most practical: that is, those places where the AirPort Base Station can get both electrical power and an Internet connection. Choose one of those locations to start.

2. **Turn on the AirPort Base Station.**
3. **Take the laptop to a computer location in the survey area.**
4. **Use AirPort Client Monitor to see what kind of signal you get at that location.**

   What you want to see is something that looks more like the exemplary display in Figure 5-10 than the one depicted in Figure 5-11, which shows a weak signal barely stronger than the noise level, and a transmission speed that fluctuates well below the base station's maximum speed.

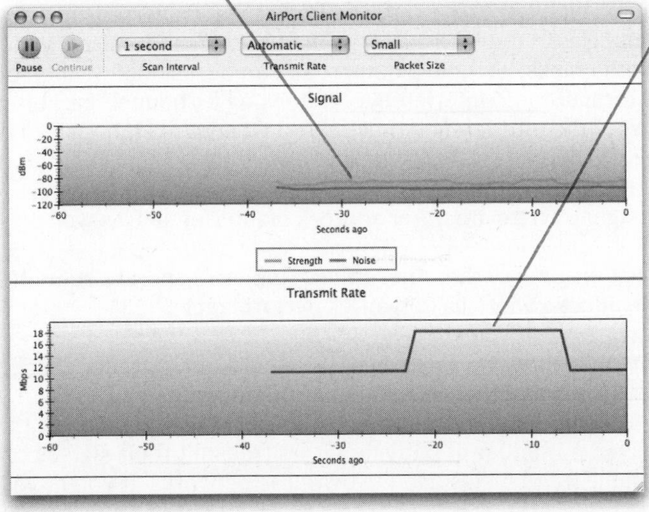

Figure 5-11: A marginal wireless signal looks something like this — or worse.

5. **Record the average signal strength, noise strength, and transmission rate on the diagram for the computer location.**

6. **Repeat Steps 4 and 5 for each computer location in the surveyed area.**

After you gather all the information you can, you have a map of where the network signal is strongest and where it is weakest in the survey area. If you get acceptable readings at all the computer locations in the area, you're done. Otherwise, go back to Step 1, choosing a different base station location, and repeat the survey.

Eventually, you'll determine one of these things:

- One of the AirPort Base Station locations you tried works satisfactorily for your purposes.
- One of the AirPort Base Station locations works well for most of the computers, and, with some rearrangement of computers in the area, you decide that you can live with it.
- No matter where you put the AirPort Base Station, you find you cannot adequately cover the area.

If that last item describes the results of your survey, you should consider getting a second AirPort Base Station and creating either a roaming network or a wireless distribution system. You can find out about both of these network-extending techniques later in this chapter.

Avoiding interference

The invisible highway that wireless signals must traverse is not always like a smooth and glassy thoroughfare — sometimes it's more like a winding road riddled with cracks, potholes, bumps, and an occasional washed-out bridge, all in the form of radio interference. "Choosing a Channel" earlier in this chapter discusses the interference that nearby Wi-Fi base stations can cause for your AirPort network, and "Seeing shadows," also in this chapter, describes some of the physical obstacles that can reduce or block your AirPort Base Station's signal. There are other sources of interference as well.

REMEMBER

Here's a list of some of the other things that can generate enough radio interference to affect a wireless network's performance:

- **Baby monitors:** Although a great boon to new parents, these walkie-talkies for the preverbal set use radio transmissions to send your baby's gurgles of joy and demands to have its diapers changed to the monitor's receiver. A number of baby monitors transmit their signals on or near the same frequencies used by wireless networks. If you have both a baby and a wireless network, check the specifications of the baby monitor before you buy and don't buy one that broadcasts on the 2.4 GHz frequency used by wireless networks.

- **Microwave ovens:** Some of the waves that a microwave oven uses to pop your popcorn, heat your coffee, and defrost your frozen entrees fall into the frequency range used by AirPort networks. Although microwave ovens are usually well shielded, some of their emissions do leak out. An AirPort Base Station located within a dozen feet or so of a microwave oven may suffer as a result. I can attest to this: Our AirPort Express base station sits about five feet away from the microwave oven in our kitchen, and, although an interior wall separates them, AirTunes goes off the air when the oven runs, making the choice between dinner and dancing a daily discussion.
- **Electric motors:** These devices don't use radio transmissions, nor are they designed to broadcast them, but the rotating coils and magnetic fields inside such motors can't help but generate some electromagnetic radiation all up and down the frequency spectrum — including the 2.4 GHz frequency range allotted to Wi-Fi networks. In short, you don't want to put your AirPort Base Station atop or beside your clothes dryer, washing machine, or refrigerator.
- **2.4-GHz cordless phones:** Yes, some cordless telephones broadcast on or near the same frequencies that AirPort Base Stations do. If you have one of these phones, expect it to degrade your AirPort network's performance every time you take or make a call. You should consider replacing such a phone with one that uses a different frequency range, such as the cordless phones that transmit at 900 MHz or those that broadcast at 5.8 GHz.
- **Television:** It's perfectly okay to watch TV and use your wireless network — my wife and I do that all the time. However, keep in mind that a television set contains a bunch of sophisticated electronic components, most of which can leak a bit of electromagnetic radiation. What's more, a television's picture tube contains an electron gun that fires subatomic particles at the surface of the screen to make it glow and to provide you with your nightly entertainment and that electron gun is, effectively, a radio transmitter. Although TVs usually don't emit much interference in the Wi-Fi broadcast range, they do emit enough interference to make it a wise decision not to place your AirPort Base Station atop or beside your television.

Upping the Signal

I'm old enough to remember the pre-cable TV days, when rabbit-ear indoor antennas crowned the tops of TVs, and to recall the fine art of twisting such an antenna into just the right arrangement to get a halfway-decent TV signal. The memory of the day that my father finally climbed a ladder to our roof to attach an outdoor antenna which looked like a modernist metal sculpture and which improved our TV reception by several orders of magnitude still fills me with nostalgic joy. So when I saw my first AirPort Base Station, I immediately wondered what the antenna looked like and whether I could get a better one. As it turns out, the antenna looks like an unassuming strip of bent metal, and, yes, you can get a better one, depending on the base station.

Adjusting antennas

When you connect to a network with a cable, your Mac's orientation doesn't much matter as far as the network connection goes: The Mac can face north, south, east, west, up, down, or sit on a diagonal, and the network connection remains just as strong. When you use a wireless network, on the other hand, your Mac's orientation can matter a great deal.

Just like the rabbit-ear antennas of yesteryear, how your Mac's internal AirPort antenna is oriented relative to your network's base station can sometimes dramatically alter your connection's performance. Just turning the Mac a few dozen degrees in one direction or another can cause the signal-strength bars on the AirPort status menu to leap up or sink down.

The AirPort antenna is what radio and electrical engineers call a *dipole antenna*, which basically means that it consists of two straightish segments connected at the center — the rabbit-ear antennas that I keep mentioning are dipole antennas, consisting of two very prominent segments. If you remember back to your high-school geometry class, you may recall that two intersecting lines, like a dipole antenna, define a plane. How the imaginary plane defined by your Mac's dipole antenna sits in relation to your AirPort Base Station's broadcast signal affects how much signal the dipole antenna receives. If that plane sits edge-on to the signal, it receives only a little signal, but if the plane faces the signal's source, it can catch much more of the signal.

Here's how some Macs have their antennas placed:

- **Mac mini:** The antenna is in the top of the case.
- **iBook:** The antenna is in the lid.
- **Titanium PowerBook:** The antenna is in the base. The metal case tends to block the signal, so Apple placed radio-transparent panels in each side of the PowerBook's base to mitigate the problem to some degree.
- **iMac G4 and iMac G5:** The antenna is behind the screen. Because the iMac G5 is nothing but screen, the placement of its antenna should come as no surprise to you.
- **G5 PowerMac:** The antenna is external, plugging into a port in the back.

REMEMBER

It doesn't help as much as you might think to know where a Mac's dipole AirPort antenna is placed: Keep in mind that radio waves can bounce off of some surfaces, so that orienting the plane of the Mac's antenna perpendicular to the AirPort Base Station doesn't always mean that you get the strongest signal that way. You may have to experiment to get the best results; the art of rabbit-ear adjustment is still with us.

Not only does a Mac's orientation with regard to the AirPort Base Station affect the strength of the network connection, so does the orientation of the base station itself. The AirPort Base Stations contain what Apple calls an *omnidirectional dipole antenna*, and although you might think that *omnidirectional* implies that the antenna broadcasts with equal strength in all directions, the antenna, in fact, doesn't. The signal radiates from the base station in a sort of flattened sphere, vaguely resembling the shape of a cream-filled doughnut, as in Figure 5-12.

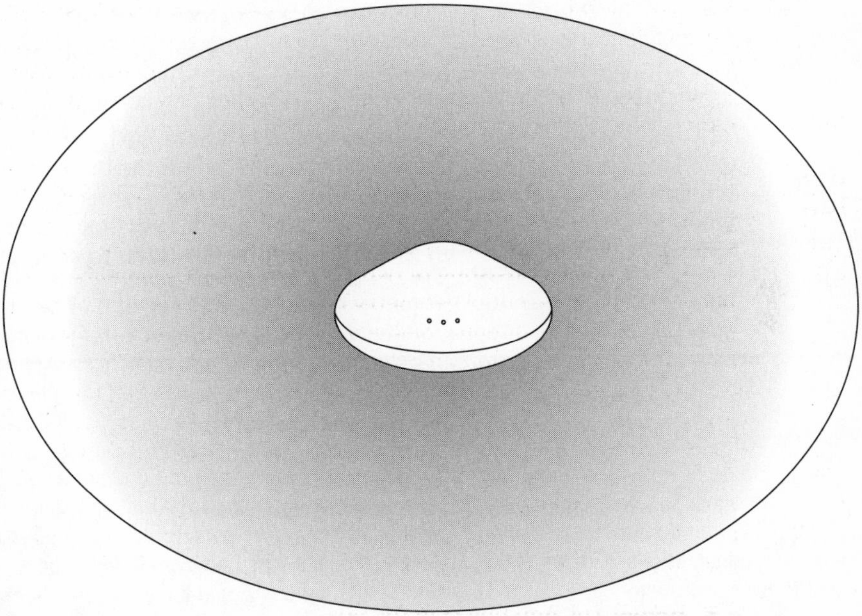

Figure 5-12: The base station broadcasts its signal in a flattened sphere.

REMEMBER

An AirPort Extreme base station sitting on its base sends out a stronger signal horizontally than it does vertically, so don't expect the base station's signal to cover adjacent floors in a building as well as it does adjacent rooms. If you mount an AirPort Extreme base station on a room's wall, using the plastic mount that comes with it, place the base station on the longer wall of the room because the signal spreads out along the wall more than it does across the room from the wall.

Using external antennas

The AirPort Extreme base station is the only one Apple has made to date that allows you to attach an external antenna. An external antenna can help your base station extend its range or direct its signal more effectively.

Two different kinds of antennas tend to be used with an AirPort Base Station:

- **Omnidirectional:** Like the AirPort's internal antenna, this kind increases the signal's range in all horizontal directions equally. Use it to increase the amount of coverage within a large room, such as an auditorium.
- **Directional:** This kind allows you to point the signal in a particular direction. Use this kind to aim a signal down a long hallway, or to relay a signal to another base station that forms part of a *Wireless Distribution System (WDS).* Wireless distribution systems are covered a little later in this chapter.

Apple's online Apple Store currently sells one of each kind of antenna, both from long-time Macintosh peripheral manufacturer Dr. Bott (`www.drbott.com`): the ExtendAIR Omni Antenna, and the ExtendAIR Direct Antenna. The first claims to increase base station range by 100 feet, and the second claims to send the signal as far as 500 feet in the direction the antenna is aimed. Dr. Bott is not alone among external-antenna vendors. QuickerTek (`www.quickertek.com`) also provides several antennas for the AirPort Base Station, as well as antennas you can attach to your laptop and desktop Macs.

Can't have just one

Among the most famous of amateur external AirPort antennas is Andrew Clapp's *cantenna*: Clapp converted a potato chip can into a very highly directional wireless network antenna and made base station history. You can find out more about his tasty project at `www.netscum.com/~clapp/wireless.html`. If you feel the urge to follow in Clapp's footsteps and embark on an antenna-tinkering adventure, or even if you just want to know more about how wireless base station antennas work, you should pay a visit to Constantin von Wentzel's "Sharing Apple Base Station Experiences" site at `www.vonwentzel.net/ABS/`. Von Wentzel has compiled detailed descriptions of the various Apple base station models accompanied by extensive internal photographs, well-researched discussions of antenna technology and theory, and step-by-step instructions for adding external antennas of varying sorts to the different AirPort Base Stations.

Adding an external antenna may not solve your wireless-connection woes. The AirPort Base Station's location, the channel on which it broadcasts, the shape and construction of the surrounding area, and the other equipment present in the area, can all affect a wireless network's performance. The antenna is just one factor in the wireless-network equation.

Adding antennas to older AirPort stations

The AirPort Extreme is not the only base station to which you can attach an external antenna if you feel determined to do so. Hardware tinkerers and hobbyists have come up with various ways to add external antennas to older AirPort Base Station models as well, and many of their experiments have been well documented on the Internet.

Adding an antenna to an older-model base station requires opening it up, which both voids the warranty, if one is still in effect, and exposes your otherwise operational base station to the risk of irreparable damage. If you don't feel willing to take on those risks, don't go down this road.

Assembling a Roaming Network

When a single AirPort Base Station can't cover the area that you want your wireless network to span — either because of interference, physical obstacles, area size, or a combination of these — you may have to bite the bullet and get one or more additional AirPort Base Stations to extend your network's range. After you have several base stations in play, you can use them to extend the network in several ways. A *roaming network* is one of those ways.

In a roaming network, as a client computer moves out of the broadcast range of one base station and into the range of another, the computer remains connected to the network. The client computer locks onto the signal of the strongest base station in the area, similar to the way a cell phone keeps its connection as it moves out of range of one cell tower, into another tower's range. Roaming networks, as their name implies, are designed with the peripatetic habits of mobile computer users in mind.

Roaming networks require that each base station in the network have a physical connection to the other. That means you can't just place the base stations anywhere you want: You must place them where they can connect to each other through a wired LAN. Roaming networks are thus well suited to situations where a wired LAN already exists, and the AirPort Base Stations are used to extend that LAN to wireless computer users. You might, for example, find roaming networks in a business setting, where a LAN already connects a suite of offices, or in a school where a LAN already connects neighboring classrooms.

A roaming network can share an Internet connection to its clients just as an individual base station can. But it doesn't have to: You can choose to establish a roaming network that remains isolated from the rest of the networked world. For example, a roaming network can connect the wireless computers toted around by salespeople at a car dealership, the wired computers used by the dealership's office staff, and the dealership's customer database and inventory tracking system. Nothing, in terms of data transmission, leaves the lot.

Setting up base stations in a roaming network

You don't need to do anything exotic with your AirPort Base Station to assemble a roaming network. Here's what you need to do to configure the AirPort Base Stations that you use to create your roaming network:

1. **Connect each AirPort Base Station to the LAN.**

   Use the LAN Ethernet port if you have a Snow or AirPort Extreme base station. For a Graphite or AirPort Express, you have only one Ethernet port, so use it. Chapter 3 describes the ports available on different model AirPort Base Stations.

2. **Open the AirPort Admin Utility.**

   You will perform the following steps for each base station that forms your roaming LAN.

3. **In the AirPort Admin Utility, in the Select a Base Station window, select a base station.**
4. **Click Configure.**
5. **Click the AirPort tab.**
6. **Give the base station a network name and network password.**

   For every base station you configure, you will use the same network name and password.

   By assigning the same network name and password to the network, the roaming computer regards the signals coming from each base station as all being from the same network. Chapter 4 tells how to set an AirPort Base Station's network name and network password.

   It should go without saying, but I'll say it anyway: When you set the network password for each AirPort Base Station, you should also make sure that each base station uses the same security protocol. If one uses WEP, they all use WEP; if one uses WPA, they all use WPA.

7. **Under the Network tab, turn off Distribute IP Addresses.**

   Figure 5-13 shows what the settings under this tab look like on an AirPort Extreme base station. Different model base stations may have some slightly different options under this tab. For example, an AirPort Extreme base station always provides a bridge between the wireless part of the network and the wired LAN, as the final sentence at the bottom of Figure 5-13 declares, and a Graphite base station has an Enable AirPort to Ethernet bridging check box that enables this bridging. In a roaming network you want to have such bridging available, so if your base station provides this check box under the AirPort Admin Utility's Network tab, put a check in it.

8. **Set the channel.**

   To avoid possible interference between adjacent AirPort Base Stations on your roaming network, try to arrange things so that neighboring base stations have channel settings several channels apart, as described in "Choosing a Channel" in this chapter. For example, if you set a base station in your roaming network to channel 1, set that station's closest neighbor to channel 4 or higher. Because stations in a roaming network tend to be spaced relatively far apart, you don't have to set their channel numbers to the optimal five channel separation described in this chapter's sidebar, "Five degrees of separation."

9. **Click Update.**
10. **Repeat Steps 3 through 9 for each base station you are configuring on the roaming network.**

These steps create a roaming network that extends a wired LAN to wireless computers.

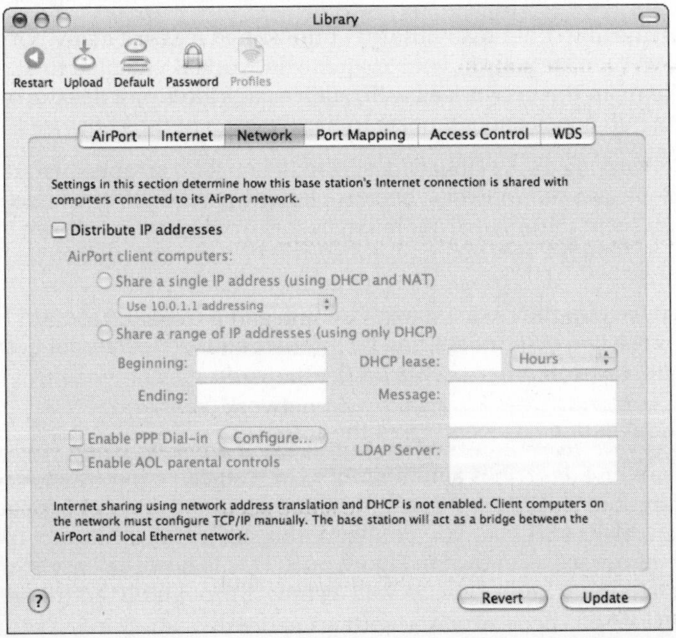

Figure 5-13: Turn off IP address distribution for a roaming network AirPort Base Station.

Building a roaming network with a DHCP-serving base station

The previous section describes how you can extend an existing wired LAN with a roaming network. You can also turn a more typical AirPort network into a roaming network by assembling a simple LAN to attach to it. In this kind of roaming network, one AirPort Base Station shares its Internet connection and distributes IP addresses with its AirPort network clients, including the other base stations.

Assembling a LAN involves acquiring cables, an Ethernet hub, and connecting the pieces:

1. **Plug an Ethernet cable into each base station.**

   You need one Ethernet cable for each base station. The Ethernet cable plugs into the AirPort Base Station's LAN port. If a base station has only one Ethernet port, use that port. Make sure that the Ethernet cables you obtain are long enough to reach from the base station to the Ethernet hub you set up in the next step.

   The AirPort Base Station that shares its Internet connection in your roaming network must be a model that has both a wide area network (WAN) and a LAN port. The other base stations in the roaming network do not need to have two Ethernet ports.

2. **Connect the other end of each Ethernet cable to an Ethernet hub.**

   You need to obtain an Ethernet hub with at least as many ports as you have base stations in your roaming network. If you plan to attach wired clients to the network as well, such as a printer or a desktop computer, the hub needs enough ports to handle each of those devices.

   Networking professionals describe the cabling arrangement described above as a *star topology*, because the Ethernet cables connected to the hub seem to radiate from it like beams from a star, as shown in Figure 5-14. Network professionals apparently have rather active imaginations.

Next, you set up the roaming network, assigning to one of the AirPort Base Stations the two tasks of sharing its Internet connection and of providing network addresses to other clients on the network:

1. **Connect one base station to the Internet.**

   Chapter 4 describes a number of ways you can connect an AirPort Base Station to the Internet. Use the method that works best for your situation. Make sure that you configure this AirPort Base Station to distribute IP addresses as shown in Figure 5-15. This is how the base station shares its Internet connection, as well as how it provides IP addresses to other devices on the network.

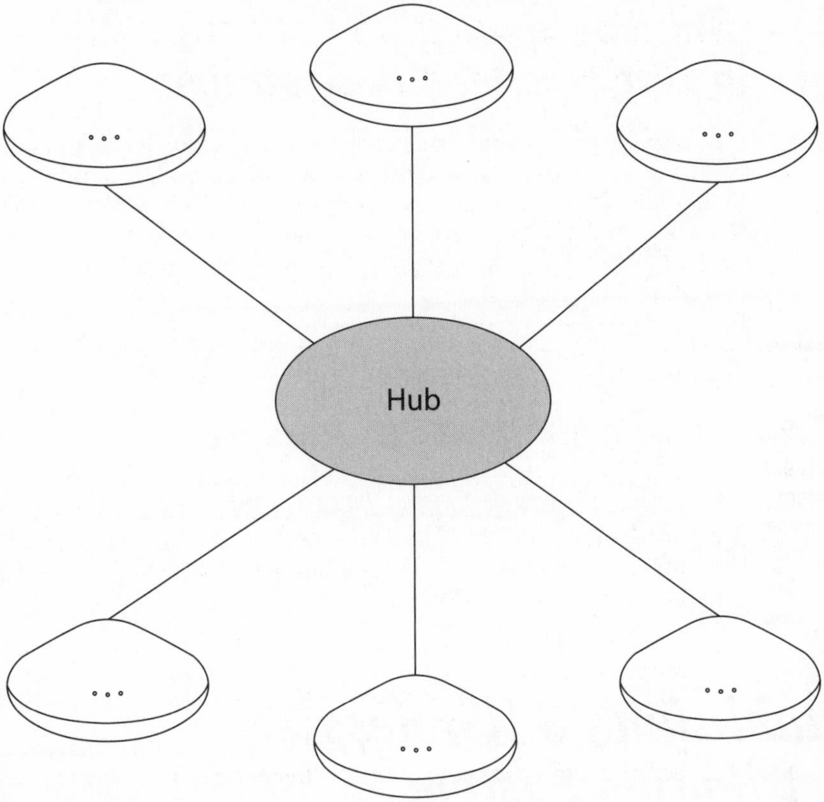

Figure 5-14: Ethernet connections to an Ethernet hub look like a star if you squint a little.

2. **Open the AirPort Admin Utility.**
3. **For each base station in the roaming network, follow Steps 3 through 9 given in the previous section, "Setting up base stations in a roaming network."**

Do not uncheck Distribute IP Addresses (Step 7 in the procedure) for the base station connected to the Internet. Do uncheck it for every other base station in the roaming network.

When you finish these steps, you have a roaming network that shares an Internet connection from one of the network's AirPort Base Stations. This same base station provides IP addresses to the rest of the network.

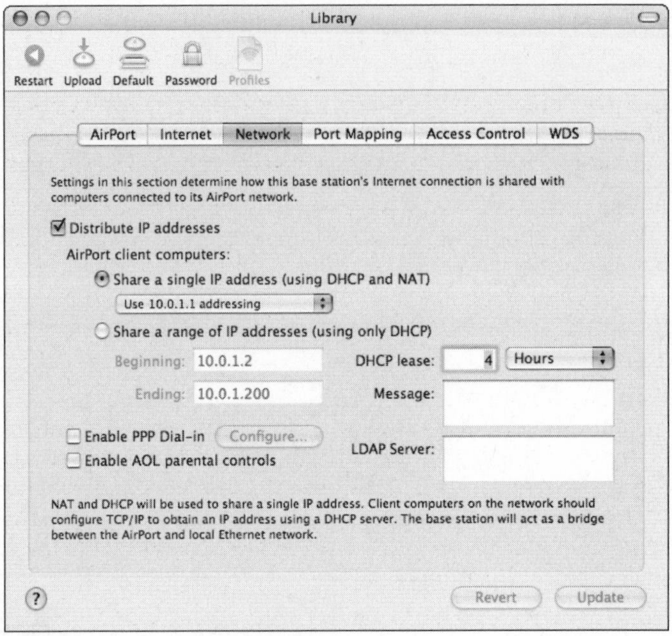

Figure 5-15: One base station shares its Internet connection with all the others.

Establishing a Wireless Distribution System

Creating a roaming network, as described in the previous section, provides one easy way to use additional AirPort Base Stations to extend the range of an AirPort network. However, roaming networks have a downside: You either have to lay down a bunch of cables and construct a wired LAN, or you have to construct the roaming network on top of an already existing wired LAN. If, on the other hand, you choose to create a *WDS (Wireless Distribution System),* you don't have to create or adapt a wired LAN to extend your AirPort network with additional AirPort Base Stations. The added base stations can extend the network's range all by themselves, without the need for any extra wiring.

REMEMBER

In a WDS, one AirPort Base Station shares its Internet connection both with the AirPort network's clients and with the other AirPort Base Stations in the WDS, each of which can forward the shared Internet connection to additional clients.

The base stations that make up the WDS can play one of three roles in the network:

- **Main:** This base station connects to the Internet. It provides network addresses to, and shares its Internet connection with, the other devices on the network. In a WDS, there is only one main base station.
- **Remote:** A remote base station receives its Internet connection from the main base station and shares that connection with any computers wirelessly connected to the remote station. The remote station also passes along the main base station's network address assignments to those computers. The remote station does not provide any network addresses itself.
- **Relay:** Like a remote base station, a relay station receives its Internet connection from the main base station and shares that connection with any remote stations connected to it, as well as with any computers wirelessly connected to the relay station. And, like a remote station, the relay passes along the main base station's network address assignments to its wireless clients, including the remote stations that the relay station serves.

Figure 5-16 illustrates the layout of an example WDS, which comprises

- One main base station connected to the Internet
- Two remote base stations connected wirelessly to the main base station
- One relay base station connected wirelessly to the main base station
- One remote base station connected wirelessly to the relay base station

All of the base stations in this WDS can provide wireless network connections to the computers on the network in addition to performing their WDS functions. In a sense, the WDS is a roaming network without the wires.

Here are a few key points to keep in mind about a WDS:

- **Each WDS base station must be within range of at least one other WDS base station — either a relay or the main station.**

  This differs from a roaming network, where the base stations can be located anywhere that the wired LAN connecting the base stations reaches.
- **All base stations in a WDS broadcast on the same wireless channel.**

  Though this may not matter in many cases, if your WDS has to coexist in the presence of a different nearby wireless network, you have fewer options when it comes to avoiding interference from the base stations on that neighboring network.

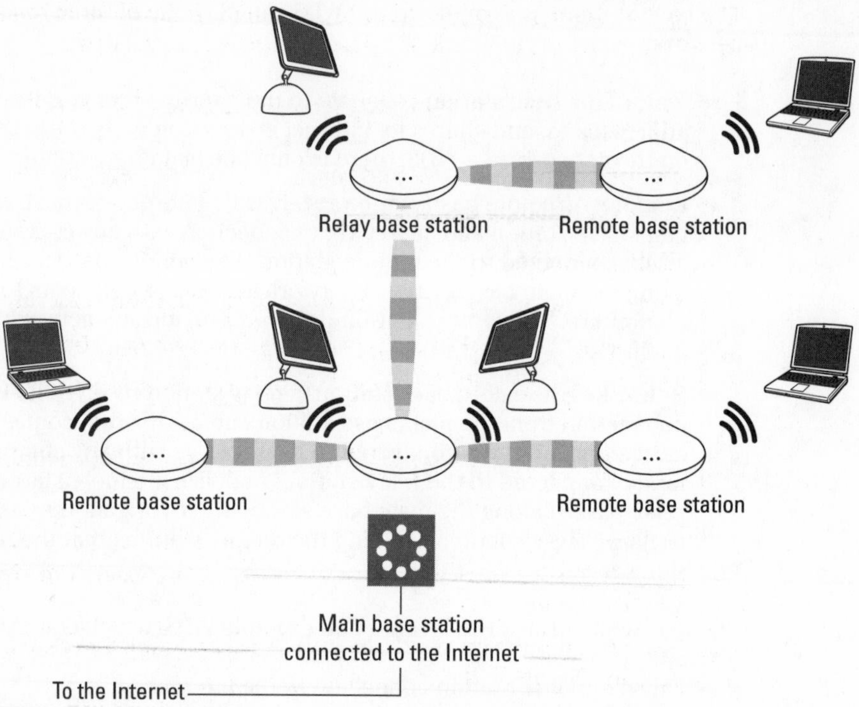

Figure 5-16: The layout of a Wireless Distribution System.

- **The more base stations involved in the WDS, the slower the wireless network.**

  TIP

  Although an AirPort Extreme or AirPort Express can transmit up to 54 megabits per second, a WDS uses some of those bits for relaying various kinds of bookkeeping information among the network's other base stations. As a consequence, the amount of data transmitted each second that's available for other uses, such as downloading a Web page on a network client, drops by a measurable amount on a WDS network. Although this slowdown often is not very noticeable, you should consider another approach to extending your wireless network if you need to use your network for high-speed or large volume data transmissions.

Setting up a WDS with a new base station

Apple has made it very easy for you to add a new base station to your AirPort network to form a WDS: Use the AirPort Setup Assistant. When you use the Assistant to set up a WDS, it both sets up the remote base station and, at the same time, modifies the settings of your current base station to turn it into the WDS's main base station.

The following steps describe how to add a new or newly reset base station to an existing AirPort network with the AirPort Setup Assistant:

1. **Make sure your current base station is powered on and functioning normally.**
2. **Power on the new base station.**
3. **Start the AirPort Setup Assistant.**
4. **Click Continue.**
5. **Click Set Up a New AirPort Base Station, and then click Continue.**
6. **When the Assistant detects your new base station, click Continue.**
7. **Click Connect to My Current Wireless Network, and then click Continue.**
8. **Click Extend the Range of My AirPort Extreme or Airport Express Network.**
9. **Click Continue.**
10. **Choose the network you want to extend from the AirPort Network pop-up menu.**
11. **Type the name you want to give your remote base station in the AirPort Name field.**

    Figure 5-17 shows what the Assistant's display looks like when you are extending a network with an AirPort Express base station.
12. **Click Continue.**

    A sheet descends from the Assistant window, listing the name of your chosen AirPort network's base station, as in Figure 5-18.

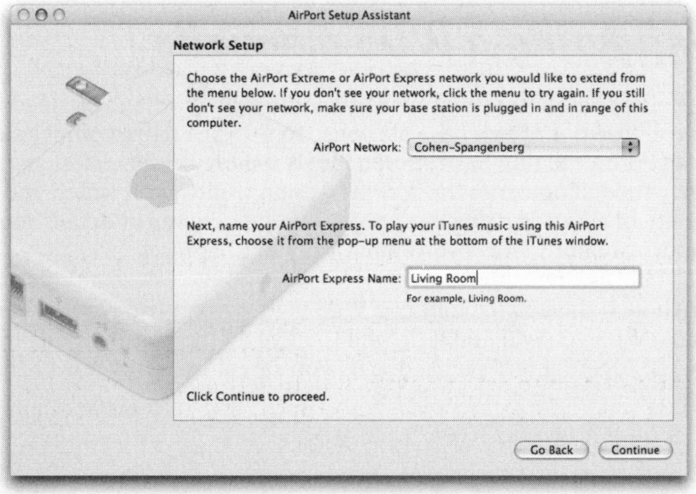

Figure 5-17: Extend a network with an AirPort Express base station.

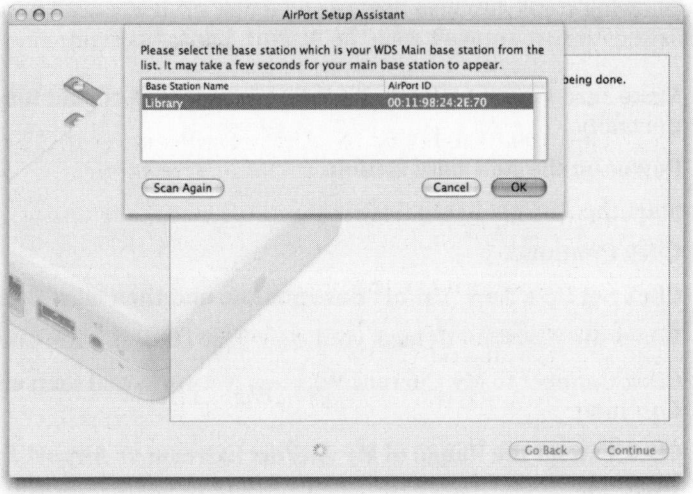

Figure 5-18: Select a station to act as the main base station.

13. **In the sheet, click the base station you want to act as the main base station and then click OK.**

    The Assistant configures the new base station to act as a remote base station and configures your network's current base station to act as a main base station. If your existing network has a network password, the Assistant asks you for it.

14. **Enter your AirPort network's password and click OK, if necessary.**

And that's it. The Assistant presents you with a summary of what it has done, and you now have a working WDS.

Configuring a WDS manually

The AirPort Setup Assistant performs quite nicely when you need to quickly create a WDS out of two base stations. To set up a more complicated WDS, or to convert an existing base station into a remote or relay station, you want to employ the full power of the AirPort Admin Utility, with which you can adjust a variety of wireless network settings to a fine degree of detail. You can find out more about the AirPort Admin Utility in Chapter 4.

TIP

Even though the whole point of a WDS is to cover a wide area, when you first set up a WDS you will find it helpful to have all the base stations involved gathered together and plugged in so you can more easily power them on and off, and, if necessary, reset them.

Here's how you can set up an AirPort Base Station to act as the main station of a WDS, and how you can configure remote stations at the same time:

1. **Power on the base stations that will comprise the WDS.**
2. **Make sure you have joined the AirPort network created by the base station you want to make the main WDS base station.**

   You can use either the AirPort status menu or the Internet Connect application to perform this step.
3. **Start the AirPort Admin Utility.**
4. **In the Select Base Station window, double-click the base station that you want to be the main WDS base station.**

   If necessary, supply the base station's password.
5. **Click the WDS tab.**
6. **Click the Enable This base Station as a WDS check box.**
7. **From the pop-up menu on the same line as the check box, choose main base station.**
8. **Click the + button beside the list box in the middle of the window.**

   A sheet appears listing nearby base stations, as in Figure 5-19.

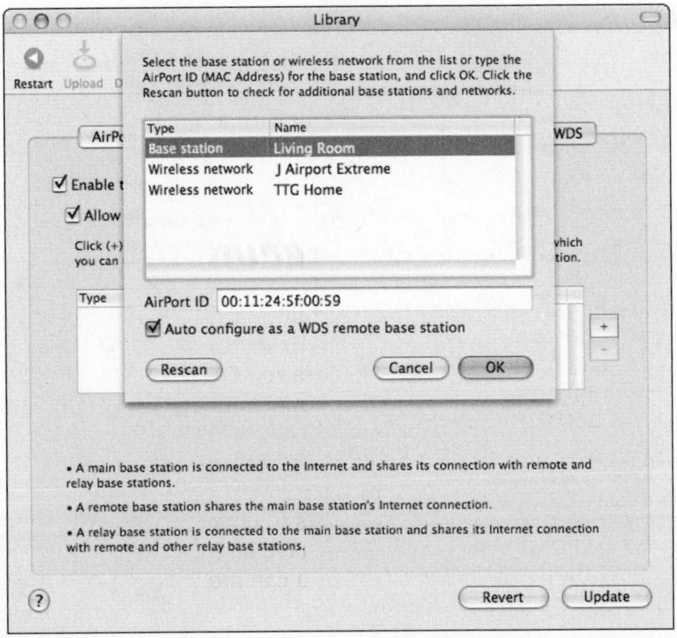

Figure 5-19: Pick the base stations that you want to be remote base stations.

9. **Select the base stations you want to make remote base stations from the list in the sheet, and then click OK.**

   The selected base stations appear in the list box in the configuration window. If you change your mind about configuring any of these base stations as remote base stations, click its entry in the list and then click the – button to remove it from the list.

10. **Click Update.**

    For each of the base stations that appear in the list, a sheet appears, as shown in Figure 5-20.

11. **For each of the sheets, click OK.**

    When you finish, the AirPort Admin Utility updates all the base stations and creates the WDS for you.

After you have set up a WDS, you may wish to add a remote base station to it. To do that, follow these steps:

1. **Power up the base station you want to make into a remote base station.**

2. **Join the AirPort network created by the base station you want to make into a remote base station.**

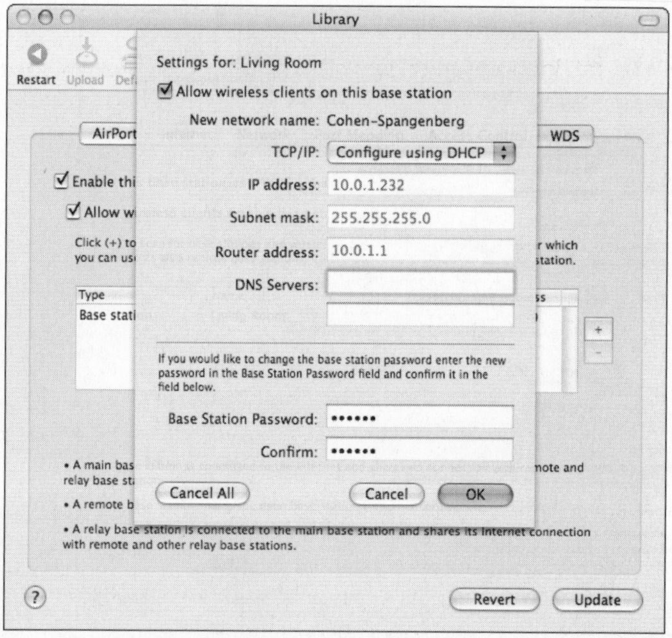

Figure 5-20: AirPort Admin Utility configures the remote stations for you.

You can use either the AirPort status menu or the Internet Connect application to perform this step.

3. **Start up the AirPort Admin Utility.**
4. **In the Select Base Station window, double-click the base station that you want as the remote WDS base station.**

   If necessary, supply the base station's password.
5. **Click the AirPort tab.**
6. **Select the channel number used by the WDS to which you want to add this station as a remote.**

   All the base stations in a WDS use the same channel number.
7. **Click the WDS tab.**
8. **Click the Enable This Base Station as a WDS check box.**
9. **From the pop-up menu on the same line as the check box, choose remote base station.**
10. **In the Main AirPort ID field, enter the AirPort ID of the main WDS base station.**

    The AirPort ID is Apple's name for the base station's MAC address. Apple prints this address on the bottom of each base station. You can also click the Select button beside the field to have the AirPort Admin Utility present a list of possible main base stations so you don't have to type the number by hand. Figure 5-21 shows you what the entries under the WDS tab look like when you choose to create a remote base station.

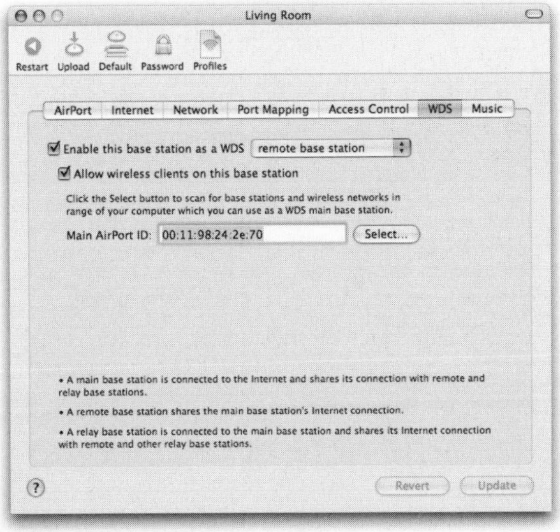

Figure 5-21: Setting up a remote base station manually.

11. Click Update.

The AirPort Admin Utility updates the remote base station and changes the settings on the main base station to recognize the addition of a remote station.

The process of setting up a relay station is almost identical to setting up a remote base station — with one crucial difference: Under AirPort Admin Utility's WDS tab, you not only have to enter the main base station's AirPort ID, but the AirPort IDs of the remote base stations that the relay base station serves.

When you select the relay base station entry from the WDS tab's pop-up menu, the other settings under the WDS tab change to look like a combination of those shown when you set up a main base station and those shown when you set up a remote base station. That is, you have a field in which to enter the main base station ID, and a list box that you use to select the remote base stations — and any relay base stations — that the new relay base station will serve.

Yes, you read that right. In a WDS you can have a main base station that serves a relay base station that in turn serves another relay base station that serves yet another relay base station that serves a remote base station. Of course, such a WDS configuration may not provide the fastest network connection in the world, but, under certain circumstances, it sure beats laying down hundreds of feet of Ethernet cable!

Chapter 6

Networking Wirelessly Without a Base Station

In This Chapter

- Introducing your Mac's inner base station
- Securing your firewall
- Setting up your Mac-hosted Wi-Fi network
- Making a computer-to-computer network
- Changing network locations

The typical Mac wireless network contains at least one AirPort Base Station, but that doesn't mean you *must* have a separate base station to create a wireless network. All you really need for a wireless network is a Mac with an Airport card. Which leads to what I'm guessing is your next question: Why would you want to create a wireless network without a base station?

Well, for one thing, it's cheaper: base stations cost money. If you can't afford both an Airport card *and* an AirPort Base Station, you can skip the base station purchase and still get a wireless network going with just a Mac and an Airport card. But even if you have an AirPort Base Station, the ability to create a wireless network without a base station can save the day if the base station you *do* have suddenly stops working for some reason (I know, because this has happened to me). And, of course, the power to set up a computer-to-computer wireless network can come in very handy in a business meeting or study group if you suddenly discover you need to exchange some critical files with your co-workers or classmates.

There's another reason, too, which is my favorite reason: Because you can.

Making Your Mac into a Wireless Base Station

It's a paradox that to turn your Airport-equipped Mac into a wireless base station you have to use wire to connect it to a network. If, however, you glance at some of the earlier chapters in this book, you'll notice that the one element of a wireless network that almost *always* seems to have a wired network connection is the AirPort network's base station. Whether it's a dial-up modem connection to an Internet Service Provider, an Ethernet cable plugging into a DSL or cable modem, or a direct connection to a corporate or school network, the AirPort Base Station gets tied down by a wire so your Mac doesn't have to. So when your Mac serves as a wireless network's base station, it sacrifices its freedom to a wired network connection so that the other Macs on the wireless network can roam free.

REMEMBER

Making your Mac into a wireless base station is not the same process as creating a wireless computer-to-computer network. You can find out about computer-to-computer networks later in this chapter in "Working and Playing with Computer-to-Computer Networks."

Setting up your Mac's wired network connection is the second of the five steps you need to turn your Mac into a base station. Here are all five, in order:

- Turning on your Mac's firewall to protect your wireless network from unwanted activity.
- Physically connecting your Mac to the Internet.
- Telling your Mac which network port to use for Internet access.
- Adjusting that network port's settings to make an Internet connection.
- Modifying your Mac's sharing preferences to create the wireless network.

When you finish preparing your Mac to play the part of an AirPort Base Station, your Mac will:

- Connect to the Internet by way of a wired connection, most commonly through its modem port or through its Ethernet port.
- Transmit and receive information through its Airport card and antenna.
- Perform a base station's networking routing functions, such as transferring data between the Internet and the wireless network's clients, using the Mac's system software.

The return of the software base station

Using a Mac as a wireless base station has been possible — on and off — since the days of System 9. Back in the old Classic Mac OS days, the sharing capability was referred to as setting up a "software base station." When Mac OS X reached the light of day, however, the software base station capability was absent, and it remained absent through the various versions of Mac OS X 10.1 while Apple worked on a more general, more powerful, more Unix-flavored solution to the problem. Software base station capability returned in Mac OS X 10.2 (Jaguar) under the name "Internet Sharing." It was refined further in Mac OS X 10.3 (Panther), and it was polished to a fine sheen in Mac OS X 10.4 (Tiger).

Today's Internet Sharing is not just for wireless sharing. Unlike the original software base station of Mac OS 9, Internet Sharing lets you use your Mac to share an Internet connection from almost any network port on your Mac through any other port. You can even do such things as share your Mac's modem connection to the Internet through your FireWire port: possibly the ultimate in fast sharing of a slow connection.

Yet even though your Mac is playing the part of a wireless base station, that task does not completely consume its attention: It still has enough processing power at the ready to do all the things that it ordinarily does for you, such as playing music, or calculating spreadsheets, or word processing your latest book on wireless networking.

If your Mac connects to the Internet through a local area network (LAN) at work or at school, stop right here and contact the network's administrator. Setting up a shared connection on such a network, aside from possibly violating any corporate or school policies, might very well cause network problems for other users.

(1) *Raising your firewall*

System Preferences → Sharing Preferences → Firewall tab

Before you do anything else, you need to open your Sharing preferences in System Preferences and start up your Mac's built-in firewall. Why? Because when your Mac acts as a base station, it usually connects directly to the Internet. And when your Mac connects directly to the Internet, nothing stands between it and all the hackers and identity thieves and other scoundrels out there who have nothing better to do than try to ruin your day.

Your Mac's firewall provides security against unwanted Internet intruders by stopping them from initiating network activities that could give them access to your Mac and to the wireless network that it creates.

Cf. 227-232 for information about the Services tab
171-173 - Internet tab

Even if you don't plan to use your Mac as a base station, it doesn't hurt to turn on your Mac's firewall. Whenever your Mac connects to any network, wireless or not, a firewall provides one extra line of defense against hacking attacks.

Figure 6-1 shows the Firewall pane, which you can find under the Firewall tab in your Mac's Sharing preferences. It provides a Start button that lets you put the pane's settings into effect.

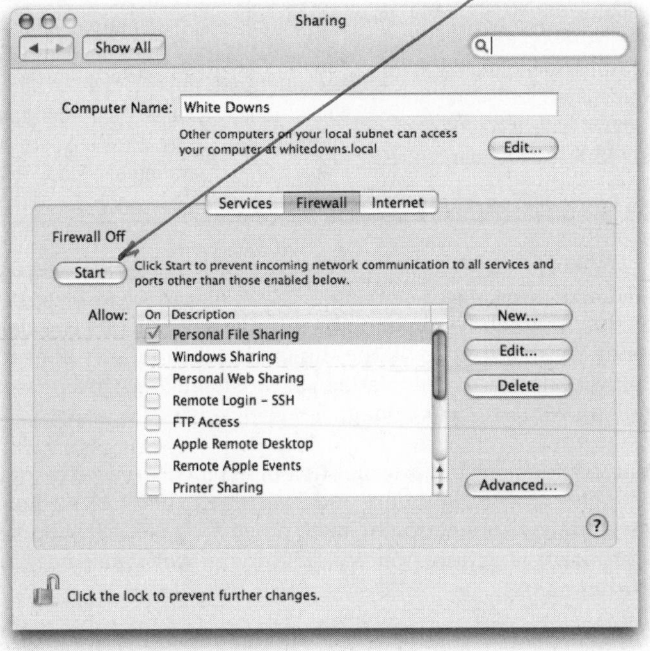

Figure 6-1: The firewall helps repel intruders when your Mac connects to the Internet.

Your Mac can provide a number of useful network services, including file sharing for Mac *and* Windows computers, traditional FTP (File Transfer Protocol) service, remote login service that lets other network users log in to your Mac, and quite a few more. The Mac's Firewall settings pane lets you choose which network services to allow and which ones to prohibit, and blocks any network connections used by the services you have prohibited. Mac OS X's default firewall settings provide the most security against unwanted network activity.

If you want to know what any of the various items in the Firewall settings list do, hover your mouse over a setting to see a help tag that explains it.

Cf. 227-232 for information about the Services tab

[System Preferences → Sharing → Firewall tab]

To turn on the firewall, perform the following four steps:

1. **Open System Preferences by clicking the System Preferences Dock icon.**

   Alternatively, you can select System Preferences from the Apple menu.

2. **Click the Sharing icon in the System Preferences window.**

   This item is located in the Internet & Network section of the window.

3. **Click the Firewall tab in the Sharing pane.**

4. **Click the Start button.**

   The firewall immediately becomes active when you click the Start button.

Wiring the connection

To wire your Mac, you need to connect your Mac physically to the device that provides your Internet connection. This device may simply be a telephone jack if you use a dial-up service to provide your Internet connection, or the device may be a cable modem or DSL modem.

WARNING!

It's a good idea to shut everything down before you start rearranging your cable connections. Although, technically, you really don't *have* to shut down your Mac — Mac OS X is designed to handle network connections coming and going at will — there is always the slight chance your Mac can get confused. More importantly, if things are turned off, you stand less chance of accidentally shorting something out as you plug and unplug cables.

To share a dial-up Internet connection, you connect your Mac's modem port to a telephone jack, using the modem cable that came with your Mac. Don't worry if you can't find where you put that cable when you unpacked your Mac: you can use just about any phone cable that has a standard square plastic RJ-11 connector on each end. If necessary, you can borrow a phone cable from one of your phones as a short-term solution — you won't want to talk to anybody anyway while you're setting up all this stuff, right?

To share an Internet connection provided by a DSL modem or cable modem, you need to use an Ethernet cable. If you want to connect your Mac directly to one of these devices, you may need to use a *cross-over* Ethernet cable. Usually, DSL and cable modems come with such a cable if they require one for a direct computer connection. If your DSL or cable modem connects to an Ethernet hub, you can connect your Mac to the same hub by using a normal *straight-through* Ethernet cable.

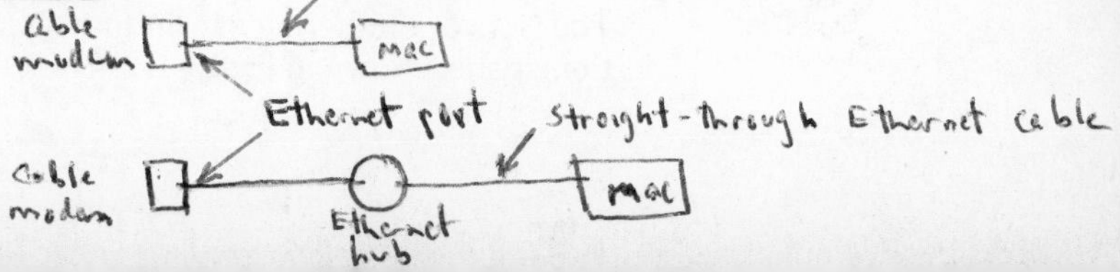

Chapter 4 describes how to connect an AirPort Base Station to a broadband modem, such as a DSL modem or a cable modem. When your Mac acts as an AirPort Base Station, the <u>cabling procedures for connecting an AirPort Base Station to a broadband modem</u> apply to the Mac.

ff. 110 – 112

If you are replacing an AirPort Base Station with your Mac, you can simply unplug the Ethernet cable from the base station and plug it into your Mac.

And that's it for the wiring. You can now restart your Mac.

③ *Putting your ports in order*

A Mac running Mac OS X is hungry for an Internet connection. It wants one so badly that it will look through every active network port that your Mac possesses in order to find one. After you've established the physical connection, you need to <u>let your Mac know which network ports to look at and which ports to ignore</u>.

Don't confuse the network ports being discussed here with the network port numbers that appear in your Mac's Firewall preferences pane. The <u>network ports</u> described in this section are <u>physical connections</u>: think of them as being <u>like the cable socket on a cable TV</u>, and think of the port numbers in the Firewall pane as being like the channels you can tune in after you connect the cable to the TV.

Your Mac's Network Preferences let you <u>choose which network ports the Mac uses</u>, and the order in which the Mac looks at them. Follow these steps to choose and arrange your Mac's network ports:

1. **Open System Preferences.**

   You can click System Preferences' Dock icon or you can select System Preferences from the Apple menu.

2. **Click the Network icon.**

   You can find this icon with the other Internet & Network icons in System Preferences' main window. When you click it, the Network preferences window appears.

   The final item on your Apple menu's Location submenu is Network Preferences. Choose it to open System Preferences and go directly to the Network preferences window.

3. **Select <u>Network Port Configurations</u> from Network preferences' <u>Show pop-up menu.</u>**

   The Network preferences window displays the network port configurations pane, shown in Figure 6-2. The checked ports on the pane's list of

[System Preferences → Network → Network Port Configurations]

ports are the ones the Mac examines for an Internet connection, and the top-to-bottom order in which network ports appear is the order in which your Mac examines them. You can click and drag the ports in the list up and down to change their order.

4. **Click and drag to the top of the list the network port whose Internet connection you intend to share.**

   The other ports in the list move down to make room for it. In Figure 6-2, the Ethernet port's position at the top of the list means that the Mac checks it first, which is what you want if you want to share an Internet connection coming from a broadband modem over an Ethernet cable. To share an Internet connection from a dial-up Internet connection, you need to check the Internal Modem check box in the network port list and then click and drag the Internal Modem port to the top of the network port list.

5. **Click Apply Now.**

   The changes you've made to the network port configurations are now in effect. You can close the Network preferences window.

When you've completed these steps, your Mac knows which network ports to examine and in which order it should examine them when it looks for an Internet connection.

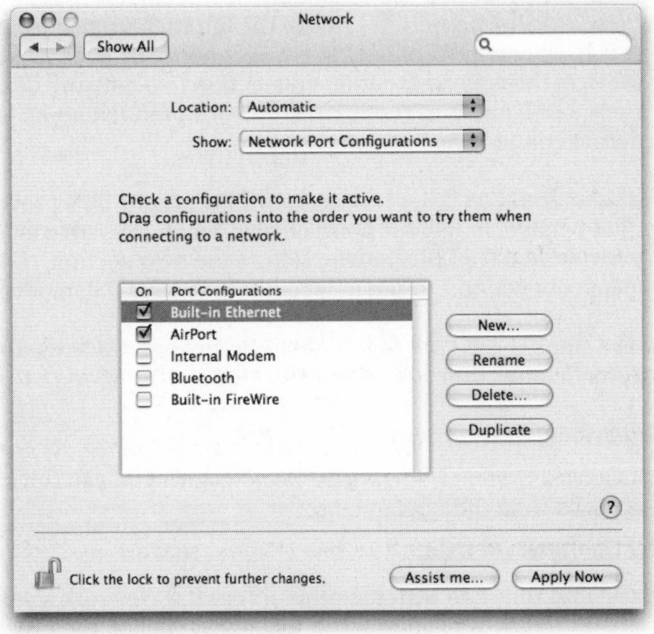

Figure 6-2: The checked network ports are the ones to which your Mac pays attention.

Configuring the Internet connection

After you've connected your cables and set your network ports in order, your next step is to tell your Mac *how* to establish the Internet connection. This step requires some information that you should have received from your ISP (*Internet Service Provider*). You enter this information into your Mac's Network preferences window.

The most common ways your Mac can connect to the Internet are through a DSL modem, a cable modem, or a dial-up modem. Follow the step-by-step instructions for the connection method that you use.

TIP

Check your ISP's connection instructions as well as those that I provide below, and, if the two sets of instructions conflict, follow those that your ISP provides.

Configuring a DSL connection, using PPPoE

DSL (*Digital Subscriber Line*) provides a high-speed Internet connection over regular copper telephone lines. DSL service usually includes a DSL modem, which you connect to a telephone jack. DSL modems also include an Ethernet port, which you connect to your Mac as described in the "Wiring the connection" section earlier in this chapter.

Many DSL service providers use a standard protocol known as PPPoE (*Point-to-Point-Protocol-over-Ethernet*) to provide the Internet connection. This protocol lets your local network connect to another network, such as your ISP's Internet service, through an Ethernet connection to a network device, which in this case is a DSL modem. Your Mac's Network preferences let you set up a PPPoE Internet connection.

REMEMBER

Keep in mind that not all DSL services use PPPoE. Some ISPs provide DSL modems that behave, from your point of view, much like cable modems do. If your ISP provides such a DSL modem, skip to the next section, "Configuring a cable modem connection," to see how to hook up your DSL modem.

Follow these steps to set your Mac's Network preferences to establish a PPPoE Internet connection:

1. **Open System Preferences.**

   You can click System Preferences' Dock icon or you can select System Preferences from the Apple menu.

2. **Click the Network icon.**

   You can find this icon with the other Internet & Network icons in the System Preferences main window. When you click it, the Network preferences window replaces the System Preferences main window.

3. **Select Built-in Ethernet from Network preferences' Show pop-up menu.**

   The Built-in Ethernet pane appears in the Network preferences window.

4. **Click the PPPoE tab.**

   The PPPoE settings appear. Most of the settings are disabled at first, but the next step takes care of that.

5. **Click the Connect Using PPPoE check box to enable it.**

   When you enable the Connect Using PPPoE check box, the fields under the PPPoE tab become active and a PPPoE Options button appears.

   Figure 6-3 shows the settings available under the PPPoE tab of the Built-in Ethernet pane after you check Connect Using PPPoE. You use these settings to specify the username and password that your ISP gave you when you ordered your DSL service. You can also set the name of the ISP and specify the DNS (*Domain Name Server*) if your ISP's setup instructions direct you to do so. Often you don't have to make these last two settings, which explains why these settings are labeled "Optional."

6. **Enter your Internet service's account name into the Account Name field.**

7. **Enter the account's password into the Password field.**

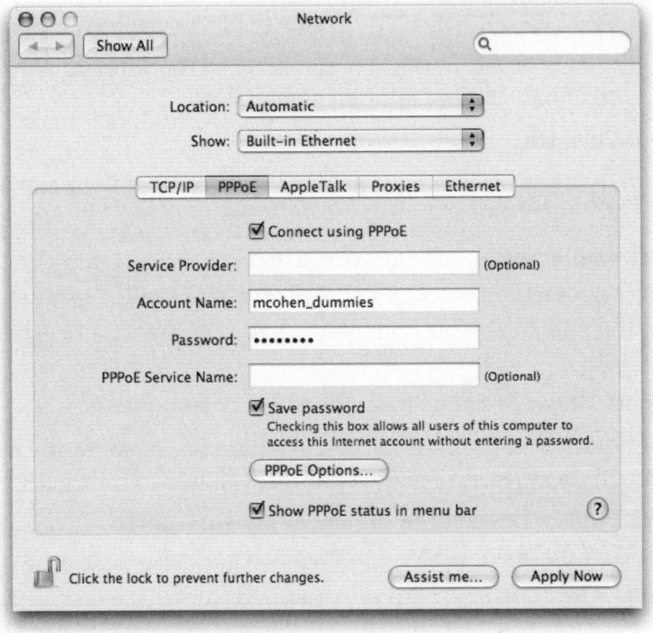

Figure 6-3: The PPPoE settings tab in the Network preferences window.

Because you are planning to share the Internet connection, you should set some of the additional connection options that the PPPoE Options button makes available.

Follow these steps to set the PPPoE options:

1. **Click the PPPoE Options button.**

   A sheet containing options descends from the top of the Network preferences window, as shown in Figure 6-4.

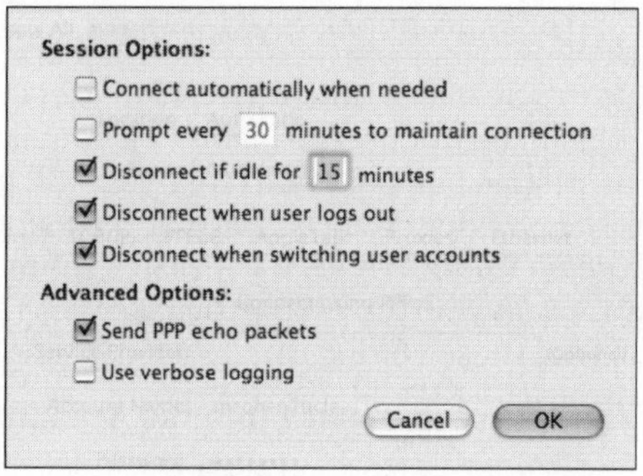

Figure 6-4: Connection options for PPPoE.

2. **Click the Connect Automatically When Needed option to enable it.**

   This option makes sure that your Mac connects to the Internet whenever any of the computers using the wireless network you are creating needs an Internet connection.

3. **Click OK.**

   The changes you have made to the PPPoE options take effect and the sheet closes.

8. **Click Apply Now.**

   This applies the settings you've made, but your Mac isn't connected yet. You still must have your Mac tell the DSL modem to establish the connection.

9. **Choose Network Status from the Show pop-up menu.**

   The Network Status pane, shown in Figure 6-5, replaces the Ethernet pane.

10. **Click Built-in Ethernet in the list of network ports.**

    This enables the Built-in Ethernet entry in the list as well as the pane's Connect button.

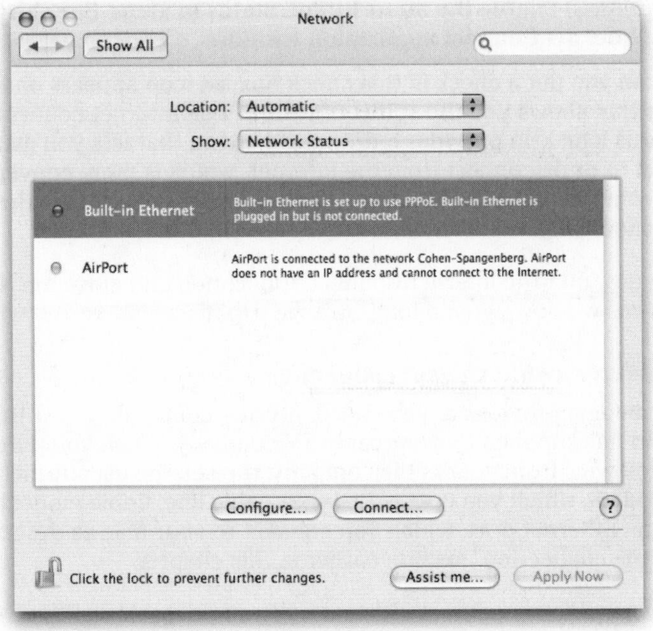

Figure 6-5: The Network Status pane tells what each of your active network ports is up to.

11. **Click Connect.**

    The Internet Connect application opens. You use this application to make the Internet connection.

12. **Click Connect in the Internet Connect window.**

    Your Mac attempts to establish a PPPoE connection. When it succeeds, the Connect button label changes to Disconnect and the connection's status appears in Internet Connect's Built-in Ethernet window, as shown in Figure 6-6.

Figure 6-6: The Internet Connect program makes the PPPoE connection for you.

13. **(Optional) Enable the Show PPPoE Status in Menu Bar check box in the Internet Connect application window.**

    When you put a check in this check box, an icon appears on the menu bar that shows you the status of your PPPoE Internet connection. The status icon also provides a drop-down menu that lets you manually connect to or disconnect from the Internet, which is more convenient than opening Internet Connect every time you want to open or close your connection.

You can now quit the Internet Connect application and close the Network Preferences window. What a long, strange trip it's been.

Configuring a cable modem connection

A cable modem provides a high-speed Internet connection over the cable television line supplied by your cable TV company. When you purchase Internet service from your cable company, the service usually includes a cable modem, which you connect to your cable line. Cable modems also include an Ethernet port, which you connect to your Mac as described in the "Wiring the connection" section earlier in this chapter.

pp. 159-160

Unlike DSL modems, cable modems usually don't require a PPPoE connection. In addition, a cable modem connection normally is always turned on, so you don't have to fiddle with settings that allow you to disconnect and connect, and you don't have to use the Internet Connect application like you do with a DSL modem that uses PPPoE.

You use your Mac's Network preferences to set up a cable modem connection. Follow these steps to set up your cable modem:

1. **Open System Preferences.**

   You can click System Preferences' Dock icon or you can select System Preferences from the Apple menu.

2. **Click the Network icon.**

   You can find this icon with the other Internet & Network icons in the System Preferences main window. When you click it, the Network preferences window replaces the System Preferences main window.

3. **Select Built-in Ethernet from Network preferences' Show pop-up menu.**

   The Built-in Ethernet pane appears in the Network preferences window.

4. **Click the TCP/IP tab.**

   Figure 6-7 shows the settings available under TCP/IP tab. The settings you need to make depend on the instructions your cable company gave you when you purchased cable modem Internet service. Usually, these instructions simply require you to choose Using DHCP from the Configure IPv4 pop-up menu, but you may also have to provide a DHCP Client ID as well.

[System preferences → Network → Built-in Ethernet]

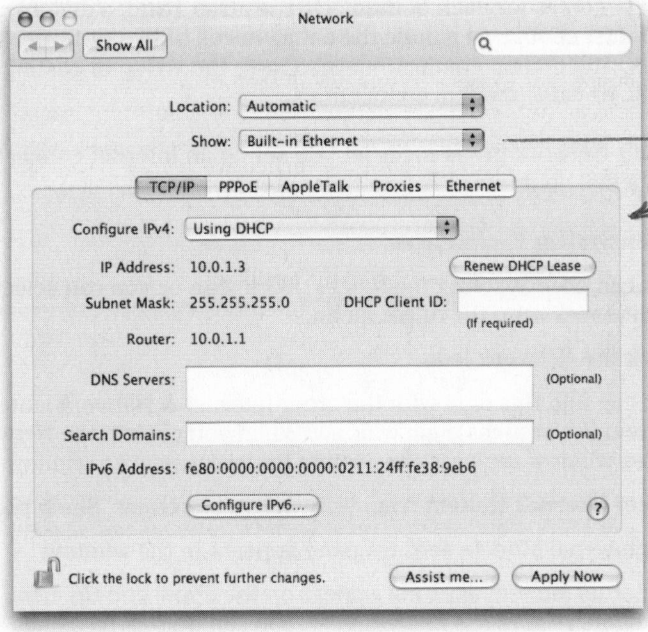

Figure 6-7: A cable modem connection usually provides an IP address by using DHCP.

5. **Enter the settings your cable company gave you.**
6. **Click Apply Now.**

   The settings take effect immediately. Your TCP/IP settings should look similar to those shown in Figure 6-7, although the numbers appearing in your TCP/IP settings probably will differ from those in the figure.

Congratulations. You should now be connected to the Internet through your cable modem. You can close the Network preferences window.

Configuring a dial-up modem connection

A dial-up modem provides a low-speed Internet connection, using regular telephone service. Here's how it works: Your Mac's modem dials the ISP's phone number and establishes a connection with a modem at the ISP, using a standard protocol known as PPP (*Point-to-Point Protocol*). The modems at each end of the connection also convert the digital information that your computer and the Internet use into the audio signals that ordinary phone connections can handle.

REMEMBER

Dial-up modem connections are usually much slower than DSL or cable modem connections — at least ten times slower and often more than that. When you share a dial-up modem connection, every user on your network shares that slow connection. The more users sharing the connection, the slower it becomes for each of them. On the other hand, a dial-up connection is usually fast enough to handle the e-mail needs of two or three simultaneous users. And dial-up connections also have the virtue of costing less than either DSL or cable modem connections.

Your Mac's Network preferences let you set up an Internet connection, using PPP and a dial-up modem. Follow these steps:

1. **Open System Preferences.**

   You can click System Preferences' Dock icon or you can select System Preferences from the Apple menu.

2. **Click the Network icon.**

   You can find this icon with the other Internet & Network icons in the System Preferences main window. When you click it, the Network preferences window replaces the System Preferences main window.

3. **Select Internal Modem from Network preferences' Show pop-up menu.**

   The Internal Modem settings pane appears in the window.

   If Internal Modem does not appear on the Show pop-up menu, you need to enable your Mac's modem port. You can do this by using Network preferences' Network Port Configurations pane, which is described in the "Putting your ports in order" section earlier in this chapter.

4. **Click the PPP tab.**

   Note that the tab may already be selected. The settings under this tab are shown in Figure 6-8. They allow you to enter the account name, the password, and the telephone number your Mac needs to establish a dial-up connection. The settings also let you provide an alternate phone number for your Mac to dial if it encounters a busy signal on the primary number.

5. **Enter the account name, the account password, and the telephone number to dial in the appropriate fields.**

   If your phone uses call waiting, you should check with your phone service to find out the prefix you need to dial to disable it, and add that prefix to the phone number that the PPP connection dials. Otherwise, an incoming call may cause the Mac to drop the Internet connection, and your clients will not be amused.

6. **Click PPP Options.**

   The sheet, shown in Figure 6-9, slides down from the top of the window. You use the Session Options at the top of this sheet to set various dialing and connection options.

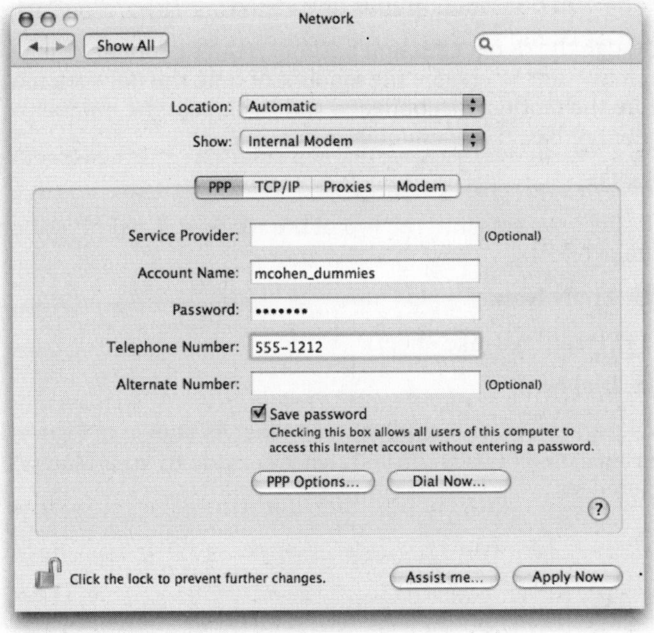

Figure 6-8: The PPP settings let your Mac dial the phone for digital access.

7. **Click to enable the check box labeled Connect Automatically When Needed.**

   This option tells your Mac to dial the phone and establish an Internet connection should any user on your wireless network require it.

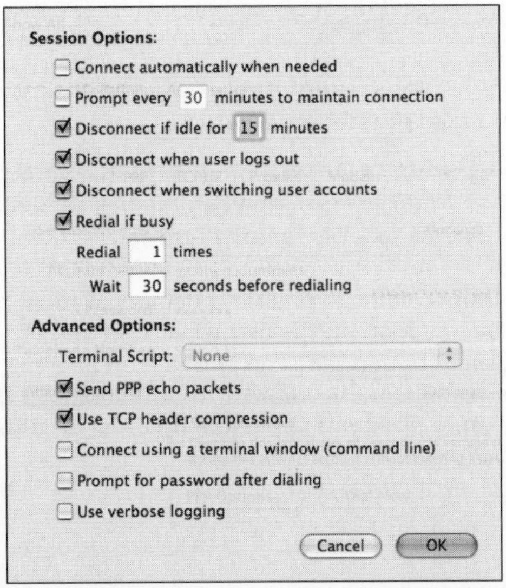

Figure 6-9: You can control a plethora of PPP options.

8. **Set the other Session options as needed.**

   Although the default Session option settings probably will work for you, you may want to change the amount of time the network must be idle before the modem disconnects, and to change the number of times the modem redials if it encounters a busy signal.

9. **Click OK.**

   The sheet retracts into the top of the window. The PPP options you have set take effect.

10. **Click Apply Now.**

    The PPP settings take effect.

11. **Click Dial Now.**

    The Internet Connect application opens. As shown in Figure 6-10, Internet Connect uses the settings you made to your Mac's Network preferences.

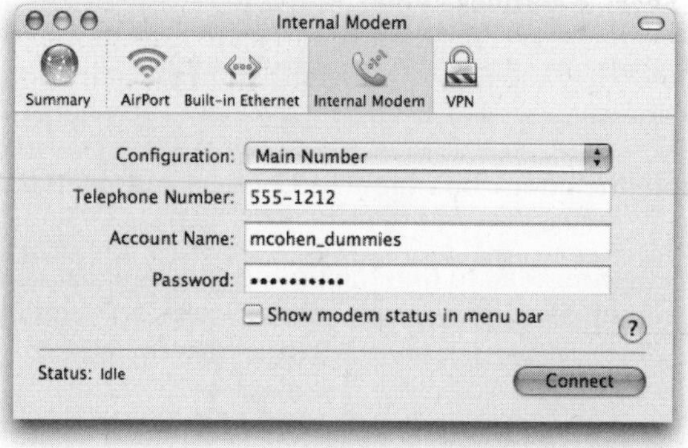

Figure 6-10: Internet Connect makes the dial-up connection for you.

12. **Click Connect.**

    Internet Connect uses the Mac's internal modem to dial your ISP and to establish a connection. When the program establishes the connection, the Connect button changes to a Disconnect button.

In the Internet Connect window, you can enable the Show Modem Status in Menu Bar check box to place an icon on the menu bar that shows you the status of your dial-up Internet connection. The status icon also provides a drop-down menu that lets you manually connect to or disconnect from the Internet, which, given the unreliability of some dial-up connections, can be a real convenience.

⑤ *Sharing the connection*

Cf. 227-232 for information about the ~~Shring~~ Services tab

Now that you've taken all the steps to prepare your Internet connection for sharing, the time has finally come to share it using your Mac's Airport card. You never thought you'd get here, did you?

To share your Internet connection with other users on your wireless network, follow these steps.

1. **Open System Preferences by clicking the System Preferences Dock icon or by selecting System Preferences from the Apple menu.**

   The System Preferences window opens. This Sharing icon is located under the window's Internet & Network heading.

2. **Click the Sharing icon in the System Preferences window.**

   The Sharing window replaces the System Preferences window.

3. **Click the Internet tab in the Sharing window.**

   The Internet Sharing settings shown in Figure 6-11 appear. You use these settings to specify the physical network port that provides your Mac's Internet connection, as well as the network ports on which you share that connection.

System Preferences → Sharing → Internet

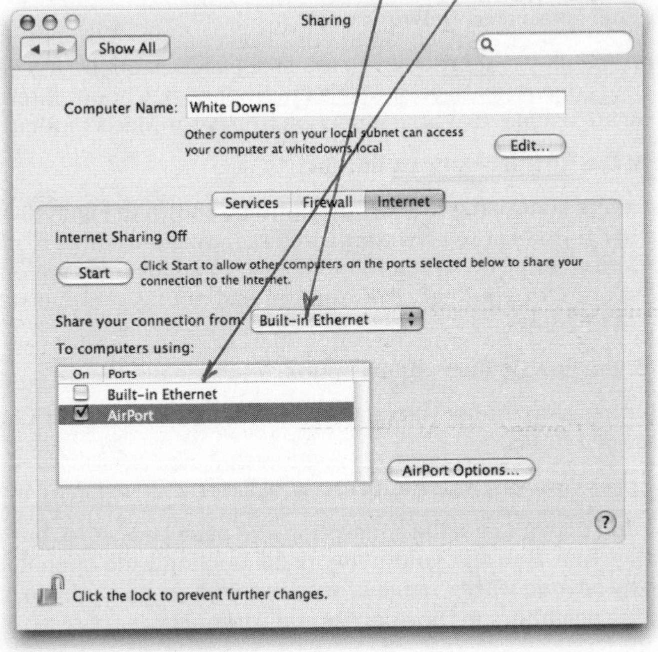

Figure 6-11: The Sharing preferences let you share your Internet connection.

4. **In the To Computers Using list, uncheck every network port except AirPort port.**

   WARNING!

   Other computers can share your Internet connection by connecting to any network port on your Mac *except* the port that your Mac uses to make the Internet connection. If the network port making the Internet connection remains checked in the To Computers Using list, it can cause all sorts of peculiar network problems. If you don't uncheck that port in the list, your Mac presents the alert sheet shown in Figure 6-12.

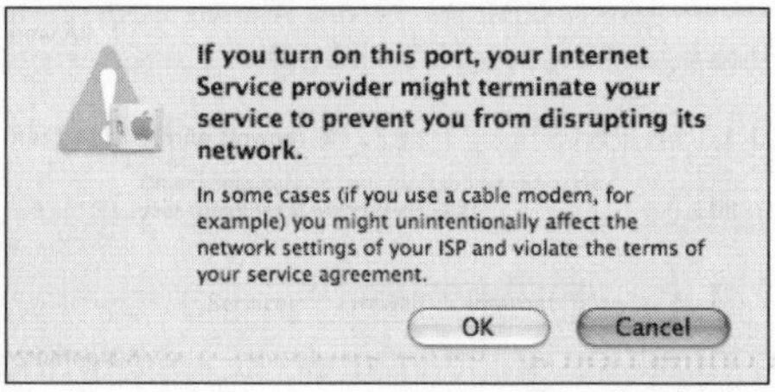

Figure 6-12: Other computers shouldn't connect to the port that provides your Internet connection.

5. **Click the Share Your Connection From pop-up menu and select your Internet-connected network port.**

   Choose Built-in Ethernet if you are using a DSL modem or a cable modem connected to your Mac's Ethernet port. Choose Internal Modem if you are using a dial-up connection from your Mac's modem port.

6. **Click the AirPort Options button.**

   The sheet containing the AirPort options, shown in Figure 6-13, appears. You use this sheet to give your wireless network a name, to choose a wireless channel on which to broadcast, to enable encryption, and to set a password for your network. You can find out more about setting channels in Chapter 5.

   pg. 119-126

7. **Click the Enable Encryption option.**

8. **Enter a password for your network in the Password and Confirm Password fields.**

   The password you enter appears as bullets.

   WARNING!

   You don't *have* to enable encryption and set a password, but you'll be leaving your Mac and your network connection wide open to exploitation by anyone within range of your network (such as those suspicious-looking neighbors in the apartment next door).

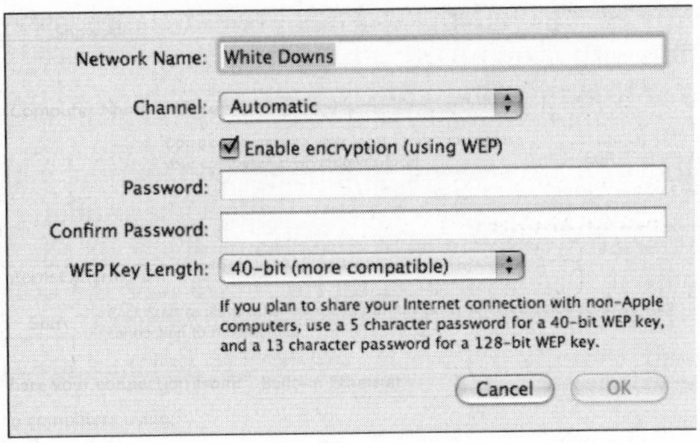

Figure 6-13: AirPort options you can set for your network's protection.

How does this differ from the wireless security options explained on pp. 91-93?

9. **Click OK.**

   The AirPort options take effect and the sheet disappears.

10. **Click Start.**

    The Sharing pane gives you one last warning, as shown in Figure 6-14. If you are sharing your connection at home and you have followed the instructions in this chapter, everything should be set properly, so it is safe to click the warning's Start button. However, if your computer is connected to a school or business network, heed the warning in this sheet and contact the network's administrator before you share your Internet connection, if for no other reason than that schools and businesses tend to have network use policies, and you don't want to unintentionally violate them.

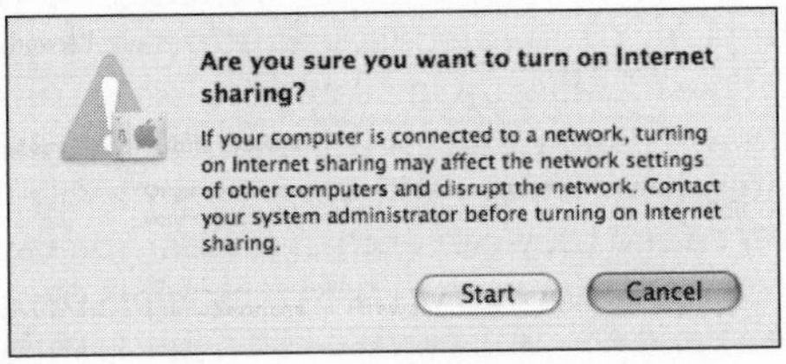

Figure 6-14: You see one last warning before your network goes on the air.

11. **Click the warning's Start button.**

    The warning shown in Figure 6-14 disappears and your Mac begins sharing its Internet connection.

12. **Select Quit from the System Preferences menu.**

    The currently displayed preferences window closes.

By following the steps above, you have turned your Mac into a wireless base station: a base station with a keyboard, a hard disk, and a screen, and that can also play *World of Warcraft.* How cool is that?

One final tip: Now that you are sharing an Internet connection, make a note to visit your Energy Saver preferences in System Preferences and set your Mac to never sleep. Otherwise, when the Mac does go to sleep, its shared connection goes to sleep with it. The users sharing the Mac's Internet connection probably won't like that very much. It is okay to allow your Mac's display and hard disk to sleep, though: neither is required to keep an Internet connection active.

Working and Playing with Computer-to-Computer Networks

You don't need to turn on Internet sharing to make your Mac into a base station. Instead, you can create a simple wireless network that consists of just your Mac and a few other computers. Apple calls this a *computer-to-computer network*. Such a network forms a closed environment, in which the participating computers can communicate with each other but remain isolated from the big wide world of the Internet.

Why would you want to do this? Here are a few examples of when such a network might come in handy:

- **To exchange files.** Granted, you can engage in file sharing on any sort of network, such as a wired LAN or an existing AirPort network. But consider the case of two salespeople meeting up in an airport lounge on a 3-hour layover and who need to exchange contact lists and PDFs of their company's latest product literature. These road warriors probably don't have a network available over which they can exchange files. You can quickly set up a computer-to-computer network — which requires only that the two computers involved have wireless cards — anywhere, without the need for an Internet connection or any other pesky infrastructure.
- **To collaborate.** Again, you can use any kind of network to engage in collaborative work — if you and your collaborators happen to have a network handy that you all can join. But when you and your study partners are monopolizing a table at an all-night diner as you feverishly try to finish your term project, you can set up your own network and get the job done.

- **To play games.** Many computer games come with network multiplayer capability. And some of these games can consume a lot of network bandwidth, so much so that many network administrators restrict their networks to more important data traffic. A computer-to-computer network confines that game-generated data to just the players involved.

REMEMBER

In short, sometimes you have to use a computer-to-computer network as a substitute network for when no other network is available, and sometimes you use a computer-to-computer network because you don't want your data passing through a more accessible network.

Creating and using a computer-to-computer network

You wouldn't find a computer-to-computer network very useful if you had to open a bunch of preference windows and change a lot of settings to set one up. Luckily, you don't have to.

You can create a computer-to-computer network by using the AirPort status menu, if you happen to have it in your menu bar. If not, you can use the Internet Connect application to create one.

Using the AirPort status menu to create the network

The AirPort status menu on your menu bar provides perhaps the most convenient way for you to create a computer-to-computer network.

TIP

If you don't have this menu on your menu bar, maybe you should think about giving up a few pixels of menu bar space to it. You can add the status menu with the Internet Connect application, or from the AirPort tab in your Mac's Network preferences window.

Follow these steps to set up a computer-to-computer network, using the AirPort status menu:

1. **Click the AirPort status menu.**

   If you have already turned on your AirPort card, the menu looks something like Figure 6-15. If your card is turned off, choose Turn AirPort On from the AirPort Status menu and then click the menu again. After all, you can't create a wireless network if you don't have your wireless network card turned on!

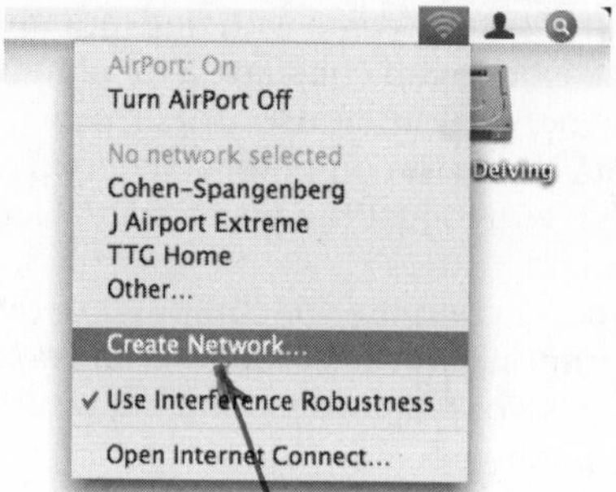

Figure 6-15: To create a computer-to-computer network, click Create Network.

2. **Click Create Network.**

   A Computer-to-Computer window appears, similar to the one in Figure 6-16. The window contains a place for you to provide a name for your network; the default name is the same name you have given your computer, but you can use something else. The window also provides a menu that you can use to select a broadcast channel. The default choice is Automatic, which allows the AirPort software to choose a channel less likely to conflict with neighboring base stations. Chapter 5 explains channels and how to choose them.

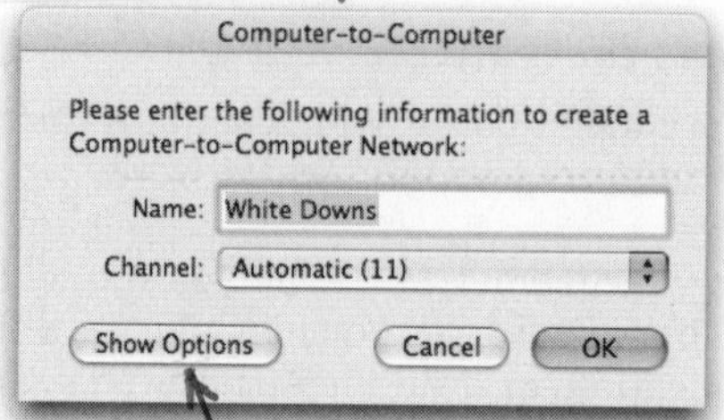

Figure 6-16: Name your network and pick a channel.

3. **Enter a name in the Name field (optional).**
4. **Click the Channel pop-up menu and select a channel (optional).**
5. **Click Show Options.**

   The window expands to provide encryption options, as shown in Figure 6-17. Although this step is also technically optional, I emphatically recommend that you always protect your wireless networks with a password. If, however, you like to live dangerously, skip ahead to Step 9.

Figure 6-17: Only the foolhardy don't bother giving their networks *some* protection.

6. **Click Enable Encryption (Using WEP).**
7. **Click the WEP Key pop-up menu and select an encryption key length.**

   The longer the WEP (*Wired Equivalent Privacy*) you choose, the longer the password. You have two choices: A 40-bit key requires you to create a password that is exactly five characters long, and a 128-bit key, which provides stronger protection, requires a 13-character password. These password length restrictions allow computers running recent versions of Windows to join your computer-to-computer network along with Mac users.

   Users of older versions of Windows may not be able to enter normal passwords with their networking software, but instead might be required to enter a password made up of *hex* digits, comprising the numerals 0 through 9 and the letters A through F. Hex is short for *hexadecimal,* and refers to the base-16 numbering system beloved of hardcore programmers. 04AF529BCA is an example of a ten-digit hex number. To accommodate these Windows users, you can create hex password keys directly: A 40-bit key requires ten hex digits, and a 128-bit key requires a rather clumsy 26-digit sequence.

8. **Make up a key and enter it twice.**

   You can't see the password as you type: the field displays bullet characters. Typing it twice keeps you from inadvertently entering something other than what you meant to type — unless you make precisely the same typos both times.

9. **Click OK.**

   The AirPort status menu changes appearance to show a tiny computer inside of the AirPort icon, indicating that you have established the computer-to-computer network.

If you skip the optional steps in this procedure, and if you don't require a network password, the process of creating a computer-to-computer network requires only three clicks. Of course, you should always protect your network with a password unless you are sure that the only computers within broadcast range of your network are computers you trust.

Creating a computer-to-computer network with Internet Connect

When it comes to setting up a computer-to-computer network, the Internet Connect application is misleadingly named: a computer-to-computer network usually doesn't connect to the Internet. But, misleading name or not, the application can set up a computer-to-computer network just as easily as the AirPort status menu can. Owners of Macs with small screens, such as the original iBooks or iMacs, often don't have the AirPort status menu showing on their menu bars to save room for other menus.

Not surprisingly, the steps you follow to set up a computer-to-computer network using the Internet Connect application resemble those you follow when you create one using the AirPort status menu:

1. **Open the Internet Connect application.**

   You can find this application in your Applications folder.

2. **Click the AirPort icon on the application window's toolbar.**

   The Internet Connect application window displays the current AirPort status. If your AirPort card is turned off, click Turn AirPort On.

3. **Click the Network pop-up menu.**

   The window looks similar to the one shown in Figure 6-18, with the menu displaying currently available AirPort networks and a choice for you to create one.

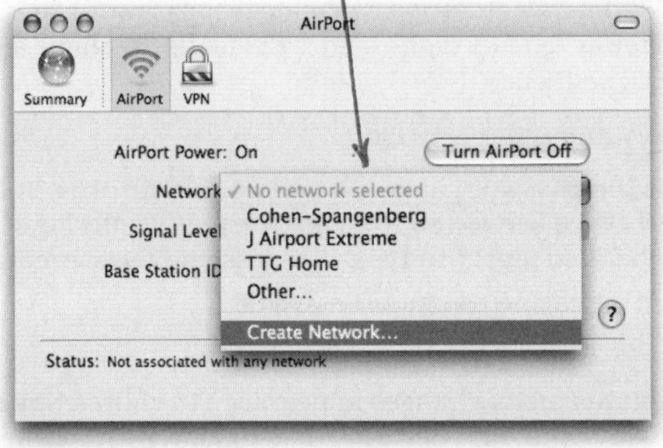

Figure 6-18: Internet Connect offers choices similar to the menu bar's AirPort status menu.

4. **Click Create Network.**

   A sheet descends from the window's title bar. The contents of this sheet are identical to those in the window displayed by the AirPort status menu, depicted earlier in Figure 6-16.

5. **Enter a name for your network in the sheet's Name field (optional).**
6. **Click the sheet's Channel pop-up menu and select a channel (optional).**
7. **Click Show Options.**
8. **Click Enable Encryption (using WEP).**
9. **Click the WEP Key pop-up menu and select an encryption key length.**

   See the description of this menu and what it means in the step-by-step instructions for creating a computer-to-computer network with the AirPort status menu, presented in "Using the AirPort status menu to create the network" immediately above.

10. **Make up a key and enter it twice.**
11. **Click OK.**

    The computer-to-computer network is now on the air.

12. **Quit the Internet Connect application (optional).**

Joining and leaving an existing computer-to-computer network

Joining an existing computer-to-computer network is even easier than setting one up. As with setting up a computer-to-computer network, you can join one by using either the AirPort status menu or the Internet Connect application.

When your Mac is within range of a computer-to-computer network, the AirPort status menu looks something like Figure 6-19, with the available computer-to-computer networks grouped together following any other wireless networks listed on the status menu. To join the network, simply select it from the menu. If the network requires a password, a dialog appears so you can provide it.

Figure 6-19: A computer-to-computer network is on the menu.

If you want to use Internet Connect instead to connect to a computer-to-computer network, you can choose the computer-to-computer network from the Network pop-up menu in the Internet Connect application's window.

When you want to disconnect from a computer-to-computer network, you can, again, do so either from the AirPort status menu or from the Network pop-up menu in Internet Connect.

Figure 6-20 shows the AirPort status menu as it appears when your Mac is connected to a computer-to-computer network. The network to which you are connected appears with a check mark beside its name. To disconnect from it, all you need to do is select Disconnect from current network. And if you don't have the AirPort status menu visible, rest assured that a similar disconnect item appears on the Network pop-up menu in Internet Connect.

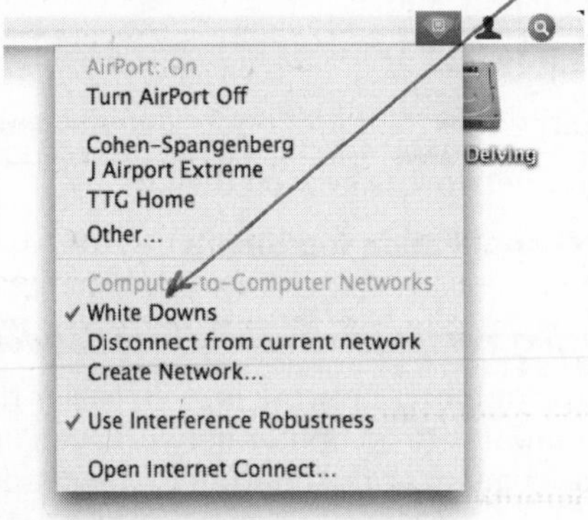

Figure 6-20: Disconnecting from a computer-to-computer network is just a click away.

Making a computer-to-computer network location

If you use AirPort to establish your Internet connection, switching to a computer-to-computer network presents no problems: just follow the steps in the previous section, and you can bid farewell to the wild world of the Internet and enter the safe confines of an isolated computer-to-computer network seamlessly. If, however, you connect to the Internet in some other way, such as through an Ethernet connection, and if you get your network address from that connection, you can still join a computer-to-computer network, but you may encounter some difficulties when you try to share files, play network-enabled games, or use collaborative network software. Creating a special *network location* for computer-to-computer network use helps you avoid such stumbling blocks.

When you create or join a computer-to-computer network, your Mac makes up a usable network address for you, one which allows other network users to identify your Mac so they can engage in various network activities with it. However, when your non-AirPort network connection provides you with a network address from a DHCP server, or if that connection requires a manually entered network address, your Mac continues to use that address even after you create or join a computer-to-computer network. If the network address that your Mac uses falls outside of the range of network addresses that the computer-to-computer network uses, no other users on the computer-to-computer will find your Mac.

A *network location* is a set of network configurations that you save under a name, so you can recall it and use it at need. For example, you can create a network location that always uses Ethernet to connect to the Internet, another network location that uses a specific AirPort network to provide the connection, a network location that uses only the modem port, and so on. Network locations come in particularly handy for laptop Macs that travel from one network environment — such as an office network, which requires a specific network configuration — to another network environment — like your home network, which might use a different network configuration.

What you need to do is relatively simple: Create a computer-to-computer network location, and configure your Network preferences for that location so that the only active network port is your AirPort card. After you've done that, every time you need to use a computer-to-computer network, you choose the computer-to-computer location from your Apple menu.

Here's how you create the computer-to-computer network location:

[System Preferences → Network → Location pop-up menu]

1. **Open System Preferences.**

   You can click the System Preferences icon in your Dock, or choose System Preferences from the Apple menu.

2. **Click Network.**

   You can do Steps 1 and 2 all at once by clicking Apple⇨Location⇨Network Preferences.

3. **Click the Location pop-up menu at the top of the Network preferences window and select New Location.**

   A sheet descends from the top of the window, as in Figure 6-21, with a field in which you can supply a location name.

4. **Enter a name for the new location and click OK.**

5. **In the Network preferences window, click the Show pop-up menu and choose Network Port Configurations.**

   The Network Port Configurations pane appears in the Network preferences window as shown in Figure 6-22.

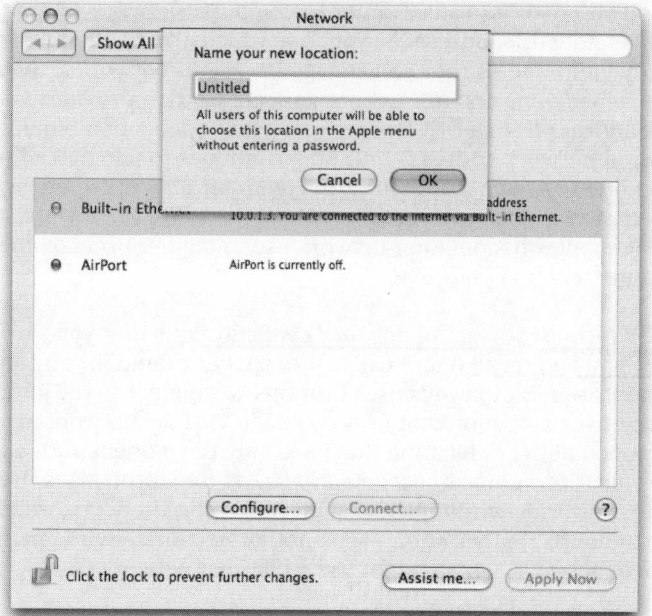

Figure 6-21: Creating a new network location.

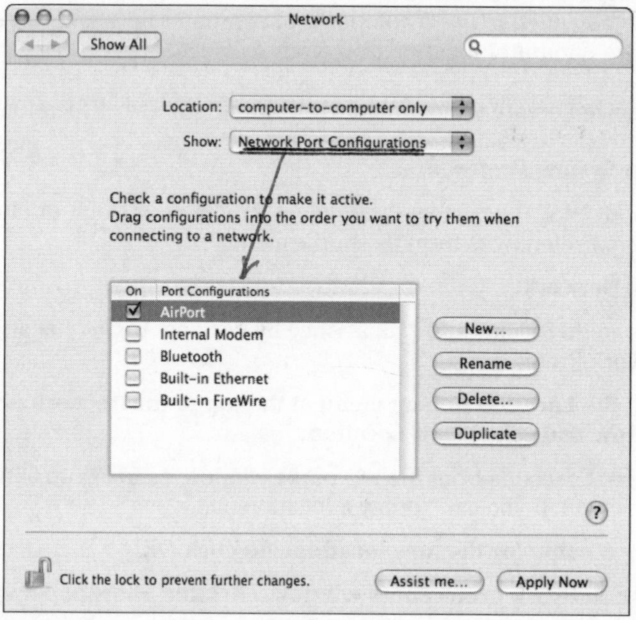

Figure 6-22: Configure your network ports here.

6. **In the port configurations list, uncheck every network port except the AirPort.**
7. **Click Apply.**
8. **Close the Network preferences window.**

Now that you've done that, you can choose this location from your Apple menu whenever you need to use computer-to-computer networking, as shown in Figure 6-23.

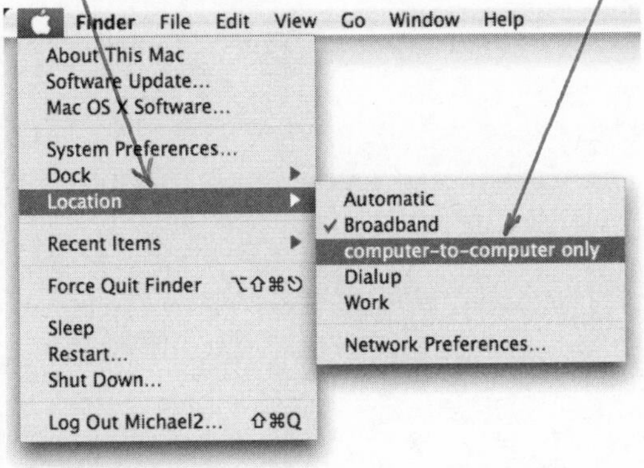

Figure 6-23: Choose a network location right from the Apple menu.

TIP

You may want to create two network locations: One that contains the locations you normally use for networking, and one that contains your computer-to-computer settings.

Part III
It's (Almost) Fun and Games

The 5th Wave By Rich Tennant

"Hey you bunch of loser jocks! This happens to be an AirPort, and I got it to go with my iBook, and you wouldn't even know what to do with it!"

In this part . . .

All singing, all dancing, all the time! Fill the airwaves with Mac-supplied music! Play with your friends across the room or around the world! Be your own multi-screen multiplex!

Turn the page and let the games begin.

Chapter 7

Streaming Music Wirelessly

In This Chapter

- Tuning in to the AirPort Express
- Controlling AirTunes remotely
- Picking other musical fruit
- Enjoying rights-protected music

Starting right around the end of the First World War, people have obtained and enjoyed entertainment wirelessly. From the early radio hobbyists' crystal-and-cat's-whisker tuners to the 21st-century condo-dwellers' digital high-definition satellite dishes, songs and speeches, plays and pictures, sports spectaculars, and cinematic masterpieces have all streamed invisibly across the skies and into the radios and televisions we keep in our homes or carry in our cars, our pockets, and our purses. It was only a matter of time before the burgeoning collections of entertaining media we amassed on our computer hard drives would cut the cord and float free — if not across the skies, then at least across our living rooms.

Apple's AirPort Express spins wireless networking technology into an audio entertainment component, and it's not the only digital device that does so. Nor is Apple's AirTunes the only way you can broadcast music from your Mac around your homes and gardens. And even when your media stays on your Mac, it doesn't have to be playing on your desktop, and you don't have to be sitting at your mouse and keyboard to surf the Web or play your favorite selections from your QuickTime movie-trailer collection on your 40-inch widescreen plasma TV.

The battle for the remote control has suddenly gotten a whole lot weirder.

Streaming Music with AirTunes

After releasing three generations of the rounded, pointy-headed Airport Base Station, Apple astonished the world — or that part of the world that pays attention to Apple's product-introduction media events — in June of 2004 by introducing the AirPort Express base station. They abandoned the previous

Airport Base Stations' friendly flying-saucer form for a shape that resembled nothing so much as a slightly oversized power adapter for the company's line of laptop computers. Although the very ordinariness of the AirPort Express's shape attracted some portion of the attention, it was Express's high-quality audio output port that attracted the lion's share of press coverage (especially among those reporters who were still trying to comprehend Apple's sudden rise to dominance in the digital music marketplace, a rise powered by the synergistic trifecta of Apple's iPod player, its iTunes software, and its online iTunes Music Store). The AirPort Express resulted from Apple's discovery, roughly five minutes before everyone else, that *Wi-Fi* was more than just a pun on *hi-fi*, and that digital wireless network technology and digital audio technology were two great tastes that tasted great together.

Looking into AirTunes

The AirPort Express, like its older and more expensive brother, the AirPort Extreme, provides wireless network connectivity using the 802.11g standard, described in Chapter 2, as well as Internet sharing. What the Express provides that its big brother doesn't is a quarter-inch (3.5mm) miniplug audio port that handles analog and digital optical audio, and that can connect to anything from a pair of cheap battery-powered speakers to a high-fidelity buff's fanciest new rig.

Through its audio port the AirPort Express outputs high-fidelity audio transmitted to it from Apple's iTunes music software running on your Mac. Apple calls this connection between iTunes and the AirPort Express *AirTunes*. After you've configured the AirPort Express and hooked it up to your home entertainment system or powered speakers, here's how you use AirTunes, assuming iTunes is already installed:

1. **Turn on the AirPort Express and your connected speakers or entertainment system.**
2. **Fire up iTunes on your Mac.**
3. **Choose the AirPort Express from iTunes' pop-up remote speakers list that appears at the bottom of the iTunes window.**
4. **Rock out.**

And that's all there is to it. The AirPort Express can play music even as it handles the normal wireless network tasks for which it is also designed, such as creating a wireless network and sharing an Internet connection with that network.

The inner AirTunes

When you beam your music from iTunes through your AirPort Express to your speakers, several different software and hardware components interact. AirTunes is Apple's name for all those components working in combination. The components include the following:

- **Mac OS X's networking software:** This monitors the state of all the network ports on your Mac. When a connection appears on one, such as when you turn on or come within range of your AirPort Express, the networking software takes note of it, and makes that information available to any other program that might be interested — such as iTunes.
- **iTunes:** This is Apple's music player and music library manager software, which can organize and play back music stored in a wide variety of compression formats. iTunes periodically checks with Mac OS X's networking software to see which AirPort Express base stations are operating within the vicinity, and adds or removes their names from its pop-up remote speakers list as needed.
- **QuickTime:** This Apple system software provides and manages the *codecs* (*co*mpression/*dec*ompression software components) that compress the music that your iTunes Library contains, that decompress music when you play it back, and that convert music from one compressed format — such as AAC, used by songs from the iTunes Music Store, as well as the popular MP3 — to other formats when necessary.
- **Apple Lossless Encoder:** iTunes, using this QuickTime codec, converts your songs from the format in which they are stored to this special high-quality format whenever it sends music wirelessly to an AirPort Express. The AirPort Express contains complementary software that converts the Apple Lossless music data that it receives into the uncompressed digital audio format that the AirPort Express's internal DAC and optical transceiver components — described next — require.
- **DAC (digital-to-analog converter):** The AirPort Express must convert the digitally encoded music it receives into an analog signal that a set of powered speakers or a nondigital stereo can play. A special DAC chip inside the AirPort Express performs this conversion.
- **Optical transceiver:** To connect to new high-fidelity systems with optical digital inputs, the AirPort Express uses this internal component to convert the digital music information from an electrical signal into an optical digital signal.
- **AirPort Express music buffer:** This is a few megabytes of RAM inside the AirPort Express that can contain a number of seconds' worth of the music sent by iTunes. The AirPort Express does not immediately convert and send the music it receives out its audio port, but waits until its music buffer contains a few seconds' worth of music data. Then, when the AirPort Express *does* start playing the music, the music can continue playing smoothly even if the AirPort Express experiences brief transmission interruptions, either from radio interference or from a burst of heavy wireless network activity.

Setting up AirTunes

Setting up AirTunes requires almost no work at all. When you configure a new AirPort Express base station with the AirPort Setup Assistant as described in Chapter 3, you don't need to do anything special: The AirPort Setup Assistant makes the necessary settings behind the scenes to activate the AirTunes feature, regardless of how you use the AirPort Express in your network. Although some of the text in the AirPort Setup Assistant's summary pane (shown in Figure 7-1) may vary depending on what you want the AirPort Express to do, the magic phrase "play iTunes music through remote speakers using AirTunes" appears somewhere in it.

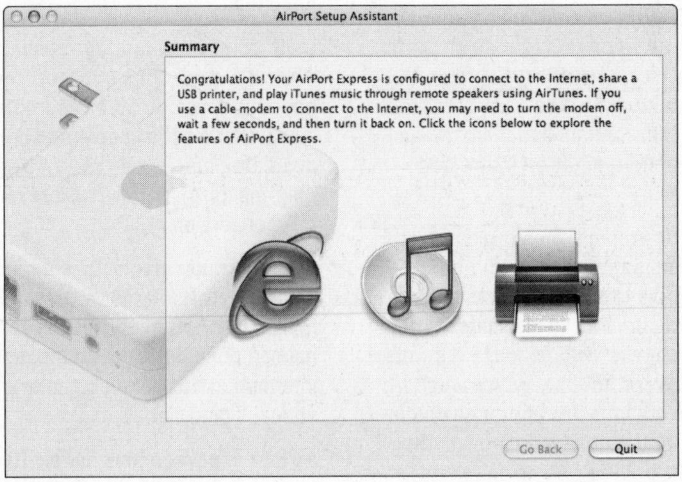

Figure 7-1: When the AirPort Setup Assistant finishes, AirTunes is ready to rock and roll.

Setting up the AirPort Express as dedicated speakers

Although the AirPort Setup Assistant always activates AirTunes, it does offer one particular setup scenario designed for using the AirPort Express primarily as a remote speaker accessory connected to your existing AirPort network. Here are the selections you make with the AirPort Setup Assistant to do that:

1. **When the Assistant asks what you want to do with the AirPort Express, choose Connect to My Current Wireless Network.**

   Figure 7-2 shows this selection, which AirPort Setup Assistant offers you once you select the AirPort Express as the device you want to set up.

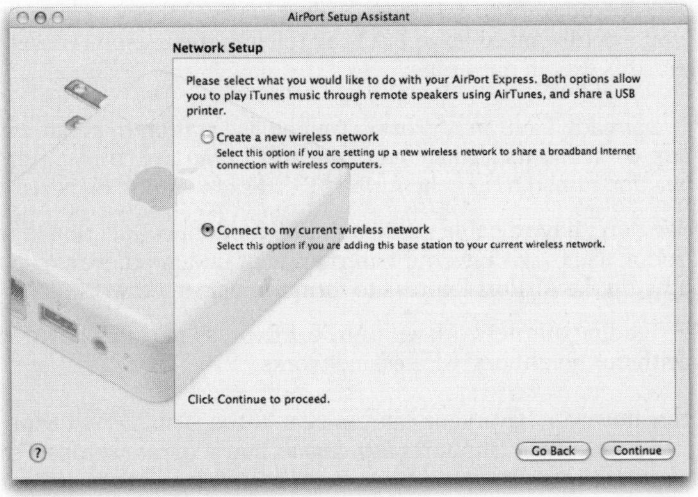

Figure 7-2: Connect to the current network to use the base station as dedicated speakers.

The Assistant asks what you want the AirPort Express to do in your existing network.

2. **Choose Join My Wireless Network.**

   This selection pane appears immediately after the one depicted in Figure 7-2, and presents the two options shown in Figure 7-3.

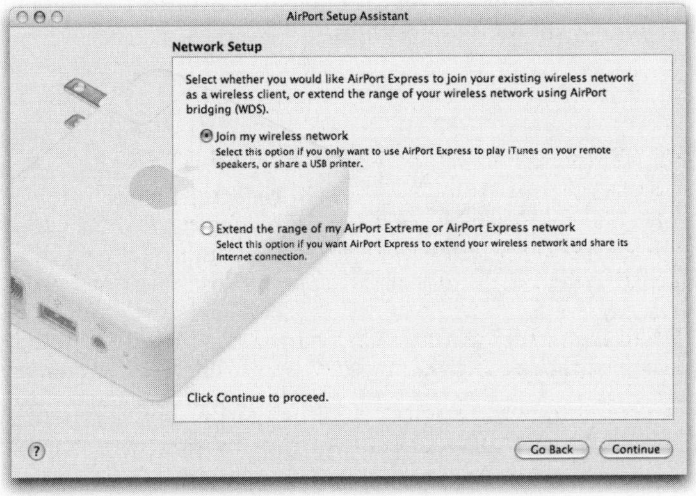

Figure 7-3: Dedicated speaker base stations join the network as a client only.

Why would you want to set up an AirPort Express this way? I can't answer why you would want to do it, but I can tell you the reasons we set up AirPort Express this way in our home:

- We already have an AirPort Extreme base station to create our wireless network, and its signal covers the area of our apartment quite well, so we don't need to use the AirPort Express to extend our existing network.
- We don't have a cable modem or DSL modem connection in our living room, where we have our entertainment unit, so there's no point in setting up the AirPort Express to form a separate network.
- Extending our network with AirPort Express possibly could interfere with our neighbors' wireless networks.

TIP

You can, however, have your cake and eat it, too. The AirPort Express, as Chapter 3 points out, supports as many as five separate setups. For example, if you ordinarily want to use the AirPort Extreme as a set of dedicated speakers, but want to use it as a main base station when you travel, do that with the AirPort Admin Utility. Simply create a second profile for the AirPort Express and modify that profile's settings to have the AirPort Express serve as a main base station, as described in Chapter 4. You can then use the AirPort Admin Utility to switch between profiles.

Recapping AirPort Admin's Music tab settings

As discussed in Chapter 4, the AirPort Admin Utility provides a Music tab when you configure an AirPort Express base station. Under this tab, you find the few settings you can select to configure the AirPort Express AirTunes feature. Figure 7-4 shows these settings.

Figure 7-4: AirPort Admin Utility provides some minimalist AirTunes settings.

The AirTunes settings you can make with the AirPort Admin Utility are

- **Enable AirTunes on this base station:** Seems obvious, and it is — check this box to turn on the AirTunes feature, and deselect to turn it off.
- **Enable AirTunes over the Ethernet port:** If you connect your AirPort Express to your *local area network (LAN)* by plugging an Ethernet cable into the base station's Ethernet port, non-AirPort-enabled Macs on the network can send music to the base station when you put a check this box. The option appears grayed out, as it does in Figure 7-4, if the AirPort Express does not have an Ethernet connection.
- **iTunes Speaker Name:** Whatever you type in this field appears in the iTunes remote speakers pop-up list when iTunes detects the presence of the AirPort Express on the network. This name does not have to be the same name as the one you assign to the base station itself.
- **iTunes Speaker Password:** Type a password in this field, and retype it in the Verify Password field, when you want to restrict use of the AirTunes feature. To remove a password after you set one, delete it from the two password fields.

When you click Update in the AirPort Admin Utility, the setting changes you have made take effect.

REMEMBER

You may never need to visit the Music tab in AirPort Admin Utility. The AirPort Setup Assistant sets up AirTunes the first time you configure your AirPort Express base station and those settings will probably suit your needs.

Connecting the AirPort Express

Hooking up your AirPort Express to your audio playback equipment is as easy, or as complicated, as you would like it to be. That is, the more sophisticated your audio playback equipment, the more choices you have, and with choice comes complexity.

REMEMBER

Keep in mind some of the advice presented in Chapter 5: Electronic components can generate radio interference that may reduce or, in extreme cases, block the Airport Base Station's wireless signal. Television sets and high-power amplifiers in particular can generate such interference. You should place your base station as far away from such equipment as practical. In addition, the higher off the floor that you place the AirPort Express, the less likely it is that its wireless signal will have to travel through various radio signal–absorbing objects, such as chairs, sofas, and coffee tables.

Miniplugs

Powered speakers provide the simplest setup. In many cases, powered speakers already come with a 3.5 mm stereo miniplug designed to plug into the headphone jack or line-out jack of devices like a portable CD player, an iPod, or your desktop Mac. Just plug the speakers into the AirPort Express audio port and turn the speakers on: connection accomplished.

RCA connecters

Some powered speakers, however, and many consumer stereo amplifiers, employ RCA connecters, which have been used on high-fidelity equipment for decades. The male RCA connector, an example of which is shown in Figure 7-5, consists of a metal post encircled by a metal crown.

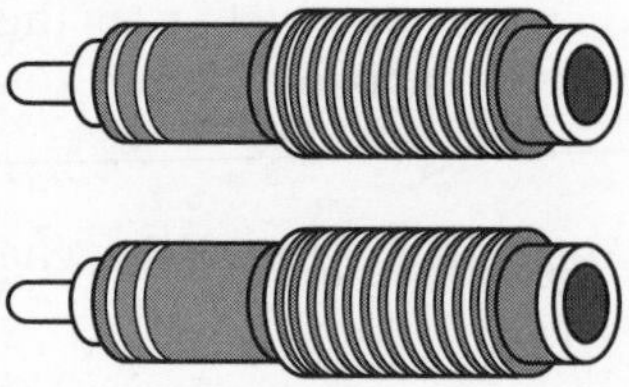

Figure 7-5: RCA plugs have been used on stereo equipment for a long time.

Most RCA connecting cables have male connectors on each end, which plug into the corresponding female RCA jacks on the back of your audio playback equipment and on the component you are connecting to it. Unfortunately, the AirPort Express does not provide RCA jacks, so you need to obtain either a cable that has RCA connectors on one end and a stereo miniplug on the other, or an adapter, such as the one shown in Figure 7-6, that has female RCA jacks on one side and a stereo miniplug on the other. Fortunately, you can find such cables and adapters in nearly all audio stores and electronics shops, as well as in most department stores that sell audio equipment.

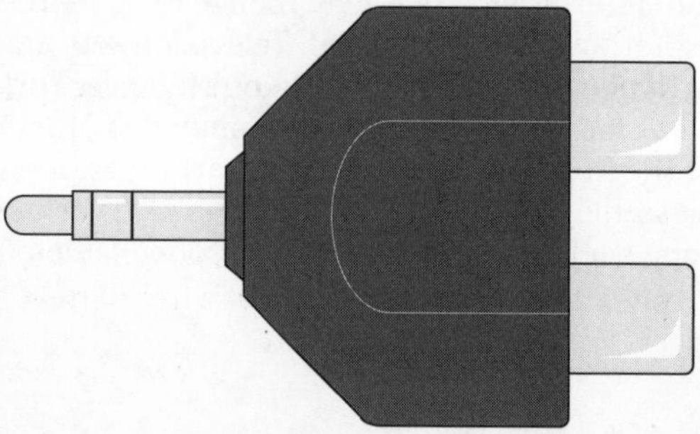

Figure 7-6: A simple RCA-to-stereo miniplug adapter.

Digital optical connections: TOSLINK and optical miniplug

Recent high-end audio equipment comes not only with RCA-style connectors but a whole bevy of other specialized connectors, including digital optical connectors. *Digital optical connections* use optical fibers instead of wires in the connecting cables, and they sport various types of optical connectors at each end of the optical fiber cable. The most common of these connectors is the TOSLINK connector, shown in Figure 7-7. The connector is named for Toshiba, the company that developed it.

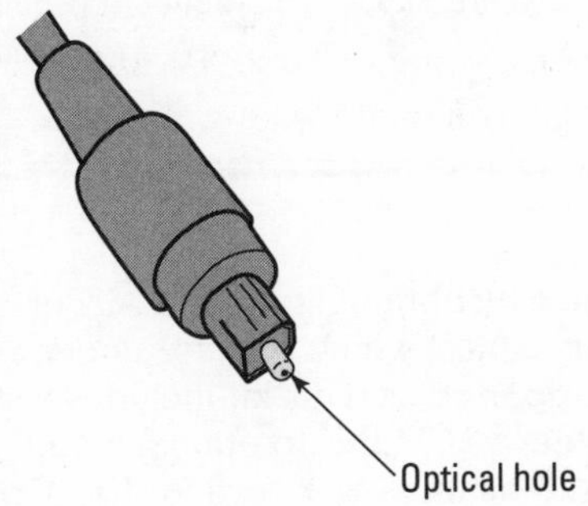

Figure 7-7: The TOSLINK digital stereo connector uses light to make music.

By the way, this connector goes by many names: TOSLINK is the registered Toshiba trademarked name, but you also may see it referred to as Toslink, TOSlink, and Tos-link, as well as by its generic name, EIAJ optical. The particular connector shape shown in Figure 7-7 also has a name: JIS F05. Furthermore, the format of the data usually delivered via the connector has its own acronymic name: SPDIF — or, sometimes, S/PDIF — which stands for Sony/Philips Digital Interconnect Format.

The AirPort Express, however, does not have a standard TOSLINK connector, but its audio port does accept an optical miniplug. This kind of plug looks almost exactly like an analog stereo miniplug, except that the optical miniplug has a small hole in its tip into which the digital optical signal from the AirPort Express shines. You have two options:

- A TOSLINK-to-optical miniplug adapter to connect a standard TOSLINK cable to your AirPort Express
- A cable that has a TOSLINK connector at one end and an optical miniplug at the other

You can obtain such adapters or cables at most high-end audio component stores, or from online vendors: Use the search terms *toslink miniplug* with Google and you should get a few thousand hits worth investigating.

Showing off your AirPort Express

Rather than have your AirPort Express base station hanging from an outlet like a sophisticated wall wart, for fewer than $25 you can obtain Griffin Technology's AirBase, a weighted, chrome-finished stand for the AirPort Express. The stand holds the base station so that its antenna is effectively oriented and so that its status light can easily be seen. The AirBase comes with an eight-foot power cable integrated into the stand, into which the AirPort Express connects after you remove the AirPort Express's detachable plug. The AirBase also has a metal loop on its back through which you can neatly route the Ethernet and audio cables that you plug into the base station. You can find out more about this product at Griffin's site at `www.griffintechnology.com`.

You can trade money for convenience by obtaining Apple's AirPort Express Stereo Connection Kit with Monster Cables, which sells for under $40 at Apple's online store: `www.store.apple.com`. This kit includes a stereo miniplug-to-left/right RCA audio cable, a TOSLINK-to-optical miniplug cable, and, as a special added bonus, an extension power cord for the AirPort Express. The cables, including the power cord, each measure at least six feet in length, so you won't have to wedge the AirPort Express into a tight space behind your audio equipment if you use the cables.

Playing music with AirTunes

With the AirPort Express configured and wired up, all you have left to do is to play that funky music. Or, rather, have iTunes play that funky music: In its current incarnation, AirTunes officially works only with Apple's iTunes, though, as you'll discover a little later in this chapter, you can overcome that limitation.

Setting up iTunes to use AirTunes requires no work because iTunes comes configured to do just that. You can, however, turn iTunes' airworthiness on or off with its audio preferences, as shown in Figure 7-8.

To see iTunes' audio preferences, do the following:

1. **Launch iTunes.**
2. **Choose iTunes ➪ Preferences or press ⌘+ , (comma).**
3. **Click Audio in the preference window's toolbar.**

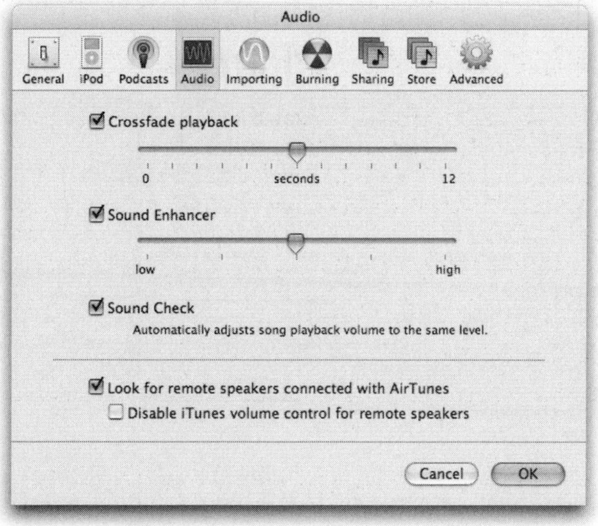

Figure 7-8: iTunes' audio preferences control its access to AirTunes.

As Figure 7-8 shows, iTunes' audio preferences offer you two AirTunes-related options:

- **Look for Remote Speakers Connected with AirTunes.** Deselect this check box only if you don't want to use AirTunes. You may want to do this if you don't have an AirPort Express base station, but it doesn't hurt to leave the option selected, as it is by default.
- **Disable iTunes Volume Control for Remote Speakers.** Check this option if you prefer to control iTunes' playback volume entirely with your audio equipment's volume controls. iTunes then transmits your music at its maximum volume setting to the AirPort Express base station.

When you set iTunes to look for remote speakers, the iTunes window gains a remote speaker pop-up menu button at its bottom right whenever iTunes detects a nearby AirTunes-compatible base station. In Figure 7-9, the button appears with the label Computer. You click this button to see a list of nearby wireless base stations offering AirTunes service, and choose the one you want to receive iTunes' musical transmissions. As described in the "Inside AirTunes" sidebar in this chapter, AirPort Express stores a few seconds of music internally before it starts playing; so, when you choose to play your music through remote speakers, you hear a slight pause as the music loads into the base station's music buffer before playback starts.

Figure 7-9: Click the Computer button to see the iTunes remote speaker list.

AirTunes plays through only one set of speakers at a time, so when you play music through a remote speaker connection, you can't simultaneously hear the music from your Mac's speakers. The remote speaker named Computer, however, refers to your Mac's speakers, and you choose this set of speakers when you want to stop transmitting music using AirTunes and play your music normally on your Mac.

If you have configured your base station to require a password before allowing AirTunes playback, as described in "Setting up AirTunes" earlier in this chapter, two things happen:

- A padlock appears beside the remote speaker name, as shown in Figure 7-10.
- When you select the password-protected remote speaker, you must satisfy a Remote Speaker Password dialog in order to play music. See Figure 7-11.

Figure 7-10: Padlocked speakers require a password before you can use them.

Figure 7-11: Enter a password to play your tunes.

You can check the Remember Password check box so you won't have to enter a password each time you use iTunes with your protected AirTunes connection.

Controlling AirTunes remotely

AirTunes' ability to waft music around your home frees the iTunes Library from the confines of your Mac, but such freedom has its price: AirTunes can't control the music after it's free. When you sit down at your Mac, set up a playlist, choose the AirTunes remote speakers, and press play, you have to go to where you've placed your AirPort Express to do your listening, and listen is all you can do when you get there. To change playlists, or pause a song, or, for that matter, even see the name of the song currently playing, you have to get back up off the sofa and amble back to your Mac. This doesn't present much of a problem if your Mac is a laptop and you have it with you on the sofa, but it does when the Mac is in your home office and your stereo is in your living room. An AirTunes remote control would come in very handy. Fortunately, although Apple doesn't supply an AirTunes remote control, others have stepped into the breach.

AirClicking AirTunes

Griffin Technology, a company that seems dedicated to providing inexpensive devices to fill small gaps in Apple's product line, fills in the AirTunes remote-control gap with its under-$40 AirClick USB. As with other Griffin offerings, the AirClick has a minimalist design that harmonizes with Apple's own aesthetics. The hardware consists of three items:

- **Receiver module:** This is a slender, white plastic–encased receiver with a USB connector and a small light that flashes when the module receives a command from the remote control. A hinge between the USB connector and the rest of the module allows it to swivel 90 degrees from the connector plug's angle.

- **Remote control:** This small, white-plastic, five-button control weighs less than an ounce, and uses a battery that Griffin says should last as long as the remote does, although the documentation does state the type of battery — CR2032 — and offers replacement instructions. The back of the control has a spring-loaded plastic clip so you can attach it to a pocket or sleeve. The buttons it provides are
    - **Play/Pause:** Starts or stops playback for programs like iTunes, iDVD, and QuickTime Player.
    - **Forward and Reverse:** These two buttons skip one track forward or backward in iTunes when pressed briefly. When held down, they scan forward or backward in the current track.
    - **Volume Up and Volume Down:** These two buttons raise and lower the playing program's volume setting. After the program's volume setting reaches either limit, continuing to hold the button down raises or lowers the Mac's system volume.
    - **USB extension cable:** You can use this 44-inch white cable to move the receiver a few feet from your Mac in order to improve reception.

Included with the hardware is a CD-ROM containing the software that allows the AirClick to communicate with various programs, including iTunes. The software installer adds the AirClick application to your user account's login items, making it available whenever you log in, but you might never notice it: It presents no windows and has no Dock icon. AirClick does add an icon to the right side of the menu bar, which, when you click it, presents the menu shown in Figure 7-12.

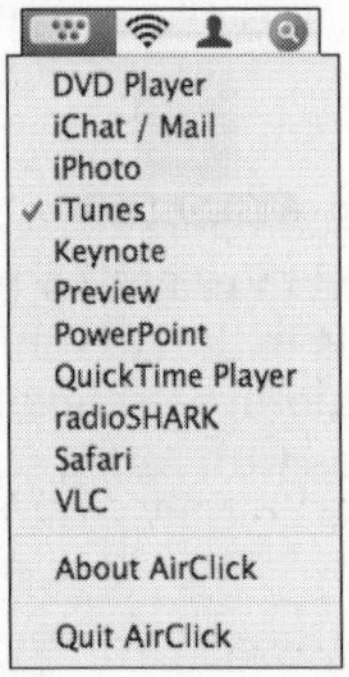

Figure 7-12: The AirClick menu is the only visible presence the program provides.

The remote control and receiver communicate using radio waves, and Griffin claims the transmission range can reach as far as 60 feet. And, like an AirPort network signal, the control's signal can pass through walls, with some attenuation: In my home, for example, the control could almost always make contact with a receiver placed 15 feet away on the far side of two intervening walls, a tiled shower, a bathroom cabinet, and a wooden bookcase.

The control does not have a lot of features, but, like the iPod Shuffle, what it does it does well, and is well worth considering as an AirTunes controller add-on. Nor should you forget that it can control programs other than iTunes: Coupled with the AirFoil application, described later in "Bypassing AirTunes with AirFoil," the AirClick becomes a rather versatile media controller.

Commanding AirTunes with Keyspan Express Remote

Keyspan, `www.keyspan.com`, has spent the last decade making peripherals for the Mac, including various models of their infrared remote control devices. One of their latest remotes, the $60 Keyspan Express Remote, looks like a shiny white version of some of their previous models, but with one big difference: It can connect directly to the USB port on the AirPort Express base station. See Figure 7-13.

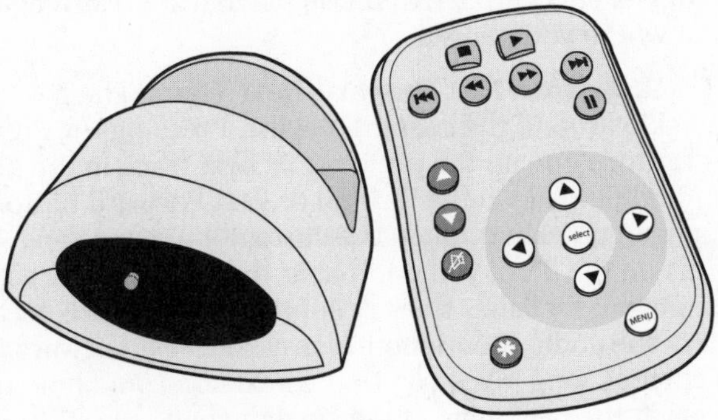

Figure 7-13: The Keyspan Express Remote's two components.

Although the control's receiver can plug into one of your Mac's USB ports and can control a vast number of programs after you load the supplied software, using the device directly with the AirPort Express and AirTunes is probably as easy as any setup process you'll encounter:

1. **Plug the receiver's USB cable into the USB port on the AirPort Express.**
2. **Launch iTunes and connect to the AirPort Express.**

   See "Playing music with AirTunes," discussed earlier, if you don't recall how to do this.
3. **Press the Play button on the Keyspan Express Remote.**

You don't need to load any special software or configure any options: The Keyspan Express Remote comes out of the box knowing how to control AirTunes through the AirPort Express.

The Keyspan Express Remote does require that your AirPort Express have firmware version 6.1.1 or later loaded into it. However, the current AirPort Express firmware version is 6.2 as I write this, so you shouldn't encounter any problems.

As you can see in Figure 7-13, the Keyspan control provides quite a few buttons — 17 of them, in fact. Although the five white buttons at the control's lower right become active only when you attach the receiver to your Mac and use the device to control other applications, the other 11 buttons give you as much control over AirTunes as any couch potato could desire:

- **Stop/Play:** These two buttons along the control's top respectively halt and start AirTunes playing. After you press the Stop button, pressing the Play button starts the track playing from the beginning. The Play button acts like a toggle: press it once to pause playback, press it again to resume from where you paused.
- **Previous Track/Rewind/Fast Forward/Next Track:** The next four buttons let you skip around the current playlist. Pressing the Previous and Next Track buttons cue up the previous or next track in the playlist. Pressing and holding down the Rewind or Fast Forward buttons moves iTunes' playhead quickly through the currently playing song. Unlike such controls on the iPod, you don't hear the song playing as you fast forward or rewind: AirTunes doesn't support that feature because of the way it buffers the audio. Continuously pressing Fast Forward beyond the end of the current song takes you to the next song, and holding down Rewind stops at the beginning of the current song.
- **Pause:** Like the Play button, this button toggles between playback and pause.
- **Volume Up/Volume Down/Mute:** You can find these three buttons along the left side of the control. The two volume buttons move the iTunes volume slider in the appropriate direction. Unlike the Griffin AirClick, they don't affect the System volume when they reach the limits of the iTunes volume control. The Mute button toggles between suppressing audio and resuming audio at the current volume setting, much like iTunes' Mute command on its Control menu.
- **Cycle:** Pressing the asterisk-labeled button, which Keyspan calls Cycle, turns on iTunes' shuffle feature if it is currently off. Pressing it again takes you to a random song in the current playlist.

Keep in mind that the Keyspan remote control uses infrared technology rather than radio technology, so you need to position the receiver within line of sight of the control when you use it. You can, however, use it across a rather large room: Keyspan claims the range of the control can reach 40 feet, which is larger than any room in which I was able to use the device.

Streaming Music Wirelessly Beyond AirTunes

AirTunes, as the previous section points out, is Apple's name for its particular blend of technologies that stream music wirelessly from your iTunes music library, through your AirPort Express base station, to an external audio system. But you don't have to use AirTunes to play sound from iTunes or any other application through your audio system. Nor do you need an AirPort Express to play music wirelessly from iTunes.

Bypassing AirTunes with AirFoil

The clever code-crafters at Rogue Amoeba, whose slogan is "Good software with a bad attitude," have produced Airfoil, a program that, like Apple's AirTunes feature, can send audio data from your Mac to an AirPort Express. Airfoil takes the sound-transmission trick a step further: Using some of the same software technology the company's Audio Hijack recording software uses, Airfoil taps into Mac OS X's sound-processing system to capture the audio produced by almost any application, which AirFoil repackages and sends to the AirPort Express.

So, although AirTunes works only with iTunes, AirFoil can send audio to your AirPort Express from just about any application, including popular media players like

- **RealPlayer**
- **Windows Media Player**
- **QuickTime Player**
- **MPlayer**

When used with iTunes, Airfoil also performs a trick that AirTunes can't: You can play your music both through your Mac's internal speakers and through the remote speakers connected to your AirPort Express at the same time.

Figure 7-14 shows Airfoil in action, doing for iTunes what AirTunes does. Notice, as well, that Airfoil detects the type of speakers connected to the AirPort Express, indicated by the Analog badge that appears in Airfoil's Remote Speakers list in the figure.

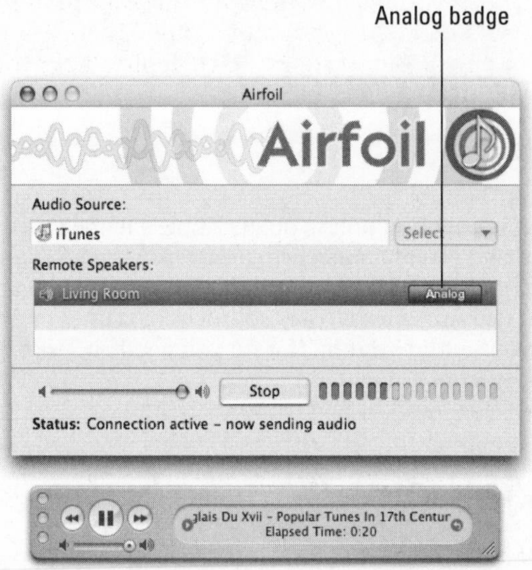

Figure 7-14: Airfoil takes AirTunes' features to a higher level.

If you have set up your AirPort Express to use AirTunes, it is already set up to use Airfoil. To play sound with Airfoil, follow these steps:

1. **Start the Airfoil program.**
2. **Click a set of remote speakers in the Remote Speakers list.**

   This list displays all the operating AirPort Express base stations within range. You can select only one set of remote speakers at a time.
3. **Click the Select pop-up menu and choose an application.**

   The application provides the audio that Airfoil plays through the remote speakers. Figure 7-15 shows the Select pop-up menu and its contents.

   You can choose an application from one of the three categories shown in the Select menu:

   - **Other Application:** Click the Select Application choice in this category to see a file selection dialog from which you can choose the application Airfoil uses as the audio source.

- **Recent Applications:** The items listed in this category are those that you have used previously with Airfoil.
- **Open Applications:** This category lists currently running applications that Airfoil can use as sound sources. When you choose an application that isn't currently running, Airfoil starts up the application for you. If you select an application that's already running, Airfoil presents the dialog shown in Figure 7-16, telling you that it must quit and then restart the selected application in order to grab its sound output.

Figure 7-15: Airfoil offers a choice of applications to use as an audio source.

Figure 7-16: Airfoil normally can't grab sound from an already running application.

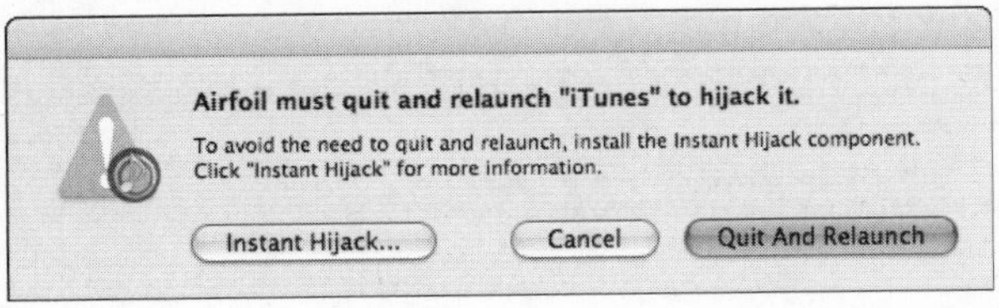

4. **Click Transmit.**

   Airfoil begins transmitting the sound produced by the selected application through the external speakers.

TIP

You can avoid the dialog that appears when you choose a running application by clicking the Instant Hijack button, which presents Airfoil's Install Extras window. From the Install Extras window, shown in Figure 7-17, you can install an Instant Hijack software component that allows Airfoil to grab the sound output from an already running application. If you plan to use Airfoil regularly, install the component. Later, if you decide you don't want the component installed, you can choose Install Extras from the Airfoil menu, and then click the Uninstall button in the Extras Installer window to remove the Instant Hijack component. And be sure to read the information presented in the Extras Installer window if you feel at all uneasy about installing the Instant Hijack component. It provides background information about the component, as well as instructions for limiting its use to a single user account on your Mac.

Airfoil provides a small set of preference check boxes you can set by choosing Airfoil ➪ Preferences, or by pressing ⌘+ , (comma). Figure 7-18 shows the rather spartan Airfoil Preferences window.

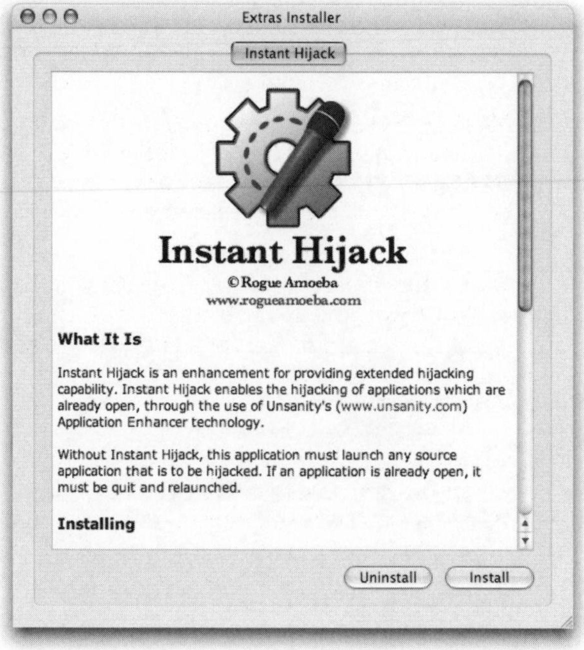

Figure 7-17: The Instant Hijack component lets Airfoil grab sound from running applications.

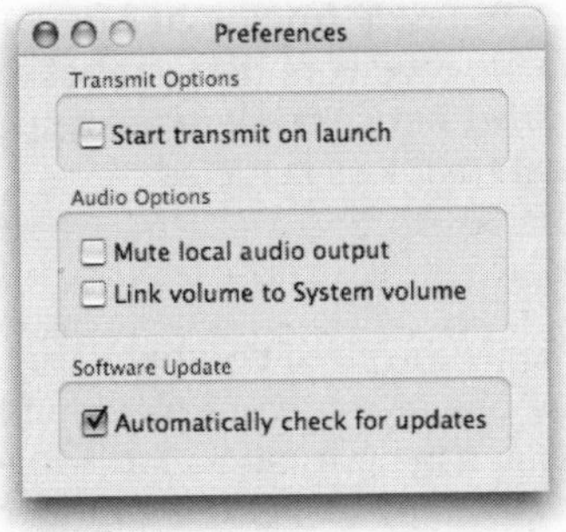

Figure 7-18: Airfoil's user preferences are few.

Airfoil provides four preference check boxes:

- **Start Transmit on Launch:** Check this setting to have Airfoil open the last application you used with it, and to route its audio immediately to the AirPort Express.
- **Mute Local Audio Output:** To have Airfoil behave more like AirTunes, check this setting, and audio sent to the AirPort Express won't play through your Mac's internal speakers.
- **Link Volume to System Volume:** When you check this setting, you can control volume through the AirPort Express with your Mac's audio volume controls, available in System Preferences or by using the volume keys on your keyboard. This setting disables the volume slider in Airfoil's main window.
- **Automatically Check for Updates:** Check this setting to have Airfoil contact the Rogue Amoeba Web site upon starting to see if any updates exist and, if they do, to give you the opportunity to download them. If you disable this setting, you can still check for updates by choosing Airfoil➪Check For Updates.

You can download Airfoil directly from Rogue Amoeba's Web site at `www.rogueamoeba.com`. The software costs $25 to register, but you can try it for free. Each time you launch an unregistered copy of the program, it plays ten minutes of high-quality audio and then introduces distortion in the audio signal as an incentive to pay the registration fee.

Using other wireless audio players

Just as AirTunes does not provide the only avenue to streaming audio from your Mac to an AirPort Express, an AirPort Express does not provide the only destination for audio that you stream wirelessly from your Mac. Several companies manufacture wireless digital music receivers that can receive and play sound sent from your Mac over an AirPort network. Here are a couple such devices you can use to take the musical road less traveled.

Rocking with the Roku SoundBridge

Roku means "six" in Japanese, and Roku is the sixth company that has made a business out of producing and selling one or more of Roku founder Anthony Wood's inventions. Roku's silver-and-black cylindrical SoundBridge is one of the latest devices developed by Wood, who also invented one of the first digital video recorders.

The SoundBridge comes in three models, which differ primarily in their appearance:

- **The M500:** The 12-inch, low-end SoundBridge provides a two-line liquid crystal display that shows two lines of text naming the currently playing song and to use for setting up the device.
- **The M1000:** This model, also 12 inches long, provides a sharper and more flexible fluorescent display that can show both text and graphics, the latter used for graphically displaying volume levels and various audio visualizers, which are described later.
- **The M2000:** The fluorescent display on this 17-inch model is the brightest and the largest of the three, with its fluorescent display providing more pixels and larger text.

All three models come with a curved rubber cradle, allowing you to place them securely on a table or shelf and to rotate their bodies for maximum display visibility. Prices range from $150 to $400, and you can purchase the device directly from the company at `www.rokulabs.com`.

Here's some of what you get for your money:

- **Both wired and wireless operation:** Beneath one of the SoundBridge's removable end caps is both a slot for the device's included AirPort-compatible 802.11g wireless network card and an Ethernet port, for purchasers who don't have a wireless network.
- **Digital and audio outputs:** Instead of the AirPort Express's multipurpose miniplug output, beneath the SoundBridge's other removable end cap are two RCA connectors for analog stereo, a TOSLINK digital optical output jack, and another RCA connector for its SPDIF coaxial port.
- **Handheld remote:** Designed for the SoundBridge, this device lets you select songs, compile playlists, set options, control volume, and even enter your network password when you first connect to your AirPort network.

- **Preprogrammed Internet radio stations:** Even when your Mac is off, the SoundBridge can access a number of Internet radio stations through your AirPort network's — or wired network's — Internet connection.
- **Visualizers:** The SoundBridge provides several displays that visually illustrate the music. Although not as stunning as the multicolor iTunes visualizer, the SoundBridge provides an attractive real-time view of the music being played, and they can be rather stunning on the largest model's spacious fluorescent display.
- **Compatibility with several commercial player formats and services:** In addition to iTunes compatibility, the SoundBridge also works with Napster, Windows Media Connect, and Rhapsody — though these last won't matter much to Mac users.

If you have an unprotected AirPort network (which you really shouldn't), the SoundBridge connects to your network with ease. However, a protected network requires that you spend some quality time punching buttons on the SoundBridge remote: You must enter the AirPort network's password character-by-character by scrolling through and picking from a character list presented on the SoundBridge display, using the remote's arrow buttons. What's more, because the SoundBridge requires you to use the AirPort network's hexadecimal-equivalent password instead of the alphabetic one, the process requires painstaking attention and patience. On the other hand, the SoundBridge remembers the password after you enter it, even when you power down the device and then turn it back on, so you have to go through this minor ordeal only once.

REMEMBER

To find out your AirPort network's hexadecimal password, use the AirPort Admin Utility. Open the base station's configuration and choose Base Station➪Equivalent Network Password. A sheet drops down with the string of *hexadecimal* digits — consisting of the numbers 0–9 and the letters A–F — that constitute the password. Write this down unless you have an incredibly good memory, or unless you have the SoundBridge sitting beside your computer screen.

The SoundBridge works directly with iTunes; you don't have to install any additional software, although you do need to have iTunes running and set up to share those playlists that you want to use with the SoundBridge. But because SoundBridge uses the same network port numbers that iTunes' sharing feature does, you don't need to create any special firewall settings to send your music to the SoundBridge.

Note, however, that the SoundBridge, like most other third-party music players, can't play any songs you have purchased from the iTunes Music Store because of *digital rights management (DRM)* restrictions.

You can use iTunes' Smart Playlist feature to collect all the unprotected music in your iTunes library. The Smart Playlist requires you to specify just one condition, as shown in Figure 7-19: Kind is not protected AAC audio file. Setting this condition creates a playlist that includes MP3s, unprotected AAC files, AIFF files — in fact, any kind of file that iTunes can play other than protected iTunes Music Store purchases.

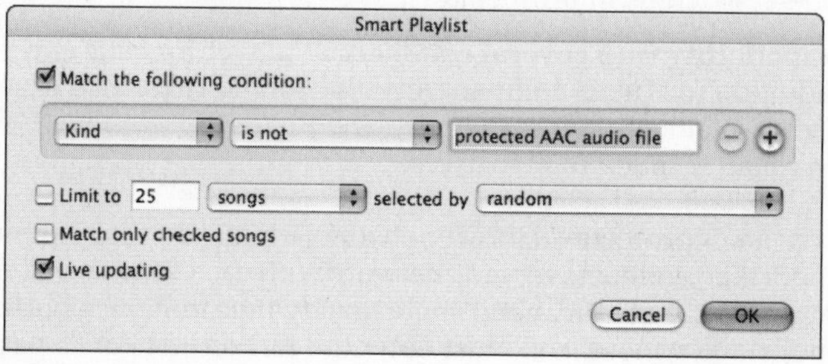

Figure 7-19: Make a smart playlist of songs you can play with Sound Bridge.

If you don't need the networking features that the AirPort Express provides, the SoundBridge is an attractive-looking and excellent-sounding addition to your collection of audio components.

Serenading with the Slim Devices Squeezebox3

The Squeezebox3 from Slim Devices provides many of the same features as its cylindrical competitor. Although the Squeezebox3 requires you to install some additional software on your Mac, unlike the more self-contained SoundBridge, this is not a bad thing: The free open-source SlimServer software, which you download from `www.slimdevices.com`, lets you tailor your wireless music experience in ways that go beyond what iTunes and either AirTunes or SoundBridge can offer by themselves.

The flat-panel unit, available either in high-gloss black or white, requires almost four inches of clearance, when set up on its integrated stand.

After you decide where you want to put Squeezebox3, here's what you get:

- **Both wired and wireless operation:** Like the SoundBridge, you can connect the Squeezebox3 to your network wirelessly, or via Ethernet cable.
- **Digital and audio outputs:** Along the back of the Squeezebox3 are an analog stereo miniplug jack, two analog RCA connectors, a TOSLINK digital optical SPDIF connecter, and an RCA digital coax SPDIF connector.
- **Handheld remote:** The Squeezebox3 remote, though petite, comes with a full range of controls, including arrow buttons to navigate the display, play, pause, reverse, forward, shuffle, repeat, volume, search, sleep, and

more. It also has a set of number buttons that also have alphabetic characters printed above them, laid out like a telephone keypad, making entry of network passwords and song titles manageable, if not completely effortless.

- **SqueezeNetwork access:** Your Mac doesn't have to be running for the Squeezebox3 to connect to various Internet radio stations, courtesy of the company's SqueezeNetwork. The device can connect to this network as long as it has an Internet connection. The SqueezeNetwork provides other goodies as well, including news headlines that scroll along the SqueezeBox3's display, an alarm clock function that provides a long list of wake-up sounds, and even a selection of natural sounds so that you can enjoy the sonic splendor of chirping crickets or a babbling brook piped into your home, using the latest digital technology.

  TIP: You should consider creating an account on the SqueezeNetwork site at `www.squeezenetwork.com` and enter your Squeezebox3's Player Identification Number. With this account you can create custom lists of favorite Internet radio stations, set your Squeezebox3's alarm clock, and add sources for headline news feeds, such as technology or business news sites.

- **Visualizers:** Although not particularly mind-blowing, the Squeezebox3's set of three visualizers shows you volume- and sound-spectrum displays much like the ones you might see on other high-end audio systems. You can log on to the SqueezeNetwork's Web site for your account to set your preferred visualizer, or you can do it from the remote control.

As you may have started to realize, the Squeezebox3's brains more than its beauty make it an attractive AirTunes replacement option. Not only does the SqueezeNetwork connection enhance Squeezebox3's intelligence beyond that of a standalone music player, but the SlimServer software that drives the Squeezebox3 from your Mac gives it another brain-boost as well.

Once you download and install the SlimServer software, it takes up residence among your System Preferences, as shown in Figure 7-20.

Open the SlimServer preference panel, as shown in Figure 7-21, and you see a deceptively simple interface:

- **Start Server:** This button lets you start up the server software. After the server starts, this button changes to Stop Server.
- **Automatically Start pop-up menu:** You can have the SlimServer software start whenever your Mac does, whenever you log in, or never start automatically at all, which requires you to click the Start Server button.
- **Web Access:** This final button, which becomes enabled when the SlimServer is running, opens the gateway to all the SlimServer's features, which appear in a control panel presented in your Web browser.

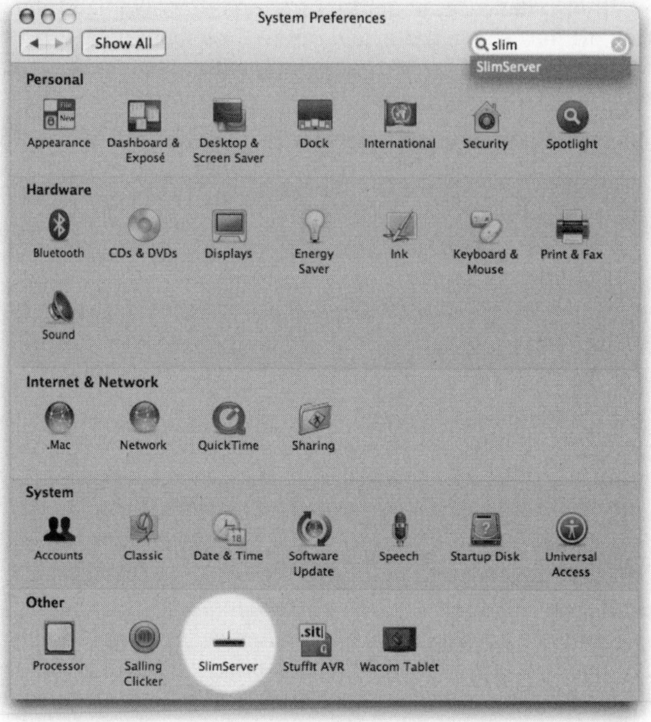

Figure 7-20: You access the SlimServer in your System Preferences.

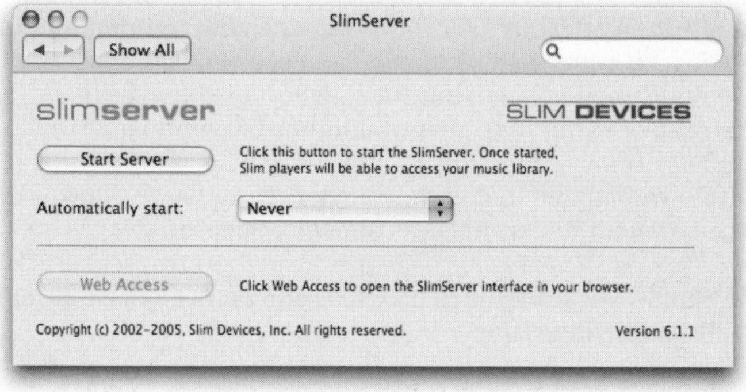

Figure 7-21: The SlimServer preference panel starts the server that feeds the Squeezebox3.

WARNING!

If your Mac's firewall, which protects your Mac from unwanted network requests, is turned off, you can simply click the SlimServer's Start Server button, click the Web Access button, and start feeding your Squeezebox3 from your own personal picnic basket of music. However, in these days of Internet hacking and thuggery, running your Mac with the firewall off is not recommended. So, if you *do* have your firewall up, you need to open a couple of ports in it to let the Squeezebox3 talk to your SlimServer.

Opening your firewall just enough to allow the Squeezebox3 to communicate with your Mac is easy, and you have to do it only once:

1. **Open System Preferences.**
2. **Click Sharing.**
3. **Click the Firewall tab.**
4. **Click New.**

   A sheet descends, presenting options for creating a new firewall opening. Figure 7-22 shows the settings you need to enter to open the firewall just enough to let the SlimServer shine through.

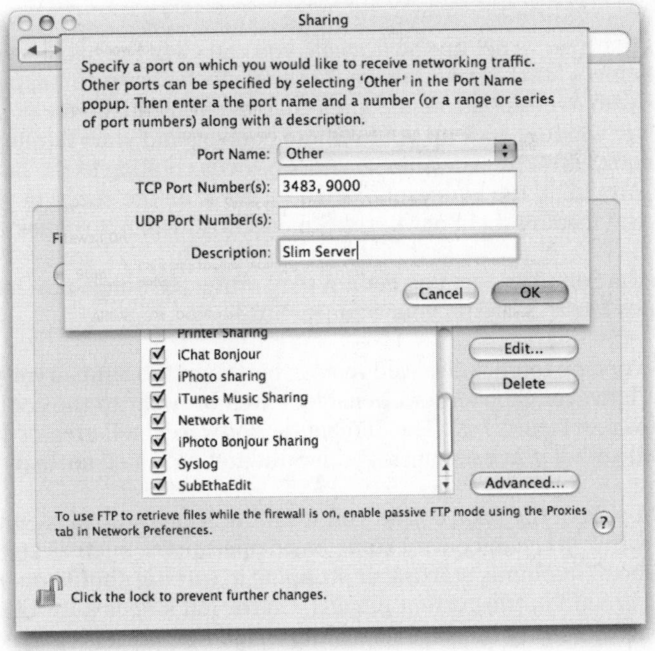

Sharing
Specify a port on which you would like to receive networking traffic. Other ports can be specified by selecting 'Other' in the Port Name popup. Then enter a the port name and a number (or a range or series of port numbers) along with a description.
Port Name: Other
TCP Port Number(s): 3483, 9000
UDP Port Number(s):
Description: Slim Server
Cancel
OK
Printer Sharing
iChat Bonjour
iPhoto sharing
iTunes Music Sharing
Network Time
iPhoto Bonjour Sharing
Syslog
SubEthaEdit
Edit...
Delete
Advanced...
To use FTP to retrieve files while the firewall is on, enable passive FTP mode using the Proxies tab in Network Preferences.
Click the lock to prevent further changes.

Figure 7-22: The firewall settings needed by the SlimServer.

5. **Click the Port Name pop-up menu and choose Other.**

   The firewall has a number of presets, but the SlimServer is not among them. Choosing Other lets you create a new firewall setting and give it a name.

6. **In the TCP Port Number(s) field, enter** 3483, 9000.

   Figure 7-22 shows how this field should look. The Squeezebox3 communicates with the SlimServer through these two TCP ports.

TCP ports are not the same thing as the physical ports (such modem or Ethernet ports) on the back of your Mac. Every packet of information that travels over your network has a TCP port number associated with it, assigned according to the kind of information it carries. The SlimServer and Squeezebox3 tag the information packets they send over the network with one of the two port numbers given here, and they pay attention only to network traffic that uses either of those two port numbers.

7. **In the Description field, enter** Slim Server.

   Actually, you can enter anything you like in this field, but you should enter something that will remind you what this new firewall setting does.

8. **Click OK.**
9. **Close the Sharing window.**

So, now that you've set up the firewall, you can click the SlimServer preference window's Start Server button. Nothing much seems to happen, except that the Web Access button becomes enabled. You can, if you like, close the SlimServer window, pick up your remote control, and start fiddling with the Squeezebox3 directly. The SlimServer continues running in the background silently, providing the Squeezebox3 with a list of all the songs in your iTunes library, and feeding them out to the Squeezebox3 when it requests one.

As with the SoundBridge, you cannot play protected songs purchased from the iTunes Music Store on your Squeezebox3.

If you do decide to click the Web Access button in the SlimServer preference window, however, SlimServer opens your Web browser to the control panel page shown in Figure 7-23. The SlimServer software itself creates this Web page, and with it you can control your Squeezebox3 in all sorts of ways.

The right side of the page shows you what the Squeezebox3 is currently up to, and with it you can control some basic operations, such as setting the Squeezebox3's volume, starting or stopping it, turning shuffle on or off, and skipping around in the current playlist. On the left side is a list of options for such tasks as setting up playlists, configuring Internet radio playback, and obtaining a wealth of online help information.

You don't need to open the SlimServer preference pane to get to this Web page. As long as the SlimServer is running in the background, you can enter the address `localhost:9000` in your Web browser to bring up the control page. You can even bookmark the page for faster access.

Playing fairly with FairPlay

The recording industry has a love-hate affair with digital music. On the one hand, the digital audio CD has almost completely replaced vinyl analog records, reducing production costs while commanding higher retail prices, a situation that can't help but warm the hearts of recording industry executives. On the other hand, anyone can easily make an exact copy of the data on audio CDs, using almost any computer that has a CD-ROM drive, and that data can be copied almost infinitely while suffering no loss of quality. This latter fact has given rise to a boom in music piracy that shows little sign of abating.

That's why before the recording industry would agree to Apple's plan for its online iTunes Music Store, Apple had to agree to create a copy-protection technology that would restrict playback of any digital music purchased online to only the purchaser of that music and a few designated others. Apple calls this technology FairPlay, and it allows the music to be played back only on five designated computers at any given time. FairPlay is what restricts songs purchased from the iTunes Music Store from playing on wireless devices like the SoundBridge and the Squeezebox3.

You can get around FairPlay's restrictions legally in two ways. One way is to simply create an audio CD containing the music, and then to recopy the music from the CD back into the Mac. Such redigitized music can be played through digital wireless devices without problem. The recording industry agreed to let FairPlay allow the round-trip redigitizing of protected music for two reasons: First, the FairPlay system keeps track of how many CDs you create from each playlist containing protected music, and it allows only a limited number of audio CDs to be made from any one playlist of protected songs. Second, the round-trip redigitizing process involves first decompressing the music and then recompressing it, which causes a slight, usually almost unnoticeable, drop in fidelity, similar to that which occurs when making an audio tape from a digital recording. Because the copy is not a perfect copy, the recording industry does not mind so very much.

The other legal way to sidestep FairPlay is to send the sound from iTunes to another device while iTunes is playing it legally on your computer. Griffin Technology, for example, manufactures a sleek little USB device called the RocketFM, which it sells for under $40. The RocketFM plugs into one of your Mac's USB ports and acts, as far as the Mac is concerned, as a set of external USB speakers. However, the RocketFM is not a set of speakers, but a small short-range FM radio transmitter that you can set to broadcast on any unused FM radio frequency. Any FM radios within that range, which is only extensive enough to cover the typical home, can tune in to the RocketFM's frequency and play the music back. The recording industry does not find this reuse of protected music troubling either, as the signal covers only a short distance, and the fidelity is only as good as a typical FM radio broadcast: with the RocketFM, you aren't using digital wireless technology, but old-fashioned analog wireless technology. Granted, the RocketFM won't help you play your protected music with a digital wireless music player, but if FM-quality music sounds good enough to you, you don't need a digital wireless music player in the first place.

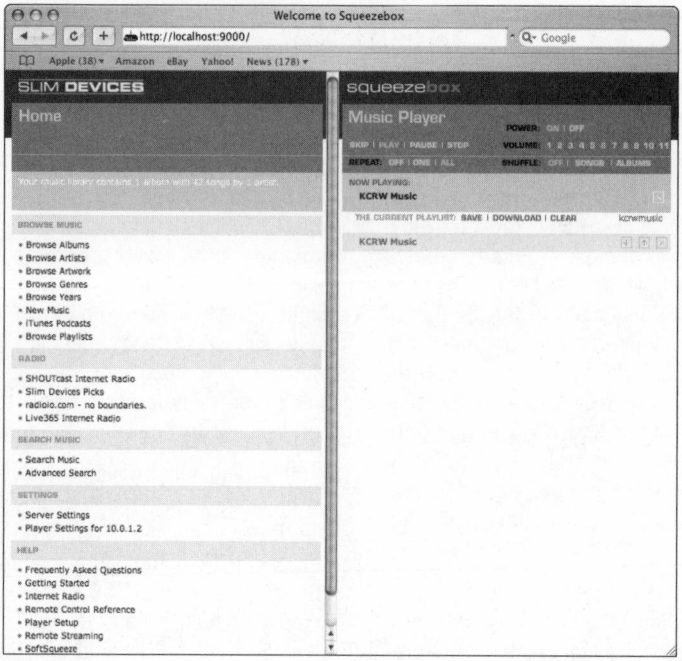

Figure 7-23: This Web page serves as your Squeezebox3's master control.

Because the Slim Devices has made the SlimServer software open source, other digital music players, including the SoundBridge, can make use of it. However, the Squeezebox3 and SlimServer were truly made for each other, and together they make a powerful addition to the digital home entertainment center.

Chapter 8

Saying Bonjour

In This Chapter

- Walking around the local network
- Tossing files through the air
- Chatting to your neighbors
- Weaving a not-so-world-wide Web
- Engaging in group-think

In Meditation XVII of his *Devotions upon Emergent Occasions*, seventeenth-century poet John Donne wrote, "No Mac is an Island." Because of a slight penmanship problem, though, that line has been misquoted ever since. Donne was right: No Mac is an island. Today's Mac hardware incorporates a variety of ports and devices designed to establish connections with other computers, and today's Mac operating system almost obsessively looks for network connections every chance it gets. Even the very first Macs, although developed in the days when personal computers really *were* personal computers, isolated from any others, had the rudiments of networking technology built into them.

And, from the beginning, Apple designed Mac networking for simplicity and ease of use. Although very complex stuff had to go on beneath the surface — and believe me, if you ever spend any time at all looking at the underlying technologies and conceptual details of modern digital communications, you'll gain a new respect for what the word *complex* can mean — Apple wanted you to have to do little more than plug in a connector to establish a network connection and find other networked devices. *Plug and Play* wasn't just a marketing phrase when it came to Apple's first networking attempts: It was a design specification.

Introducing the technology formerly known as Rendezvous

When Apple released Bonjour in 2002, the software bore the name *Rendezvous,* a name alluding to its ability to let various devices meet and greet each other on a LAN. The Rendezvous name, however, turned out to be a trademark owned by another company. As part of Apple's legal settlement with that company, Rendezvous was renamed Bonjour when Apple released Mac OS X 10.4.

And Bonjour has yet another name: *ZeroConf,* short for *zero-configuration networking,* a networking protocol developed by Apple's Stuart Cheshire and released as an open standard soon after its debut. By releasing it as an open standard, Apple has made Bonjour much more attractive to software and hardware developers who might want to integrate it into their products. And many developers do want to integrate it into their products because Bonjour allows them to build networking products that require no user configuration: They just work. When you see a product that its producer says uses Bonjour, or Rendezvous, or ZeroConf, they are all referring to the same thing, just with different names.

A couple of decades later, almost everything about how the Mac networks has changed completely, yet Apple's goal has remained the same: to make networking as simple and painless as possible for you, the user. Apple has dubbed its latest spin on painless Mac networking *Bonjour*. In this chapter, you get a chance to say hello to it.

And I was just kidding earlier. In *Devotions Upon Emergent Occasions,* John Donne didn't really write, "No Mac is an Island," but "No man is an *Iland*." So don't send me any letters about it, okay? Thanks.

Communicating Locally

Bonjour is one of those technologies that is a lot harder to describe than it is to use. In fact, the whole point of Bonjour is that you don't need to do anything to use it: Usually, it just does its job, which is to make some common networking tasks easier. It does this by helping your Mac find network resources without making you deal with addresses or special network configurations.

Bonjour deals with local networking only: local printers, local file servers, and local computers connected to a local network. This makes it ideal for working with devices on the wireless local network that your AirPort Base Station creates.

Taking a closer look at Bonjour

Bonjour does many of the things that Apple's groundbreaking AppleTalk networking protocols did several decades ago, but Bonjour does them in ways that integrate seamlessly into the vast array of interconnected networks that form the basis of today's Internet. You don't absolutely need to know what these things are, though, and if you don't want to, just skip the following list. However, when you use a Bonjour-friendly product, having a basic idea of what Bonjour does might come in handy, and skimming the following list might help you better understand the "Using Bonjour on a mixed network" section in this chapter.

The AppleTalk-like things that Bonjour does include

- **Address assignment:** Every device attached to a network has to have a numerical address so that other devices can find and communicate with it. For example, an AirPort Base Station connected to the Internet usually gets its network address from the *Internet service provider (ISP)* that provides the Internet connection, and the base station, in turn, assigns addresses to the devices on the wireless network that it creates, using its internal DHCP server. See Chapter 4 for more about the DHCP server inside your AirPort Base Station.

  Bonjour, however, provides a way for devices to choose their own network addresses, following certain rules that keep two or more devices on the same network from choosing the same address. This is especially useful for networkable devices like printers that don't have keyboards on which you can enter network addresses. Using Bonjour, the printer can choose its own address without your having to enter anything on the keyboard that the printer doesn't have.

- **Name resolution:** A computer has no problem using the long strings of numbers that make up a network address. People, however, tend to find names easier to use. Take the numeric address of the Web site for this book's publisher: 208.215.179.146 — easy for a computer to remember, but not as easy as `www.wiley.com` is for you to remember. Bonjour offers devices a way to associate a name with a device's numeric address on a local network, and to avoid conflicts when two or more devices on that network want to use the same name. When you browse your network in the Finder and see a name instead of a numeric address in the list of servers, for example, Bonjour probably supplied the name.

- **Service discovery:** Each of the devices connected to a network can provide services to other devices. For example, some devices may offer file-sharing services, some may offer printing services, and some may offer media-sharing services such as iTunes music sharing. Knowing a device's address or name doesn't necessarily help you know its services. Bonjour, though, includes a mechanism that devices can use to announce their services to the rest of the network, and for devices to request the names and addresses of the devices that offer particular services.

What's all this talk about AppleTalk?

Bonjour might be considered AppleTalk, the NeXT Generation: The problems that Bonjour solves are the same problems that Apple's original AppleTalk networking protocols addressed in the mid-1980s. Back when AppleTalk was being born, the Internet was still a research project used mostly by academic and military researchers, and the Internet's underlying TCP/IP communication protocols, which now form the basis for most of today's networks, were just one possible set of protocols that network developers might consider using — at the time, it was one of the more expensive ones to implement.

Apple developed AppleTalk to provide simple plug-and-play networking in support of two early Macintosh-related projects: the never-completed Macintosh Office, which was a file-server and resource-sharing product, and the Apple LaserWriter. This latter, of course, did see the light of day and, in fact, spawned the desktop publishing revolution. Because the original LaserWriter was an expensive item — the first ones cost rather more than a Macintosh computer itself — Apple needed a way to connect it to more than one Mac at a time, so that a small office of Macs could share it economically. Sharing meant networking. However, networking meant, as it still means today, assigning addresses to networked devices, identifying services that the devices offer, and providing names for those devices so that mere mortals can tell which device is which.

AppleTalk handled these tasks by combining protocols like Apple's *Name Binding Protocol (NBP)* and *AppleTalk Address Resolution Protocol* (AARP), among others. With AppleTalk, a LaserWriter could be connected to the AppleTalk network and all the Macs on that network could detect it and print to it. All the nasty details that comprise networking were hidden from the user. As far as a Mac user was concerned, when a LaserWriter was connected to the network, its name immediately appeared in the *Chooser*, a small program in the Mac operating system, so named because it was how users chose the printers and, later, file servers, they wanted to use. Plug and play. Click and use.

AppleTalk was so popular that AppleTalk networking was built into many laser printers besides those sold by Apple. Even today many printers still support it. And even with the advent of TCP/IP-based networks and the explosion of the Internet, AppleTalk still survives, modified so that its protocols can ride piggyback-style atop modern TCP/IP networks. Although slowly falling by the wayside, AppleTalk is still with us, and may be for several years to come.

None of the networking tasks that Bonjour handles depend on the others. For example, although your Mac may have its network address assigned by your AirPort's DHCP server and not by Bonjour, Bonjour can still provide your Mac's network name. Bonjour does not serve as an alternative to other network technologies, but as a collaborator with them.

Using Bonjour on a mixed network

Networks come in a variety of shapes and sizes, ranging from the very simplest, such as a single Mac and a printer, to the great big grandmother network of all, the Internet, which interconnects huge numbers of networks all over the world. Very few networks exist in a completely isolated state: If you have set up an AirPort network, chances are that it not only connects your local devices to each other, but to the Internet as well.

Sometimes when you mix several networks together, things don't go as you might expect. Although Bonjour is designed to work well in a mixed network environment — as it should, given that most networks *are* mixed — it does have certain limits and issues that you might need to address, or, at least, understand.

Communicating within subnets

As I mentioned near the beginning of this chapter, Bonjour deals with local networks only. But what does that mean, exactly? The answer has to do with *subnets.* A subnet is just what it sounds like: a smaller piece of a larger network.

What defines a subnet is the range of valid network addresses within it. A network routing device like an AirPort Base Station handles only the data traveling to network addresses within its subnet, and passes data going to addresses beyond its subnet to some higher authority, such as an ISP's router.

How is that range of network addresses that make up a subnet established? Consider Figure 8-1, which shows the TCP/IP settings for a Mac on an AirPort network. That Mac's address, like all TCP/IP addresses, consists of four numbers, separated by periods: In Figure 8-1, the Mac's IP address is 10.0.1.7. Next, consider the *Subnet Mask* setting shown in the figure: 255.255.255.0. This setting tells the network router (the AirPort Base Station in this case) to ignore all but the very last of the four sets of numbers that make up the TCP/IP address. To oversimplify, each 255 in the subnet mask more or less tells the router to ignore, or *mask,* the corresponding portion of a TCP/IP address and to pay attention only to the portion that corresponds to the 0 in the subnet mask. As a result, the AirPort network can have only 255 valid addresses on it; or, to put it another way, the AirPort's router deals with a subnet comprising no more than 255 addresses.

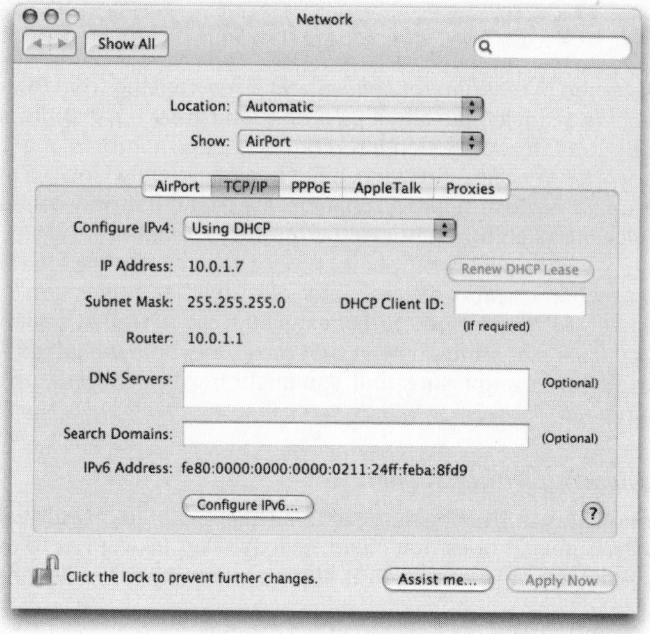

Figure 8-1: The TCP/IP settings pane showing the subnet assignment.

Bonjour works only within the local subnet for very good reasons, and they have to do with the services that Bonjour provides:

- **Address assignment:** Remember that Bonjour allows devices to choose their own addresses, and makes sure that two devices don't choose the same address. If Bonjour had to check each proposed address against the more than four billion possible TCP/IP addresses on the Internet, it would take a very long time to spot and resolve address conflicts. Restricting its checks just to the local subnet speeds things up enormously.
- **Name resolution:** Bonjour lets devices select their own human-readable names and handles conflicts. Handling only a few dozen names on a subnet is a lot easier for Bonjour to do than examining the names assigned to every device on the Internet.
- **Service discovery:** Bonjour-enabled devices announce the services they offer to other network devices. Confining those announcements to the local subnet avoids the network congestion that would ensue if, say, every network printer announced its existence to every Internet-connected computer on the planet.

As you can see, in the vast majority of cases, limiting Bonjour to your subnet avoids all sorts of problems. Sometimes, however, that limitation can chafe. For example, iTunes' music-sharing feature, discussed in Chapter 9, uses Bonjour to allow other computers on your network to listen to your music collection. But if you want to share the music collection on your home computer in Lompoc with, say, your sister who lives in Philadelphia, you can't — at least, not very easily.

Communicating between subnets

Using Bonjour across subnets is possible, but doing so is not for the fainthearted. I won't go into all the details here, because, quite frankly, they're complex, exacting, and require specialized network expertise. If you do want to attempt it, the key is to set up a special Bonjour forwarding system using a free program like Network Beacon, shown in Figure 8-2.

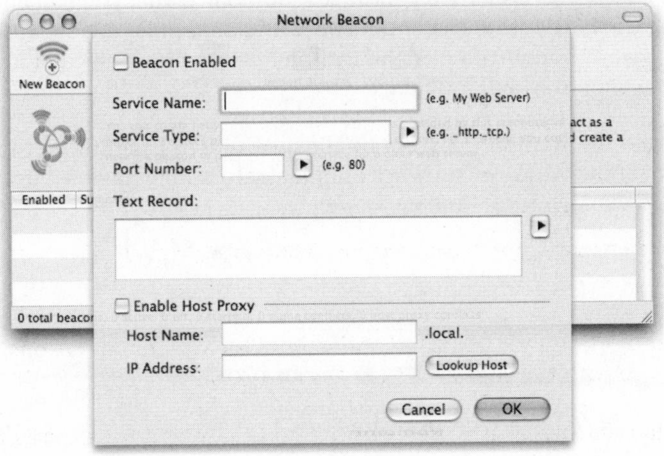

Figure 8-2: Bonjour can escape your local subnet with Network Beacon, if you dare.

By applying the right settings in Network Beacon, adjusting other settings in your firewall, and optionally issuing some commands using Mac OS X's Terminal application, you can advertise various Bonjour services across subnets. You can obtain Network Beacon from `www.chaoticsoftware.com` and you can find various instructions for using it by doing some Internet searching: For example, the search string **network beacon itunes sharing** should yield a few useful hits for iTunes music sharing.

Avoiding locally broken Windows

If you use an AirPort network in a business or school setting, you may find that the network also connects to a Windows network, which offers file servers and other resources that you may need to use. However, because of the way that Bonjour assigns names to network devices, coupled with the way that some Windows networks are configured, you may encounter problems getting from your AirPort network to those resources on the Windows network. You can solve this problem, though, without having to give up the convenience of Bonjour.

Here's what causes the problem: When Bonjour creates a name for a network device, it appends *.local* to the name of that device, establishing a network domain similar to the .com, .org, .net, and other domains used in Internet addresses. Of course, .local does not exist as a real top-level domain on the Internet like .com and its friends do. The .local domain is simply a convention that Bonjour uses to label the subnet on which it operates. In Figure 8-3, for example, look right below the Computer Name field in the Sharing window and you can see that the Mac has the Bonjour name *whitedowns.local*. That means that when other devices on the network want to find this Mac, they will look in the .local network domain created by Bonjour for a device named *whitedowns*.

By the way, if you don't like your Mac's Bonjour name, you can change it by clicking the Edit button to get the sheet depicted in Figure 8-4. However, notice that although you can change the Mac's name, you can't alter the .local domain to which it belongs.

Unfortunately, Bonjour may not be the only one to call dibs on the .local domain. Some Windows network administrators may have set up their networks to use the .local domain as well.

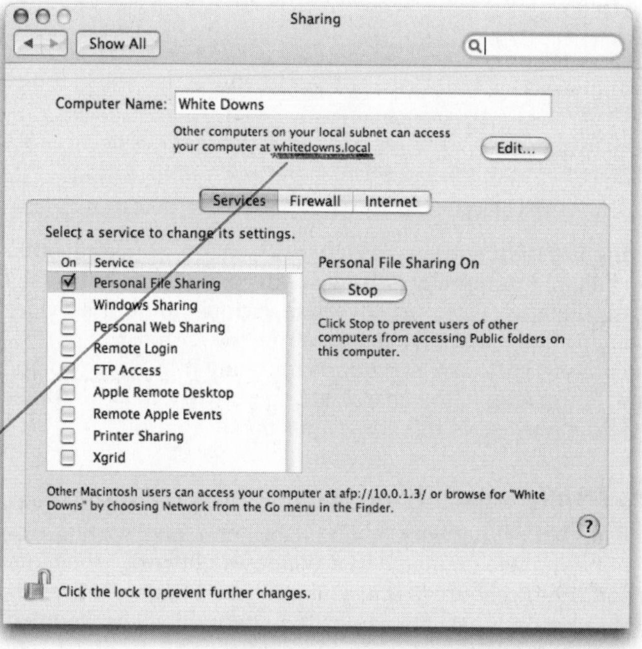

Cf. p.238

Figure 8-3: The computer's Bonjour name appears below the Computer Name field.

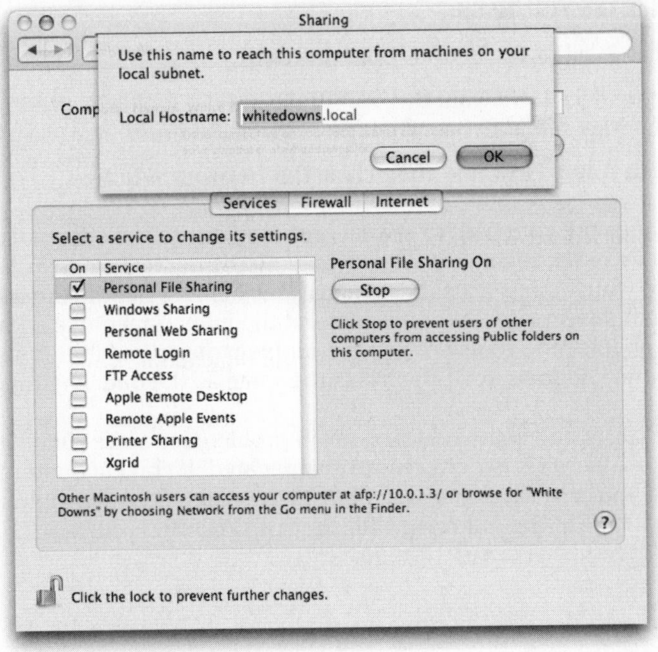

Figure 8-4: You can change the name, but not the domain, in the Sharing window.

They don't have to: In a Windows network, there is nothing special about the .local domain, and there's no technical reason for a Windows network to apply that domain name to itself. However, in a now-outdated Microsoft technical document that explains how to set up a Windows network, the examples use the .local domain name, and many Windows network administrators have followed those examples quite literally. As an unfortunate consequence, neither Macs nor Windows clients that use Bonjour — Apple has made Bonjour available for Windows, too — can find any resources on a Windows network that uses the .local domain.

The easiest way to solve the problem, of course, is to change the domain name used by the Windows network, but that may cause other network administration problems.

Mac users running Mac OS X 10.4 or later can solve the problem another way — by altering their network preferences:

1. **Open System Preferences.**
2. **Click Network.**
3. **Choose AirPort from the Show pop-up menu.**

4. **Click the TCP/IP tab.**
5. **Type** local **in the Search Domains field.**

   Figure 8-5 shows what the TCP/IP tab should look like after this step. Your Mac's IP address, of course, may be different.
6. **Click Apply Now and then close the Network window.**

TECHNICAL STUFF

These steps tell your Mac to search both Bonjour's .local domain, and then to search any other .local domain provided by the nearest *Domain Name Server (DNS)* to which it has access. Domain Name Servers handle the translation between numeric Internet addresses and the human-readable addresses most normal people tend to use. When the problem described in this section occurs, the Windows server's DNS is the same as the one your Mac uses.

TIP

Unfortunately, solving the .local domain problem for Macs running an earlier version of Mac OS X is rather more complicated. If you encounter this problem, and you aren't running Mac OS X 10.4 or later, consult Apple's instructions available at the following URL: `http://docs.info.apple.com/article.html?artnum=107800`.

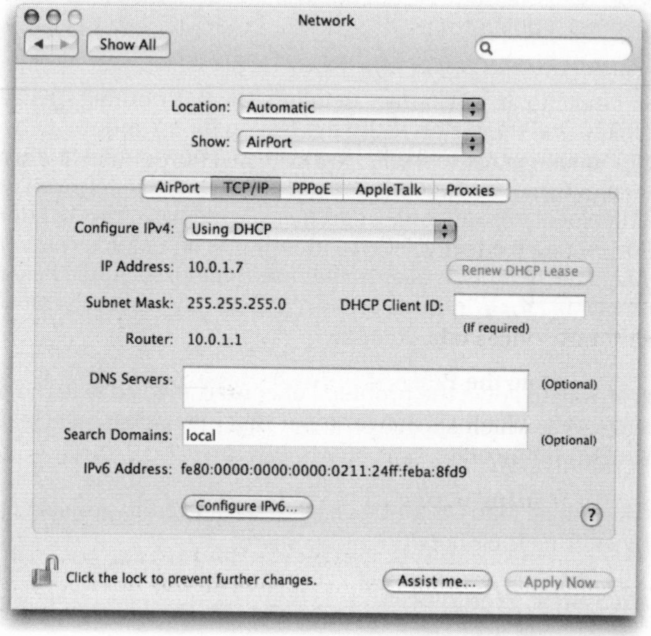

Network
Show All
Location: Automatic
Show: AirPort
AirPort
TCP/IP
PPPoE
AppleTalk
Proxies
Configure IPv4: Using DHCP
IP Address: 10.0.1.7
Renew DHCP Lease
Subnet Mask: 255.255.255.0
DHCP Client ID:
(If required)
Router: 10.0.1.1
DNS Servers:
(Optional)
Search Domains: local
(Optional)
IPv6 Address: fe80:0000:0000:0000:0211:24ff:feba:8fd9
Configure IPv6...
Click the lock to prevent further changes.
Assist me...
Apply Now

Figure 8-5: Tell the Mac to search any *local* domains it may find.

Networking in Plain Sight

Marketing professionals usually try to promote a technology based upon its features, so I admit to feeling a twinge of sympathy for the marketing people at Apple who have to try to explain Bonjour: Its principal feature is that it's featureless. It simply sits in the background, doing its job, making your networking life easier. In many cases, you won't even see it unless you look very hard. But it's there, often silently lending a hand as you exchange files, share personal Web sites, chat, and work with your network neighbors.

Passing files around

File sharing has to rank among the earliest of networking tasks developed, and it was among the earliest of the networking tricks that the first Macs learned to perform. Over the years, the networking technologies that have made Mac file sharing possible have changed several times, but the user experience has altered much less: You turn file sharing on and drag files from one place to another using the Finder. What is Bonjour's role in this? It's subtle, and mostly behind the scenes. As I describe how you share files over your network, I'll point out where Bonjour comes in. Don't blink, because you might miss it.

Cf. 157-159 for information about the Firewall tab

Cf. 171-173 for information about the the Internet tab

Using Personal File Sharing

In order for someone else on your AirPort network to copy files to or from your Mac, you have to turn on file sharing. Here's how you do that:

1. **Open System Preferences.**

   System Preferences → Sharing → Services

2. **Click Sharing.**

3. **Click the Services tab.**

4. **Click to enable the Personal File Sharing check box.**

   In Figure 8-3, which appears a little earlier in this chapter, you can see how the Sharing preferences window should look after this step.

5. **Close the Sharing window.**

Although Bonjour is mentioned nowhere in this process, your Mac uses Bonjour to announce to other clients on your AirPort network that the Mac is now shared and that network users can access any files on it for which they have permission.

When other clients on your AirPort network enable file sharing, you can access files on their Macs like this:

1. **In the Finder, choose Go⇨Network.**

   You can also click the Network icon in the sidebar of any Finder window (if you have the sidebar showing). The Network window that appears, such as the one shown in Figure 8-6, displays items on your network to which you can connect. For example, the icon selected in Figure 8-6, which happens to be my trusty old iBook that goes by the name of *MEC's iBook,* is connected to my AirPort network and has sharing enabled. Bonjour has made the shared nature of my iBook known to the rest of the network so that other network clients know they can connect to it.

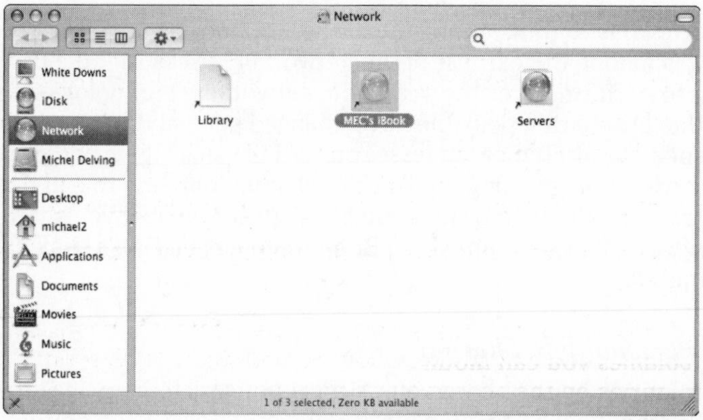

Figure 8-6: Bonjour supplies the names of shared computers to other network clients.

TECHNICAL STUFF

Personal File Sharing uses the *Apple Filing Protocol (AFP),* which predates Bonjour. Bonjour now uses AFP, but it usually only works between Macs on a network. If you wish to share files with Windows users, Mac OS X provides Windows Sharing, which both Mac and Windows network clients can use.

2. **Double-click the icon of the shared Mac.**

   An authentication dialog, like the one shown in Figure 8-7, appears, giving you the choice of connecting to the shared Mac as a registered user or as a guest.

   • **Registered user:** Enter a username and password that matches a user account on the shared Mac. Connecting as a registered user gives you access to all files for which the user account has access.

   • **Guest:** This option gives you access to only the files in the Public folders that exist in each user account's Home directory on the shared Mac. In addition, Guest access restricts you to putting files only in those Public folders' Drop Box folders.

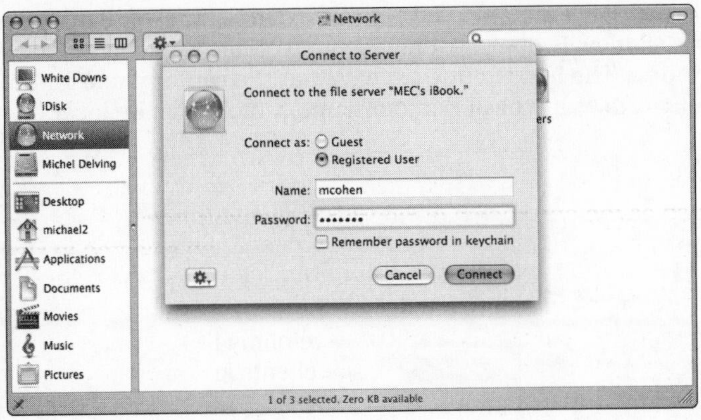

Figure 8-7: You can connect as a guest or as a registered user.

To keep your Mac secure, don't give your username and password away, except to highly trusted individuals. Providing your username and password to other users allows them to do anything with files on your Mac that you can do, including deleting them or replacing them. Instead, you can put any files that you wish to share in your Home directory's Public folder, where network clients can pick them up by using a guest connection.

3. **Enter a username and password, or click Guest, and then click Connect.**

   As shown in Figure 8-8, another dialog appears, offering you a selection of *volumes* you can mount. These volumes may include both specific directories on the shared Mac's hard disk as well any disks connected to that Mac.

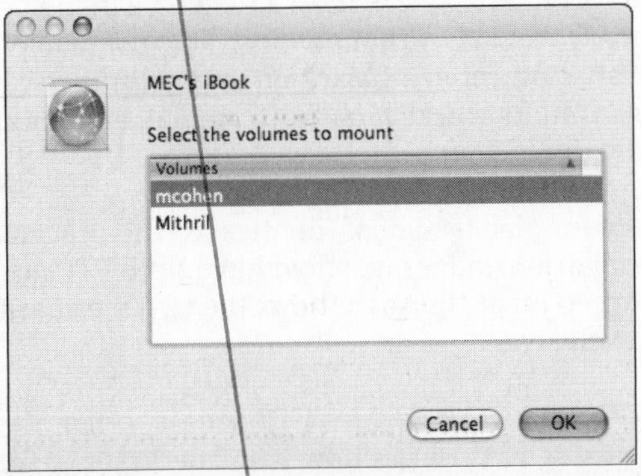

Figure 8-8: Both shared directories and disks are called *volumes.*

4. **Select the items you want to access, and then click OK.**

   The shared items appear on your Desktop in the Finder, just as though they were on your own Mac. However, Bonjour actually is helping out behind the scenes. This becomes clear when you choose File➪Get Info for one of the shared volumes. For example, in Figure 8-9, the mcohen Info window is open for the mcohen volume selected in the Finder window. The Info window reveals that the server name for the mcohen volume uses a typical Bonjour name, which ends in .local.

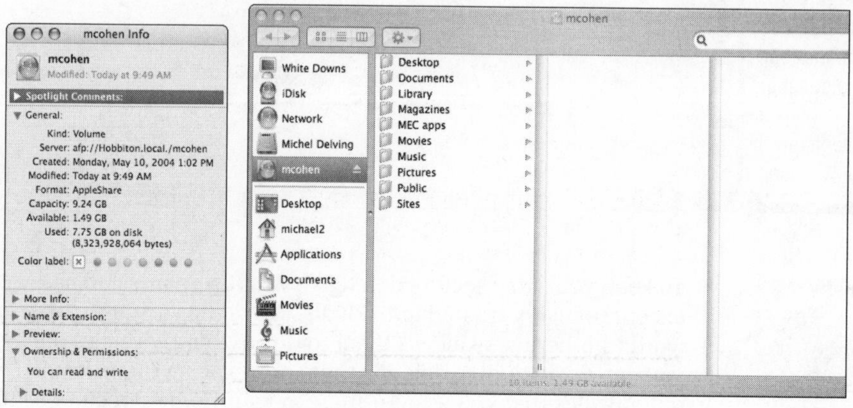

Figure 8-9: Bonjour reveals its presence in the Info window's Server name information.

As you can see, Mac OS X uses Bonjour to provide the same ease of use that Apple's earlier, proprietary, file-sharing technologies provided in previous versions of the Mac OS.

Enabling and using FTP

There's an even older file-sharing technology that Bonjour assists with on Mac OS X: *FTP* (*File Transfer Protocol*), one of the first Internet protocols to be invented back in the early 1970s, more than a dozen years before the first Mac said "Hello" in January of 1984. Although most Mac users may never need to use FTP for file sharing between Macs on an AirPort network, it can come in handy for programmers working in the Terminal, and for similar technically inclined individuals.

Enabling FTP access on your Mac is almost identical to the process you use to turn on Personal File Sharing. In fact, you can have both FTP access and Personal File Sharing turned on at the same time: the two file-sharing services are not mutually exclusive.

After a Mac on your AirPort network enables FTP access, other Macs can access files on that Mac using FTP. Here's how you can do that using the Finder:

1. **Choose Go⇨Connect to Server, or press ⌘+K.**

   The Connect to Server window appears, as shown in Figure 8-10.

2. **In the window's Server Address field, enter** ftp:// **followed by the Bonjour name of the Mac you want to access.**

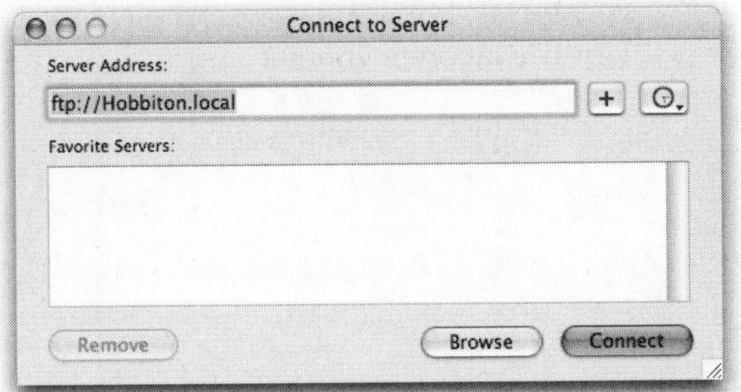

Figure 8-10: Using Connect to Server in the Finder to gain FTP access.

Figure 8-10 shows what I enter to access my iBook, which has the Bonjour name *Hobbiton.local.* Note that unlike some other Finder operations, the server name is case sensitive: *Hobbiton.local* is not the same name as *hobbiton.local.*

REMEMBER: You can find out the Bonjour name of a Mac by looking at its Sharing window in System Preferences, as shown in Figure 8-3 earlier in this chapter. The Bonjour name appears in the fine print below the Computer Name in the Sharing window.

3. **Click Connect.**

   A small Connecting to Server window appears as your Mac attempts to make the connection. When the connection is established, a dialog prompts for a name and a password, as shown in Figure 8-11.

Figure 8-11: FTP access to another Mac requires a name and a password.

4. **Enter a username and password and click OK.**

   A Finder window shows you the files on the connected Mac belonging to the user account name you entered, as shown in Figure 8-12.

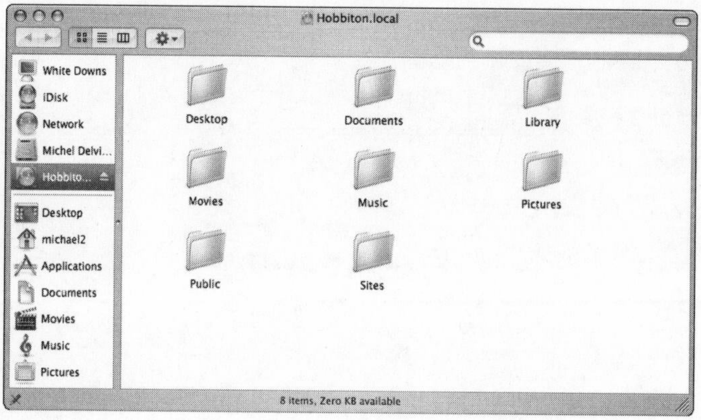

Figure 8-12: A Mac connected to another Mac using FTP access in the Finder.

FTP access does have a couple of limitations and quirks, however:

- **No anonymous or guest access:** Unlike Personal File Sharing, FTP access to another Mac does not allow guest access, or, for those of you familiar with FTP in other contexts, *anonymous* access. You must enter a username and password that matches a user account on the Mac to which you want to connect. And that name and password provides access only to the Home directory of the account.

  TIP

  As I warned in the "Using Personal File Sharing" section earlier, giving away your username and password to other network users can put your files at risk. However, you can create a user account on your Mac other than the one that you normally use, and place any files you wish to share with FTP in that account's Home directory. You can then give other network users the username and password for that account. Network clients connecting to your Mac can access only the Home directory for that account, leaving your other files secure.

- **Finder access is read-only:** If you look at the lower left of the Finder window shown in Figure 8-12, you can see the crossed-out pencil icon indicating that you cannot add any files to or delete files from the window. You can only copy files from it to your own Mac.

This last limitation, however, applies only to Finder access using FTP. If you use a dedicated FTP program, such as the open-source CyberDuck, which you can download from `cyberduck.ch`, you can both read and write files on a Mac that has FTP access enabled. CyberDuck, in fact, is Bonjour-savvy, having a special Bonjour button you can click that shows you the local Macs you can connect to using FTP, as shown in Figure 8-13.

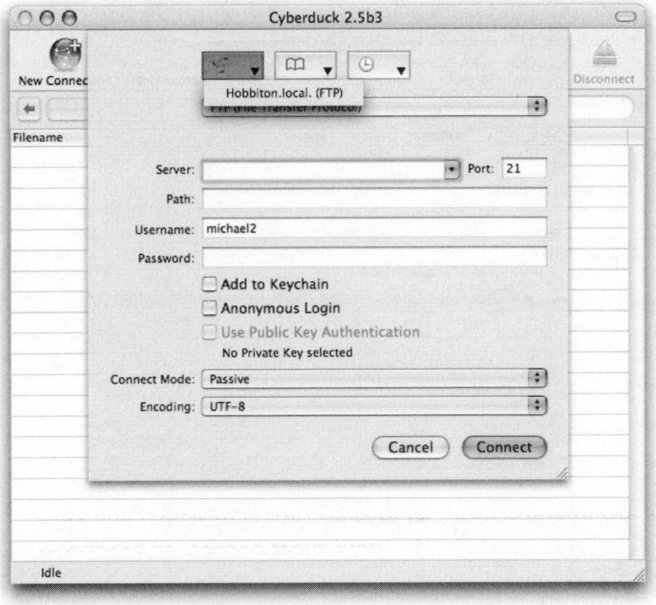

Figure 8-13: Some FTP programs, like CyberDuck, can say *bonjour* to Bonjour.

Using iChat as an intercom

One Apple program that more explicitly makes use of Bonjour is iChat, Apple's instant messaging client. In addition to providing text, audio, and video chatting using AOL's Instant Messenger service or with Jabber servers, iChat provides messaging over a local network using Bonjour. Because your AirPort network operates over a relatively limited area, using iChat with Bonjour is a lot like having an intercom that lets you send text, audio, or video to nearby workers, or family, or neighbors.

Setting up iChat to use Bonjour is very simple, because iChat enables Bonjour messaging by default. However, if you happen to find yourself working on a Mac where iChat's Bonjour messaging is not enabled, a quick trip to iChat's preferences can fix things right up:

1. **Open iChat.**
2. **Choose iChat⇨Preferences, or press ⌘+, (comma).**

   iChat's preferences window opens.
3. **In the iChat preferences window's toolbar, click the Accounts icon.**

   The Accounts settings appear in the preferences window. You can see the current iChat accounts listed along the left side of the window.
4. **Click the Bonjour account in the panel on the window's left.**

The Bonjour-specific settings appear in the right half of the window, as shown in Figure 8-14.

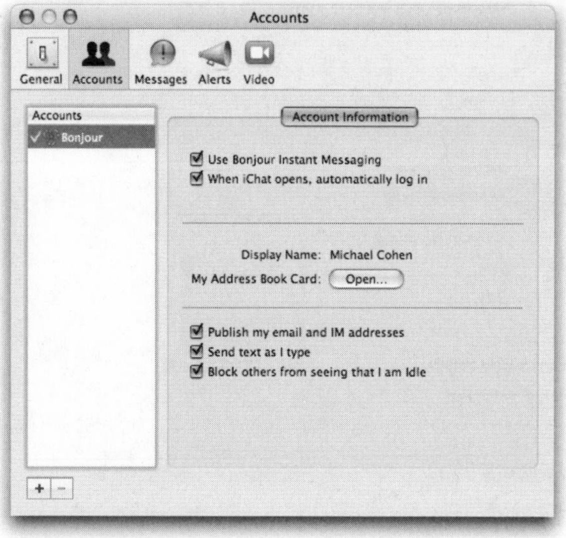

Figure 8-14: You can enable Bonjour-based messaging in iChat's preferences.

5. **Click Use Bonjour Instant Messaging and then close the preferences window.**

You're now ready to use Bonjour messaging in iChat — that is, if you have your firewall set up to let you do so. It's easy enough to check on that, and to change it if it isn't:

1. **Open System Preferences.**
2. **Click Sharing.**
3. **In the Sharing window, click the Firewall tab.**
4. **Scroll down the Allow list until you see iChat Bonjour; if it is unchecked, click to enable it.**

   Figure 8-15 shows the Firewall's Allow list with the iChat Bonjour setting checked.
5. **Close the Sharing window.**

   And now you're ready to use Bonjour in iChat — if you weren't set up to use it already.

If you're running an older version of Mac OS X that doesn't already provide an iChat Bonjour or iChat Rendezvous setting in its Firewall, you can set it up manually by creating a new Firewall setting. You need to open up port numbers 5297 and 5298 to let Bonjour iChat packets through the Firewall.

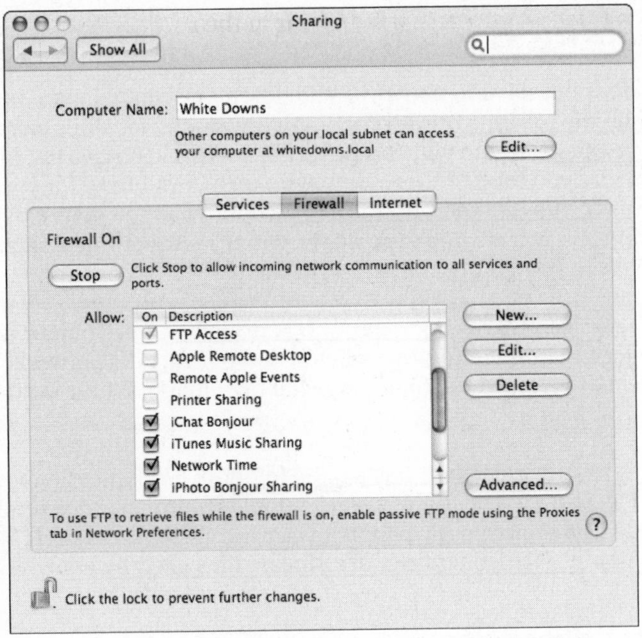

Figure 8-15: You may need to enable iChat Bonjour in your Firewall settings.

To chat using Bonjour, just choose Window⇨Bonjour, or press ⌘+2. iChat's Bonjour window opens up, as shown in Figure 8-16. Unlike iChat's Buddy list window, you don't need to add usernames to your Bonjour window. In fact, you can't: iChat uses Bonjour to automatically display every iChat Bonjour user on your AirPort network.

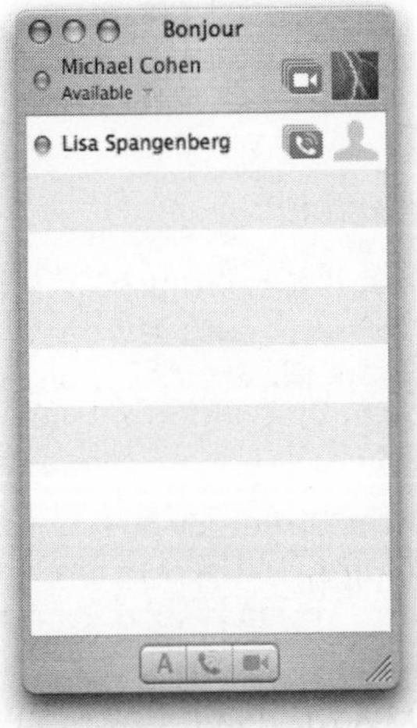

Figure 8-16: iChat uses Bonjour to show other iChat users on your AirPort network.

Publishing a local Web site

The Web does not have to be World Wide: Bonjour can help you publish Web sites for just the folk on your AirPort network. Such Web sites can act as a local notice board, where you can post the electronic equivalent of vacation schedules that you might otherwise pin up on a cork board for your office mates, or as the virtual refrigerator door where you tape your kids' weekly list of chores, or as a staging area where you try out your new small-business Web site before you go live to the rest of the world.

REMEMBER

When you use Mac OS X's Personal Web Sharing feature, Bonjour advertises the presence of your site to other clients on your AirPort network. Safari, Apple's default Web browser, has a special bookmark just for local Bonjour-advertised Web sites.

Every Mac OS X user account has a Sites folder in its Home directory, as shown in Figure 8-17. The folder comes supplied with an index.html file that you can use as a template for building your own Web pages in HTML, and an images folder into which you can drag image files used by your Web pages.

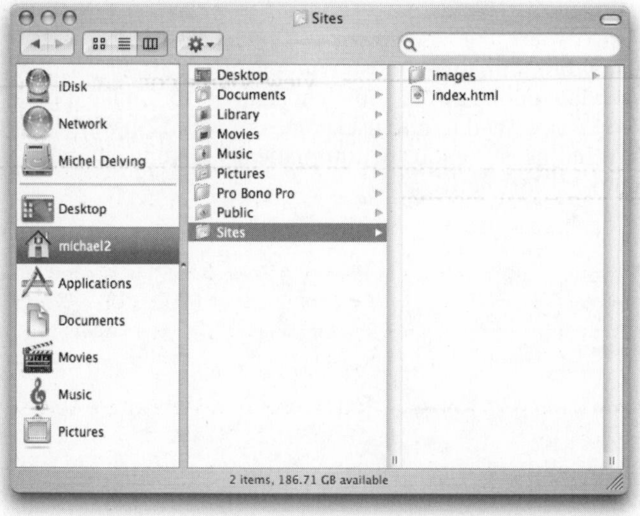

Figure 8-17: You build your Bonjour Web site in your Sites folder.

TIP

Double-click the `index.html` file in your Sites directory to have Safari open the file and show you Apple's explanation of how to use the Sites directory and Mac OS X's Personal Web Sharing feature. You may also want to copy both this file and the images folder to a safe place before playing around in this folder, so you can always return things to the way they were if you ever need to.

Although this book doesn't attempt to teach you about HTML and Web site design — you can find several other Dummies books to help you with that — you may find the following points helpful:

- **index.html is where you start:** The `index.html` file in your Sites folder is your home page, the starting point of your Personal Web Sharing site.

  If your Sites folder does not contain a file named `index.html`, other network users may get an error message when they enter the URL for your site. Make sure to name the file that contains your home page `index.html`.

- **HTML is the language of the Web:** HTML stands for *Hypertext Markup Language,* and consists of instructions that tell your Web browser how to display a Web page. Although the complete set of HTML instructions is quite extensive, you can create respectable-looking Web pages using only half a dozen or so of those instructions.

- **HTML files are text files:** You can use almost any text editor to create an HTML file. These files consist entirely of text with no special formatting. The formatted text you see when the Web page is displayed in your browser is created by the HTML instructions that are interspersed throughout the text. The instructions that format the text that you see on a Web page, that display images on that page, and that create the links you can click on that page are themselves text.

  You can use your Web browser's View Source command to see the combination of both the displayed text and the text that makes up a Web page's HTML instructions. In Safari, for example, choose View➪View Source, or press ⌘+option+U, to see a window showing the combination of text and HTML instructions that produce the Web page you are currently viewing.

- **The images displayed on a Web page are actually stored in separate files:** HTML instructions in the text file that makes up a Web page tell the Web browser which image files to display, where to find those files, and how to lay them out on the page.

- **You don't need to know HTML to create a Web page:** Many word processors can save your documents as HTML, and some of the programs you may already own, such as iPhoto, can create Web sites comprising both HTML files and image files for you.

The files that you store in your Home directory's Sites folder comprise the Web site you publish when you use Personal Web Sharing.

To turn on Personal Web Sharing, perform the following steps:

1. **Open System Preferences.**
2. **Click Sharing.**
3. **Click the Services tab.**

4. **Click the Personal Web Sharing check box.**

   Figure 8-18 shows how the Sharing preferences window should look after this step.

5. **Close the Sharing window.**

After you turn on Personal Web Sharing, Bonjour swings into action, letting the rest of your AirPort network know that your Mac is now a Web server.

WARNING!

Personal Web Sharing publishes the Sites folder in every Home directory on the Mac. If several users share your Mac, each of their Web sites becomes available on the AirPort network when you turn on Personal Web Sharing. You should check with other users of your Mac before you turn this feature on.

Although the Sharing window shown in Figure 8-18 lets you know the URL for your site on the network, which you can tell to other network clients, Mac users running the Safari Web browser don't need to bother remembering the URL: Safari gets that information directly from Bonjour and creates a bookmark for your site whenever you have Personal Web Sharing enabled.

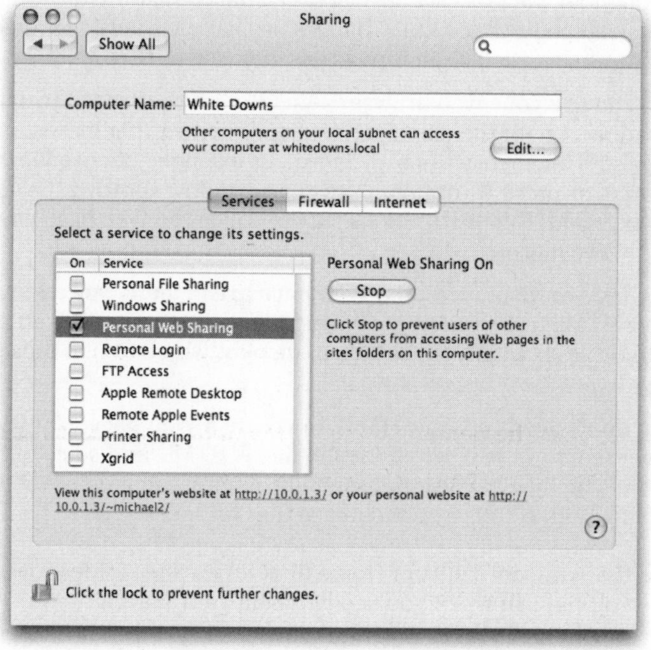

Figure 8-18: Personal Web Sharing turns your Mac into a Web server.

Cf. p. 224

To see a Bonjour-advertised Web site in Safari, do the following:

1. **Open Safari.**
2. **Choose Show All Bookmarks, or press ⌘+option+B.**

   Safari shows its Bookmarks collection.
3. **In the Collections column, click Bonjour.**

   Figure 8-19 shows Safari's bookmarks page with the Bonjour collection selected. The Bookmark pane lists the names of all network users that have Personal Web Sharing enabled.

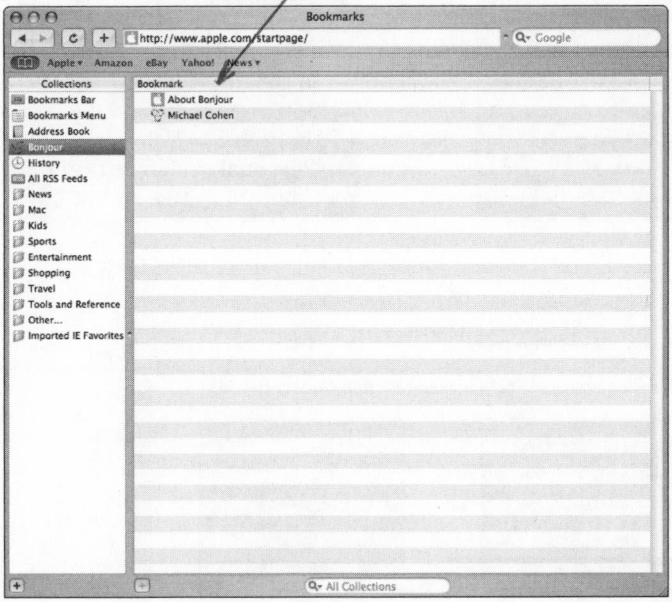

Figure 8-19: Bonjour tells Safari which network users have enabled Web sharing.

4. **Double-click the name of a user listed in the Bookmarks pane.**

   The `index.html` page in that user's Sites folder opens in Safari, as shown in Figure 8-20.

TIP

You don't need to use Safari to see a personally shared Web page. Mac users who prefer other browsers, as well as Windows users who have Apple's free Bonjour software installed, can use the same Bonjour-created URL to see the page that Safari displays; see Figure 8-21. Of course, other browsers probably won't automatically generate a bookmark for the site.

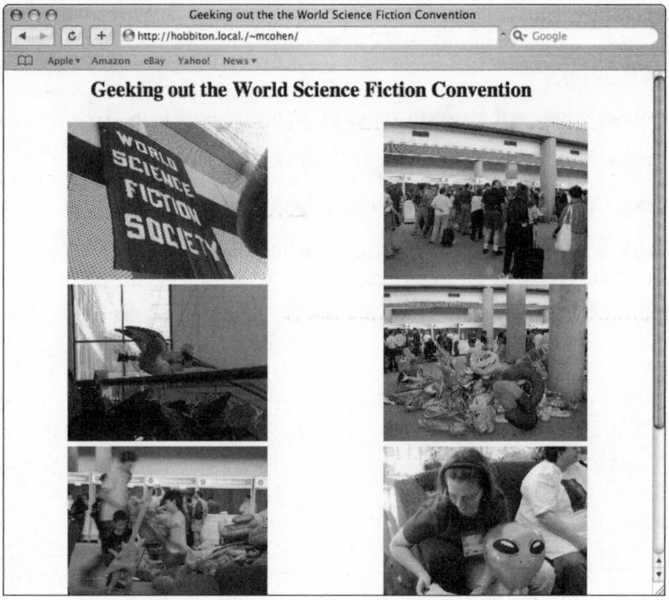

Figure 8-20: Safari displays a personal Web page, courtesy of Bonjour.

Figure 8-21: With Bonjour running, any browser can use a Bonjour-created Web address.

Collaborating with Bonjour

Bonjour extends the range of your Mac in some interesting ways. Using Bonjour's ability to announce and discover services on your network, a program can reach out and connect with other computers on your network that are running the same program. When that happens, opportunities for collaboration arise. Two simple examples are given here.

Writing in a group

One collaborative program that has become quite popular is SubEthaEdit, produced by a German company, TheCodingMonkeys, and available for free noncommercial use from `www.codingmonkeys.de/subethaedit/`. The program is simply a text editor, designed to produce nonformatted text quickly and easily. What makes it more than just a simple typing tool is its underlying collaboration engine, which lets several people simultaneously edit the same document, and which changes on each person's screen in real time.

Figure 8-22 shows SubEthaEdit in action: In the figure, two people collaborate on taking meeting notes. The text created by each person shows up highlighted with a separate color. The collaborators' names, their assigned highlight colors, and their cursors' current positions appear in the panel to the right.

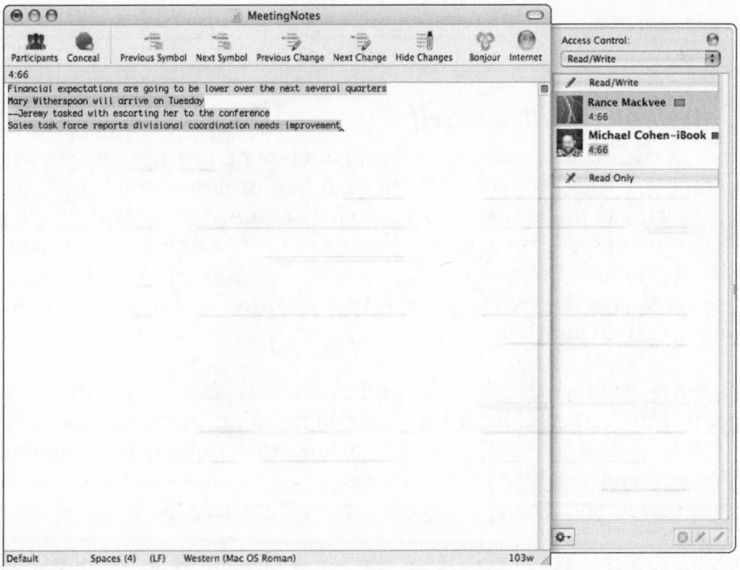

Figure 8-22: Four hands tackle one project with SubEthaEdit.

When other network users open their copies of SubEthaEdit, they can see a list of open, shared SubEthaEdit documents on the network in the program's Bonjour window, as shown in Figure 8-23. A pop-up menu allows SubEthaEdit users to either view the shared document or join in a collaborative editing session. Collaborators can also make use of Bonjour to engage in iChats with each other as they work on the shared document.

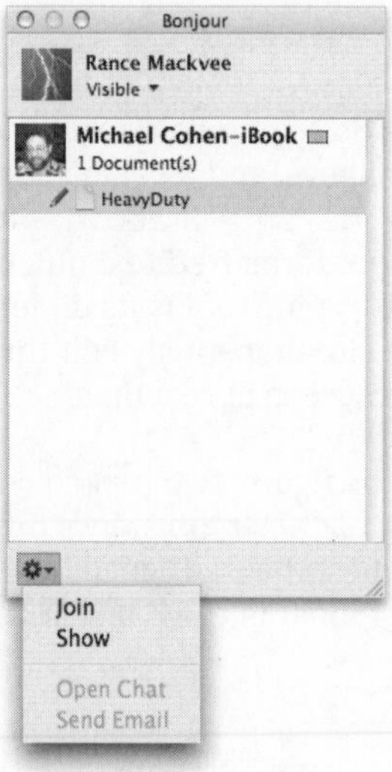

Figure 8-23: Using SubEthaEdit to join a work in progress.

Collaborating with yourself

Just as two people can work on one document, one person can work on two computers with the help of Bonjour. A free program from abyssoft.com, Teleport, lets the mouse and keyboard on one Mac control another Mac. Delivered in the form of a preferences pane that you drag into your main Library folder's PreferencePanes folder, Teleport lets you figuratively place the screens of the two Macs next to each other and drag your mouse from one screen to another.

Figure 8-24 shows the Teleport preference pane in action. A thumbnail representing the screen of each Teleport-running Mac on the network appears above the line. Drag a screen thumbnail next to the thumbnail representing your Mac's screen — labeled White Downs in the figure — and you can then drag your cursor onto the other Mac's screen anywhere the two screens touch.

When your mouse cursor is on the other Mac's screen, your Mac shows an indicator pointing to where your mouse currently is, as shown in Figure 8-25. Any mouse clicks you make, and anything you type on your Mac's keyboard, take effect on the connected Mac when your mouse is on its screen.

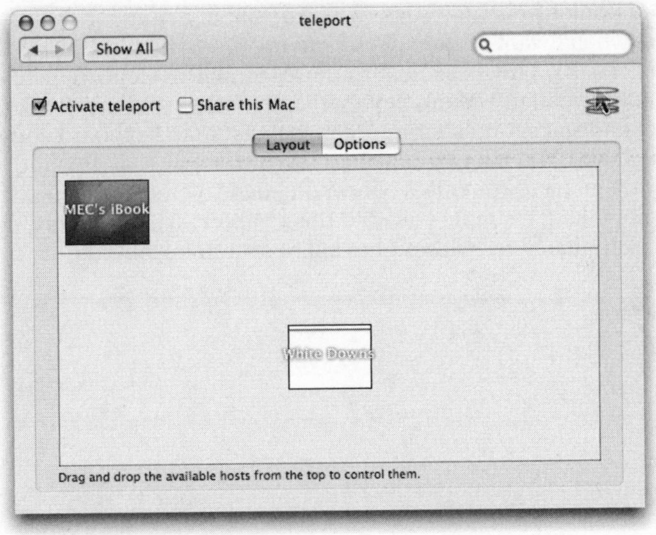

Figure 8-24: You can move your mouse over the network with Bonjour and Teleport.

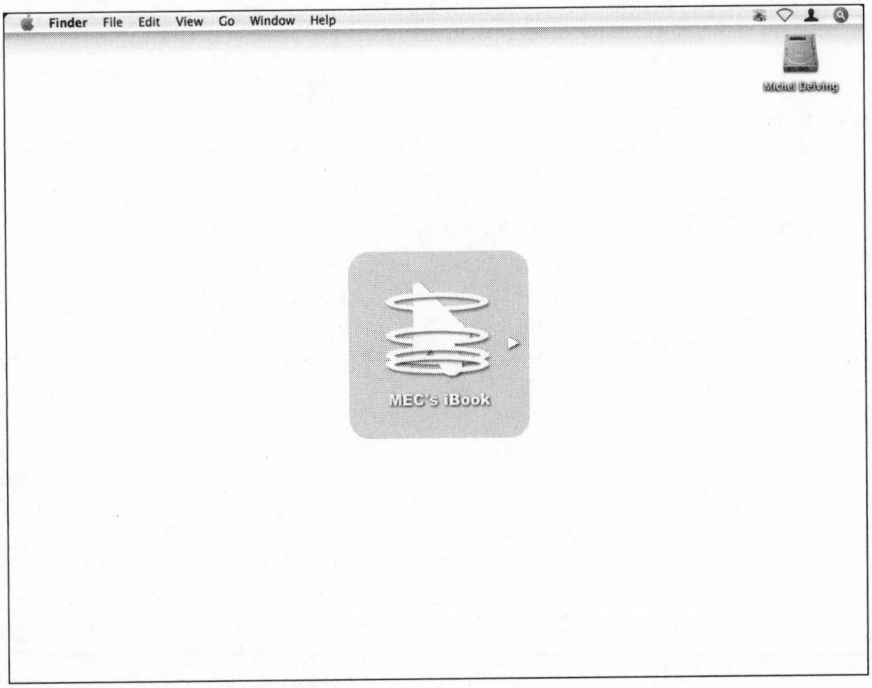

Figure 8-25: Where, oh, where has my mouse pointer gone? Follow the arrow.

This little Bonjour-enabled trick can be useful in a number of situations. For example, when I want to copy files from my iBook to my desktop iMac, I often place my iBook on my desk next to the iMac. With Teleport, instead of having to switch between my iMac's keyboard and mouse and my iBook's keyboard and trackpad as I set up sharing, I can use my iMac keyboard and mouse on both machines. You may also be interested to know that Teleport came in very handy just a few minutes ago when I had to create the SubEthaEdit group note-taking example I used in this chapter: With Teleport, I was able type on both machines without having to switch keyboards.

Chapter 9

Entertaining Your Guests

In This Chapter

- Opening your iTunes music library
- Obeying the law
- Hanging your pictures on your guests' walls
- Building your neighborhood Bijou
- Decking out your game room

For some time now, Apple has referred to the Mac as a *digital hub*. This marketing metaphor places your Mac at the center of a wagon wheel with spokes radiating from it. What each of those spokes connects to, however, depends on whom you ask — and what they want to sell you.

Some people imagine the spokes as connecting to a cool collection of digital devices, like iPods, cameras, MIDI keyboards, and digital camcorders. Others imagine the spokes connecting to the multimedia applications that manage and manipulate your media, like iTunes, iMovie, GarageBand, and iDVD.

When you put your Mac on your AirPort network, the digital hub's spokes can connect to something else: people. People who can enjoy the stuff that you put on your Mac with your cameras and keyboards and that you polished and perfected with your multimedia applications.

All of these variations on the meaning of *digital hub* can happily coexist: One hub pulls the media in, one hub rolls it about, and one hub spreads it out. This chapter's about that third hub.

So get rolling.

Sharing iTunes

In Chapter 7 you can find a lot of ways to play music from your iTunes library remotely using the AirPort Express or various third-party music players. But there's another way to play your iTunes music remotely: through another Mac — or even a Windows computer — connected to your AirPort network. Since the introduction of iTunes version 4, iTunes has included a music-sharing feature that allows others on your network to see and play songs from your music library on their computers.

Setting up music sharing

Setting up sharing in iTunes requires nothing more than configuring a few preferences. After you enable sharing, iTunes uses Bonjour — described in Chapter 8 — to announce to the rest of your network that your music library is available.

To turn on music sharing in iTunes, follow these steps:

1. **Open iTunes.**
2. **Choose iTunes⇨Preferences or press ⌘+, (comma).**

   The iTunes preferences window opens.
3. **In the preference window's toolbar, click Sharing.**

   The Sharing preference pane appears. This pane provides settings for both sharing music and using other network users' shared music.
4. **Click Share my music.**

   The setting options beneath the Share my music check box become active, as shown in Figure 9-1.
5. **Click one of the following items:**
   - **Share Entire Library:** Click this button to make your entire music library available to others on your network. Users connecting to your shared music can see all of your playlists as well as songs that don't belong to any playlist, and can play any of your unprotected songs, as well as protected songs for which their computers are authorized.
   - **Share Selected Playlists:** Click this button to make only the playlists that you've checked in the list below this button available to others on the network.

   Press ⌘ when you click a playlist to select or deselect all of the playlists in the list.

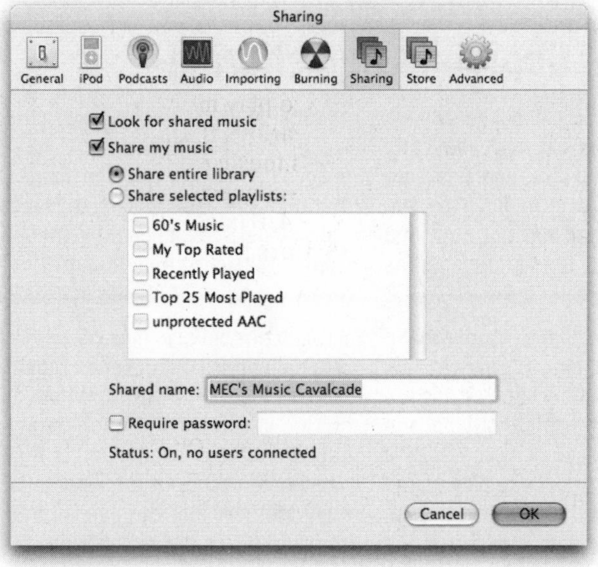

Figure 9-1: Use iTunes' Sharing preference to make your music available to the network.

6. **Enter a name for your shared music in the Shared Name field.**

   The name you enter here appears in iTune's Source panel for users who have connected to your shared music.

7. **Optionally, click Require password, and then enter a password.**

   When you require a password, other network users can see the name of your shared music collection in iTunes' Source panel, but they cannot open that music list until they enter the password you have specified.

8. **Click OK.**

   iTunes provides a reminder dialog, shown in Figure 9-2, pointing out that music sharing is for personal use only. See the "Copyright and sharing" sidebar elsewhere in this chapter for more about what this message means.

Figure 9-2: Sharing music between friends is OK, but commercial use is another matter.

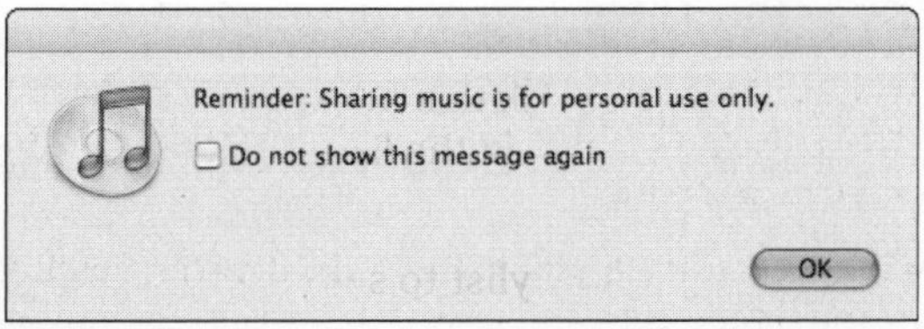

Copyright and sharing

When musical artists or music publishers own a song's copyright, it means just that: They, and they alone, own the right to copy and distribute it. Using this right, they can grant licenses that authorize who else may, and may not, copy and distribute the song, and the circumstances under which they may do so.

When you buy a CD or a tape, you purchase only the physical CD or tape. Copyright law grants certain rights as to how you can use the music recorded on the physical CD or tape, and the law imposes certain restrictions as well. You have, in short, an *implicit* license to play and to copy the recording within limits. For example, you do not have the right to mass-produce copies of the tape or CD and to sell them or give them away, or the right to distribute digital copies of the music indiscriminately to large numbers of people.

An *explicit* license governs how you can use music purchased from the iTunes Music Store. In setting up the iTunes Music Store, Apple negotiated with music copyright owners to obtain licenses that give Apple the right, in turn, to license music to you. The license that Apple grants to you has to follow the limits the copyright holder has imposed. When you open an iTunes Music Store account, you agree to follow the terms of Apple's license.

So, it comes down to this: When you buy a song or an album from the iTunes Music Store, you haven't bought the music. What you *have* purchased is a license to listen to that music in a non-commercial setting and to make a limited number of copies of it for personal use. Both copyright law and the license you have purchased govern the acceptable ways you can use the music.

This law and license are one reason a limited number of network users can share your music at any given time, and it is why network users have to authorize their computers to play your purchased music from the iTunes Music Store.

After you have completed these steps, other network users can play music from your iTunes collection on their computers — within certain limits, which are described in the next section.

TIP

You can tell how many network users are sharing your music collection by opening up iTunes' preferences again and clicking the Sharing icon: The number of connected users appears in the Sharing preference pane, as you can see near the bottom of Figure 9-1.

Using music sharing

Setting up iTunes to share your music takes only a few seconds. Setting it up to play other network users' music collections takes even less time.

Follow these steps to set up iTunes to play shared music:

1. **Open iTunes.**
2. **Choose iTunes⇨Preferences or press ⌘+, (comma).**
3. **In the preference window's toolbar, click Sharing.**
4. **Click Look for Shared Music.**
5. **Click OK.**

After you have enabled Look for Shared Music, iTunes displays an item representing a shared music source in its Source pane whenever it detects any shared music on the network. In fact, you will often see shared music sources appear in, and disappear from, your Source pane as other network users start and stop sharing their music. The detection and display of shared music sources does not cost iTunes much in terms of either time or performance, because iTunes does not fetch a list of shared songs until you open a music source.

Click a shared music source, and iTunes fetches that source's list of songs and playlists from the computer sharing it, as shown in Figure 9-3. You may also have to enter a password if the person sharing the music requires one. Downloading the list of shared music to your Mac can take anywhere from a few seconds to a minute or more if the number of shared songs is extensive. After iTunes has downloaded the list, however, you can browse the shared music list and play songs from it as quickly as you can from music on your own Mac.

Figure 9-3: iTunes fetches the music list from a shared music collection.

TIP

iTunes uses the color blue to indicate songs and playlists that are shared, but if you happen to be among the ten percent of the population who are color-blind, you can still tell if a song is shared by selecting it and choosing File⇨Get Info. The Summary tab in the information window for the song uses the word *remote* in the Kind field, as shown in Figure 9-4.

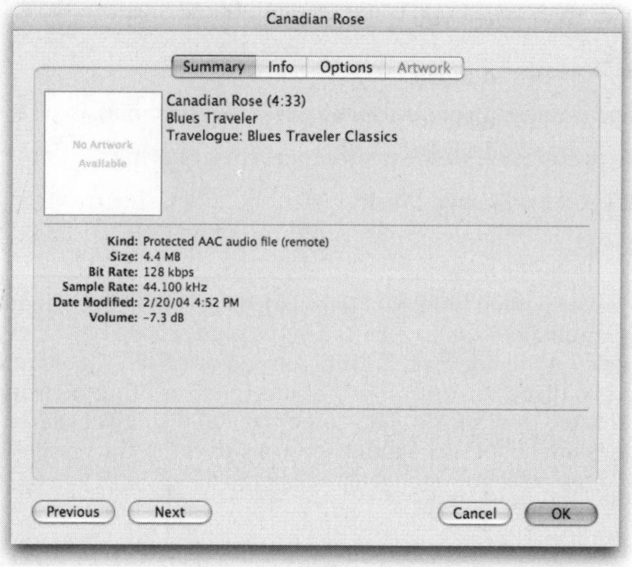

Figure 9-4: A remote song is a shared song.

Music sharing comes with certain limitations:

- **iTunes must be running for you to share your music.** This may seem obvious, but it needs to be said.
- **You can share music with only five computers at any one time.** In part, Apple's license agreements with the music industry require this, but also, streaming your music to many computers can slow down both the network and your computer.
- **You can share music only over your local network.** This poses no problems when you want to share music over your AirPort network, but don't expect to access the songs on your home Mac from your office unless you take special measures that Apple does not support. See Chapter 8 for more about extending the reach of Bonjour, the network technology that underlies music sharing.
- **You can't copy songs from other computers to your own library.**
- **You must be authorized to play shared songs purchased from the iTunes Music Store.** Again, Apple's license agreements with the music industry stipulate this. Keep in mind that the person who purchased the song can authorize as many as five other computers to play it, which just happens to match the number of computers that can access a shared music source at any one time.
- **You cannot share Audible spoken word content.** Audible, Inc. supplies the audio books sold by the iTunes Music Store, and Apple's license agreement with them prohibits sharing.
- **You cannot share QuickTime sound files.**

Follow these steps when you no longer want to share someone else's music:

1. **Select the shared music source.**
2. **Click the eject music icon that appears in the lower right of the iTunes window, visible in Figure 9-3.**

A small eject icon also appears in the Source panel beside each shared music icon; you can click that icon as well to eject the shared music. When you eject a shared music item, the item itself remains in the Source panel, but you can no longer see its contents until you open it and reload the music list.

If you have trouble accessing music from others, or if others on your network have trouble accessing your music, check your Firewall settings in the Sharing window in System Preferences. When the Firewall is enabled, you must enable iTunes Music Sharing in the Firewall's Allow list. For Windows users, or Mac users running older versions of Mac OS X, you may need to create a new Firewall setting that allows network connections on port 3689.

Posting Your Photos

Photo sharing with iPhoto resembles music sharing with iTunes. This similarity did not arise by chance: Both programs form part of Apple's iLife suite of media programs and exhibit a family resemblance.

To share your iPhoto picture collection, follow these steps:

1. **Open iPhoto.**
2. **Choose iPhoto⇨Preferences, or press ⌘+, (comma).**

   The iPhoto preferences window opens.
3. **In the preference window's toolbar, click Sharing.**

   The Sharing preference pane appears. This pane provides settings for both sharing photos and using photos shared by others on your network.
4. **Click Share My Photos.**

   This setting works very much like the music-sharing setting in iTunes, described in the previous section of this chapter. When you enable photo sharing, the options beneath the Share My Photos check box become active, as shown in Figure 9-5.
5. **Click one of the following items:**
   - **Share Entire Library:** Click this button to make your entire photo library available to others on your network. Users connecting to your shared music can see all of your photos and albums.

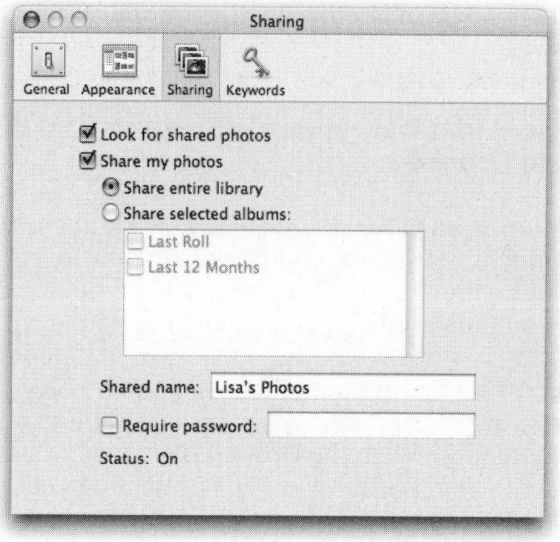

Figure 9-5: Use iPhoto's Sharing preference to share your photos over your network.

- **Share Selected Albums:** Click this button to share only the albums that you've checked in the list that appears below this button.

  TIP: As with iTunes playlists, press ⌘ when you click an album to select or deselect all of the albums in the list.

6. **Enter a name for your shared photos in the Shared Name field.**

   The name you enter here appears in iPhoto's Source panel for users who have connected to your shared photo collection.

7. **Optionally, click Require Password and enter a password.**

   When you require a password, other network users can see the name of your shared photo collection in iPhoto's Source panel, but they cannot open that photo collection until they enter the password you have specified.

The moment you check the Share My Photos check box, your photo collection appears in the Source panel of other iPhoto users on your network — if they have set up iPhoto to look for shared photos, that is.

Follow these steps to set up iPhoto to look for shared photo collections:

1. **Open iPhoto.**
2. **Choose iPhoto⇨Preferences, or press ⌘+, (comma).**
3. **In the preference window's toolbar, click Sharing.**
4. **Click Look for shared photos.**
5. **Click OK.**

Like iTunes does with music sharing, iPhoto's sharing displays photo collections in your Source panel as soon as they appear on the network, but it doesn't display the contents of those collections until you click on the shared collection. When you do, iPhoto fetches the photos from the collection and displays them in your copy of iPhoto. And, again as with iTunes, you may have to enter a password first if the owner of the shared collection requires one.

REMEMBER

It often takes some time for iPhoto to transfer images across the network, so be patient, especially if you have connected to a large photo library. The program, however, does reduce the waiting time considerably by sending smaller versions of the photos first: iPhoto only fetches the full-size version when you select a photo from a shared collection and magnify it with the size control. That said, a photo collection containing many hundreds of photos may take a minute or so to completely appear in iPhoto's viewing area when you open it across your AirPort network.

After you open a shared photo collection, you can view, print, or e-mail photos from it. Figure 9-6 shows a shared photo collection in iPhoto, including its shared albums. In the Source panel, a shared photo collection has a small blue icon that looks like a stack of photos. When you want to disconnect from a shared collection, click the eject icon beside the collection's name.

Figure 9-6: Browse another network user's photos with iPhoto sharing.

Keep in mind the following limits, and one startling fact, about shared photo collections:

REMEMBER

- **You cannot edit photos shared from another Mac:** Although the Edit button remains enabled in the iPhoto toolbar, when you select a shared photo and click the button, the only iPhoto editing tool that works is the size control.
- **Shared smart albums, slideshows, and iPhoto books appear as ordinary albums:** Although your own smart albums, slideshows, and iPhoto books have distinctive icons in your Source list, they look like and behave as ordinary albums when you use these items from someone else's Mac.
- **You cannot create slideshows or books from shared photos:** When you select photos from someone else's Mac, the iPhoto toolbar controls for creating slideshows or books become disabled.
- **iPhoto must be running to share photos:** This stands to reason, of course. It also means that when you quit iPhoto, you have to disconnect any users sharing your photo collections, as shown in Figure 9-7.

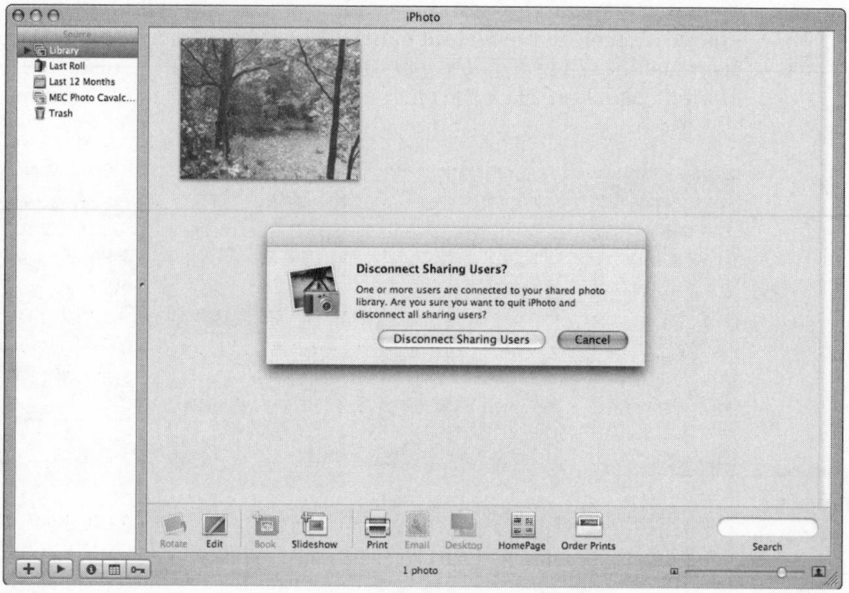

Figure 9-7: Quitting iPhoto disconnects network users who are sharing your photos.

- **Contrary to Apple's iPhoto Help, you *can* copy photos from a shared collection to your own library:** Drag a photo from a shared collection to your photo library or to an album, and iPhoto copies it across the network to your Mac.

TIP

iPhoto uses Bonjour, described in Chapter 8, to enable network sharing of photos. If your Mac's Firewall is enabled, you must make sure to enable iPhoto Bonjour Sharing in the Firewall's Allow list for others to view your photos. Mac users running versions of Mac OS X prior to 10.4 may need to create a new Firewall setting that allows network connections on port 8770.

Showing Movies

You can create QuickTime movies in many ways with the programs Apple provides with your Mac. For example, you can make QuickTime movies from your slideshows using iPhoto; and, of course, iMovie lets you make sophisticated movies quite easily. Although you can share photos with iPhoto and music with iTunes over your AirPort network, Apple has not provided so simple a way of sharing your movies. However, it has put most of the pieces in place for you to do it yourself.

The key piece is Personal Web Sharing, described in Chapter 8. When you turn on Personal Web Sharing, you can share anything in your Sites folder with others on your AirPort network. All they need is a browser, or some other application that uses Web protocols.

Using QuickTime Player to play shared movies

One application that uses Web protocols is QuickTime Player. You can open movies by giving QuickTime Player a URL that points to a movie located anywhere on the Web — including your own AirPort network. As Chapter 8 explains, Bonjour creates Web addresses for each user account's Sites folder on your Mac when you turn on Personal Web Sharing.

The URL that Bonjour creates incorporates both the Mac's Bonjour name and the name of the user account. You can construct the URL of any document in a user's Sites folder like this, substituting the stuff inside of the pointy brackets with the appropriate information:

```
http://<Mac's Bonjour name>/~<user account name>/<path to the document>
```

For example, if your Mac has the name *PlayHouse*, and your user account is *MaryJane*, and you have put a movie named *mybroadwayaudition.mov* in your Sites folder, that movie's URL is

```
http://playhouse.local/~MaryJane/mybroadwayaudition.mov
```

To see a movie inside of a shared Sites folder, anyone on your network need only do the following:

1. **Open QuickTime Player.**
2. **Choose File⇨Open URL.**
3. **In the Open URL dialog, enter the Bonjour address of the movie.**

TIP

You can use iChat to send the URL to others on your network so they can copy and paste it into QuickTime Player. Also, you don't need to use QuickTime Player to view a QuickTime movie. Most Web browsers can play QuickTime movies directly because QuickTime includes a browser plug-in just for that purpose. Enter the movie's URL into the browser to see a blank page with the QuickTime movie floating in the middle of it.

Making a movie Web page

Playing a QuickTime movie in either QuickTime Player or your browser works well, but sometimes you want additional information to go along with the movie. For that, you need to embed the movie in a Web page. Though this sounds complicated, it really doesn't have to be: Using Apple's free TextEdit program, and one free third-party program, you can create a movie page without writing a single line of HTML yourself.

REMEMBER

Before you start, you should download Francis Gorgé's free program, PAGEotX from `www.qtbridge.com/pageot/pageot.html`. This program generates the HTML you need to include in your Web page in order to display the movie.

To create a movie page, you must perform the following general steps:

1. **Modify TextEdit's preferences to display HTML as plain text so you can edit it.**
2. **Create a formatted page in TextEdit for displaying the movie, and save the page as an HTML document.**
3. **Create the HTML code to display the page using a free program.**
4. **Paste that code into the HTML document.**

I've presented each of these steps as separate procedures below to avoid creating one enormous step-by-step procedure. Actually, the process looks far more complex than it is: After you're familiar with it, making a Web page following these steps usually ends up taking only a couple of minutes to complete.

Modifying TextEdit preferences

You need to adjust one of TextEdit's preferences so you can deal directly with HTML. Although you won't have to write any HTML to make your movie page, you will have to copy HTML from the PAGEotX program and paste it into TextEdit.

To adjust TextEdit's preferences properly, do the following:

1. **Open TextEdit.**
2. **Choose TextEdit⇨Preferences, or press ⌘+, (comma).**

3. **In the Preferences window, click the Open and Save tab.**
4. **Under the When opening a file heading, click Ignore Rich Text Commands in HTML Files.**

   Figure 9-8 shows how the preferences should look.

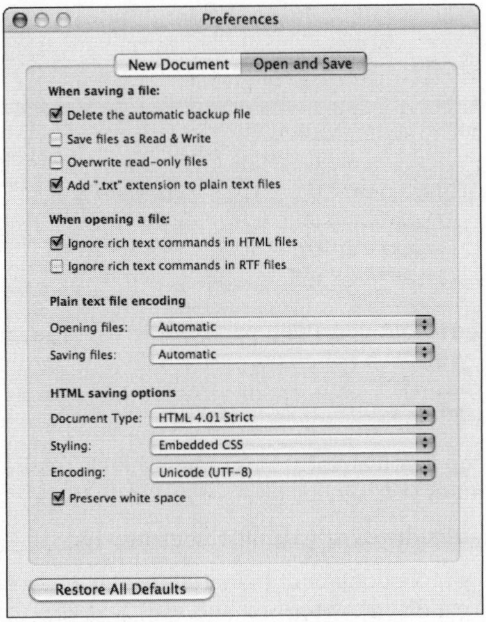

Figure 9-8: Set TextEdit's preferences for editing HTML.

5. **Close the Preferences window.**

   Now, whenever you open an HTML document with TextEdit, you see the actual HTML code rather than the formatted text that the HTML code creates.

6. **Create a folder inside of your Sites folder.**

   This folder will hold the QuickTime movie and, eventually, the Web page that displays it. Figure 9-9 shows a Sites folder containing a birdmovie folder, into which the QuickTime movie birdmovie.mov has been placed. You should set up your movie folder similarly. In the next set of steps, you add a file named `index.html` to this folder.

Mac OS X's Personal Web Sharing software displays a folder's `index.html` file when a browser requests a folder by name, but does not specify which document in that folder to view.

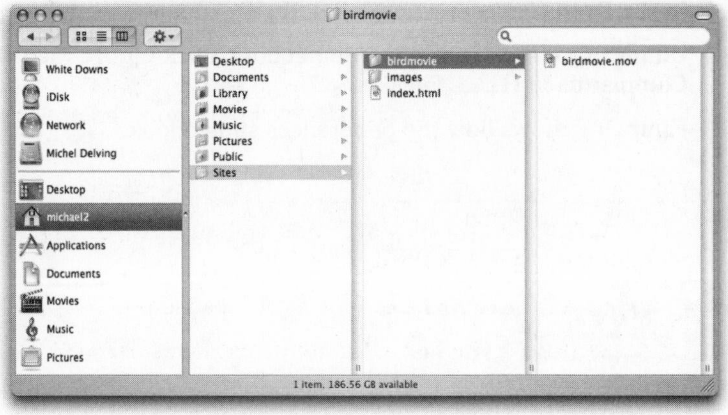

Figure 9-9: Create a movie folder in your Sites folder to hold your movie and Web page.

Creating a Web page

Now that you have a place to put your Web page, you create the `index.html` document that becomes your Web page:

1. **In TextEdit, choose File➪New.**

   You can omit this step if you already have a new, blank TextEdit document window in front of you.

2. **Write the text that introduces or explains your movie.**

   Figure 9-10 shows a simple document I created for this purpose, using TextEdit's formatting tools to center text and vary text size.

Figure 9-10: Create a document to frame your movie.

3. **Add a separate line anywhere in the document that indicates where the movie should be placed.**

   Keep the text simple and distinctive so you can spot it in its HTML form later. In Figure 9-10, for example, I used the simple yet serviceable "INSERT MOVIE HERE."

4. **Choose File⇨Save As.**

5. **In the Save As sheet, navigate to the folder inside of your Sites folder that contains the QuickTime movie.**

6. **In the sheet's Save As field, type the name** index.

7. **Click the sheet's File Format pop-up menu and choose HTML.**

   Figure 9-11 shows the choices on the pop-up menu. Note also that the Hide Extension check box is unchecked, so the file will actually be saved as `index.html`.

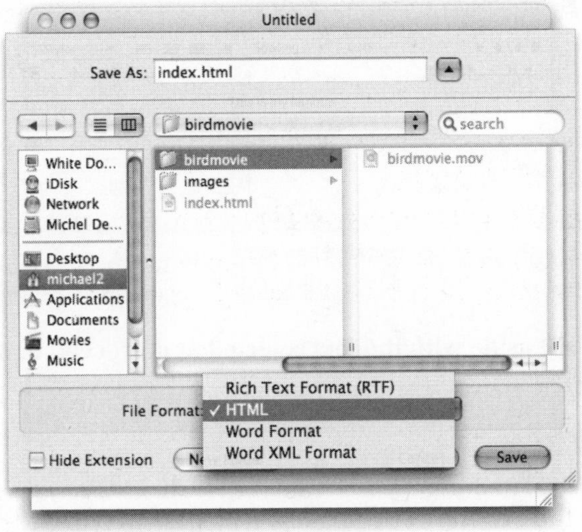

Figure 9-11: Save the file as HTML, and name it *index.html.*

8. **Click Save.**

9. **Close the document window.**

10. **Choose File⇨Open.**

    The Open dialog appears.

11. **Navigate to the document you just saved, select it, and click Open.**

    Because of how you adjusted TextEdit's preferences, the document opens as an HTML text document. Somewhere in the mass of HTML code you should be able to see the phrase that indicates where the movie should go.

12. Select the text that indicates the movie's placement.

Figure 9-12 illustrates what my document looked like when I performed this step. You can see the code `<p class="p1">` and `</p>` that brackets the selected text: In HTML, `<p>` tags mark individual paragraphs. By replacing the selected text between these two tags with the movie-displaying HTML, as you do in the next section, the movie will appear in its own paragraph on the Web page.

```
<!DOCTYPE html PUBLIC "-//W3C//DTD HTML 4.01//EN" "http://www.w3.org/TR/
html4/strict.dtd">
<html>
<head>
  <meta http-equiv="Content-Type" content="text/html; charset=UTF-8">
  <meta http-equiv="Content-Style-Type" content="text/css">
  <title></title>
  <meta name="Generator" content="Cocoa HTML Writer">
  <meta name="CocoaVersion" content="824.1">
  <style type="text/css">
    p.p1 {margin: 0.0px 0.0px 0.0px 0.0px; text-align: center; font: 12.0px
Helvetica}
    p.p2 {margin: 0.0px 0.0px 0.0px 0.0px; text-align: center; font: 19.0px
Helvetica}
    p.p3 {margin: 0.0px 0.0px 0.0px 0.0px; text-align: center; font: 12.0px
Helvetica; min-height: 14.0px}
  </style>
</head>
<body>
<p class="p1">This is my famous<span class="Apple-converted-space">¬†</
span></p>
<p class="p2">Hummingbird movie</p>
<p class="p3"><br></p>
<p class="p1">INSERT MOVIE HERE</p>
<p class="p3"><br></p>
<p class="p3"><br></p>
<p class="p1">Copyright ¬© 2005 Michael E. Cohen</p>
</body>
</html>
```

Figure 9-12: Select the text that shows where the movie will play.

13. Leave TextEdit aside with the text selected and proceed to the next section's steps.

Creating the HTML code

After creating a formatted page, you create the HTML instructions that place the movie where the currently selected text is in your Web page. For that, you use the free PAGEotX program described earlier.

To create the HTML code that displays the movie on the Web page with PAGEotX, follow these steps:

1. **Open PAGEotX.**
2. **Choose File⇨Open Movie, or press ⌘+O.**

   The Open dialog appears.
3. **Navigate to the movie you placed in a folder inside of your Sites folder, select it, and click Open.**

   The PAGEot window displays the name of the movie file at the top and fills in the movie's physical dimensions in its Spatial pane.

Why the movie code looks so darn complicated

If you think that the HTML instructions that place a QuickTime movie on a Web page look rather complex, you're right. Both historical and practical reasons have made it so.

Because of a change that Microsoft made a few years ago to how its Internet Explorer browser works, the original, relatively simple method for placing QuickTime movies on a Web page stopped working in the Internet Explorer browser, though it continued to work in other browsers. The solution requires including both the Internet Explorer–specific instructions and the older instructions on the Web page. Internet Explorer ignores the older instructions and other browsers ignore the Internet Explorer instructions. To put it more technically, you have to wrap an `embed` tag and its attributes, which constitutes the older QuickTime display method, inside of an `object` tag and its associated `param` tags, which is the newer display method.

The instructions also look complex because you can specify many options for presenting a QuickTime movie, such as the size of the movie pane, whether the movie has a controller, whether the movie plays automatically or not, and so on. Specifying each of those options twice, in both Microsoft and non-Microsoft formats can result in some forbidding-looking code. PAGEotX's sole purpose is to create these HTML movie-embedding instructions so you don't have to type in that complex mass of redundant code yourself.

4. **At the bottom of the PAGEot window, click Code and Options.**

   A window displays the HTML code that PAGEotX has created. Figure 9-13 shows the PAGEot window, the PAGEot Code & Options window, and the TextEdit window that contains the HTML code for your Web page lurking behind the two front windows.

5. **Optionally, click to disable AutoPlay in the Control pane of the PAGEot window.**

   Turning off this option causes the movie not to play on the Web page until the user clicks the movie's play button. When you adjust options in the main PAGEot window, you can see the changes reflected in the code pane of the PAGEot Code & Options window.

   If you are curious, you can find what all of the HTML code means, and an explanation of the QuickTime options that PAGEotX controls, by visiting Apple's QuickTime tutorials at `www.apple.com/quicktime/tutorials/embed.html`.

6. **Select all the text in the PAGEot Code & Options window and then choose Edit⇨Copy, or press ⌘+C.**

   This places PAGEotX's HTML code on your Clipboard for the final stage of the page-building process.

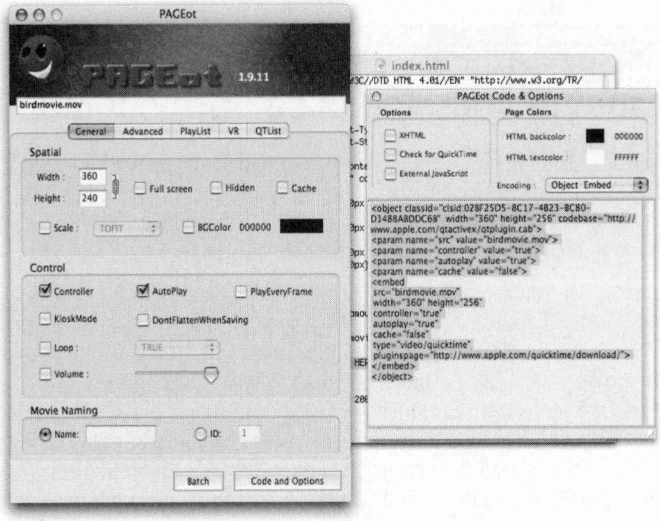

Figure 9-13: PAGEotX writes the HTML code for displaying QuickTime on the Web.

Pasting the HTML code

Now that you have some movie-displaying HTML code, courtesy of PAGEotX, you insert it in your HTML page.

Perform the following steps:

1. **Switch back to the TextEdit application.**

   If you have been following each of these procedures in sequence, it should still be running.

2. **Make sure that the text indicating where the movie should appear is still selected.**

   If it isn't, simply select it again.

3. **Choose Edit➪Paste, or press ⌘+V.**

   This replaces the selected text with the code you copied to your Clipboard from PAGEotX in the previous procedure. Figure 9-14 shows what the TextEdit window should look like after you perform this step.

4. **Choose File➪Save, or press ⌘+S.**
5. **Quit TextEdit and PAGEotX.**

Your Web page for displaying the movie is now in the right folder. As long as your Mac has Personal Web Sharing enabled, clients on your network can open the page in their browsers by entering the Bonjour-supplied URL for the page, as described in "Using QuickTime Player to play shared movies" earlier in this chapter.

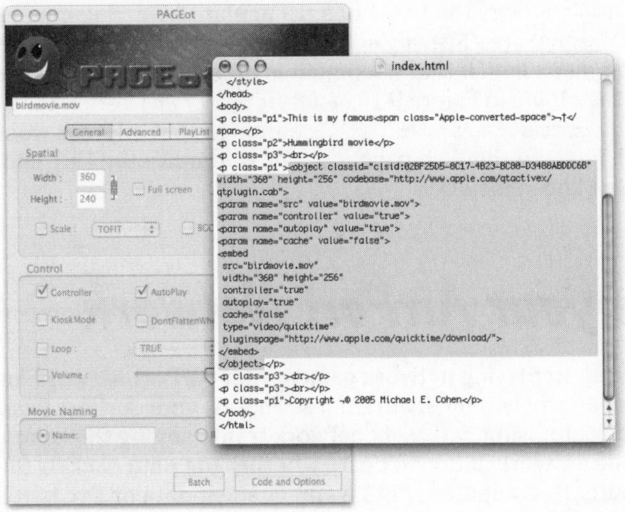

Figure 9-14: Paste the code from PAGEotX into your HTML document.

Figure 9-15 shows Safari presenting a movie page created by following this procedure. In the figure, the page is in a folder named *birdmovie*, which is inside of the Sites folder belonging to the user account *michael2*, which, in turn, is on a Mac with the Bonjour name *whitedowns.local*. The URL that brings up the movie page is `http://whitedowns.local/~michael2/birdmovie/`.

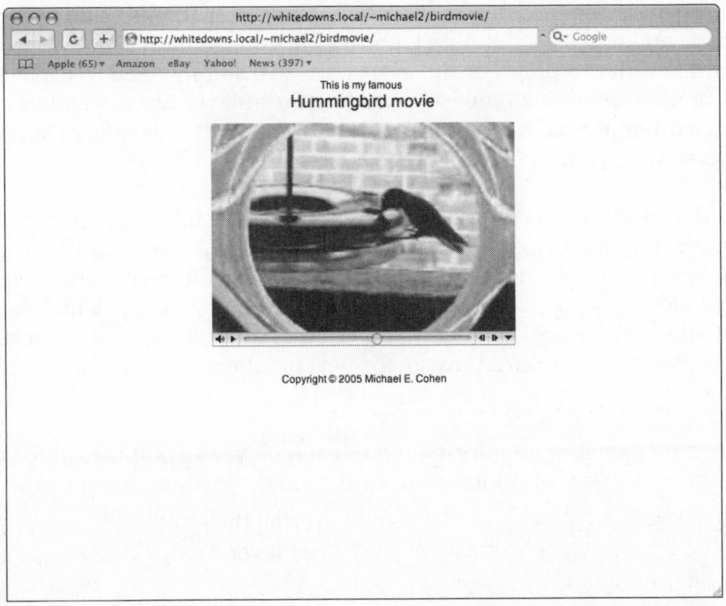

Figure 9-15: Safari displays a movie inside the Web page you created.

When you don't specify the name of a particular HTML document in a URL, the Mac's Personal Web Sharing software assumes you want to see the document named *index.html* if that document is present. That's why the URL shown for the movie in Figure 9-15 is not `http://whitedowns.local/~michael2/birdmovie/index.html`, although that address would work just as well: I named the Web page *index.html* to make the URL shorter and easier to type and remember.

Gaming on Your AirPort Network

When it comes to playing network games, your AirPort network functions more or less as any wired network does. That is, your AirPort Base Station, like a wired router, simply directs network traffic among the computers connected to the network. It doesn't care whether the data packets its shunts from computer to computer contain spreadsheet data or Orc battle plans. However, some games may require specific configurations on any network, wired or not. Between your Mac's Network preferences and Firewall preferences, and the AirPort Base Station's capabilities, you can accommodate such configurations.

Playing locally

Playing games locally across your AirPort network, as opposed to playing games over the Internet, ranks among the easier things to do with your network, and that's as it should be: After all, you shouldn't have to work too hard in order to play. For the most part, you simply need to make sure that your Mac and the computers used by other players can communicate with each other according to the game's needs, and that usually means configuring each machine's firewall settings properly.

As described in several other places in this book, information that travels across a network does so in small chunks, called *packets*, and each packet includes both the *address* of the machine it needs to reach, and a *port number* that identifies the packet to the programs on that machine which need to pay attention to the packet. Your Mac's Firewall software acts as a gatekeeper, letting only those packets through for port numbers you have approved as being okay to use.

Network games send information to each player's computer across the network, and tags that information with specific port numbers, so the game software on each computer can obtain just the game information and ignore the other information packets traveling across the network.

Different games tend to use different port numbers than the port numbers commonly used for things like mail, Web browsing, and file sharing. If you have your Mac's Firewall enabled, chances are good that the Firewall blocks those port numbers: generally, your Firewall blocks all port numbers except those you have specifically approved. You have two choices:

WARNING!

- **Turn off your Firewall:** This makes all of the tens of thousands of port numbers on your Mac open to anyone on your network. On a normal AirPort network, where the base station provides the Internet connection, this does not present as great a risk as it otherwise might: The AirPort Base Station itself acts as a firewall, protecting your Mac from most unsolicited access attempts from the Internet. Nonetheless, turning off your Firewall does make you more vulnerable to unwelcome intrusions than you otherwise might be. It's like leaving the front door to your home unlocked: you never know who might come in and what they might do.
- **Approve the port numbers that the game requires:** This requires a trip to your Firewall settings in System Preferences, and creating an entry for each port your game software needs to use. Though somewhat more time consuming to set up than simply turning off the Firewall, it doesn't take much more time and leaves your Mac better protected.

To set up your Mac's Firewall for local network gaming, do the following:

1. **Open System Preferences.**
2. **Click Sharing.**

   The Sharing preferences appear.
3. **Click the Firewall tab.**
4. **Do one of the following steps:**
   - **Click Stop to turn off the Firewall, and close the Sharing window.**

     If you see a Start button instead of a Stop button, it means that the Firewall is already off. In either case, you can skip to the last step. All of your Mac's network ports are available.
   - **Click New, and continue to the next step.**

     A sheet appears with controls you can use to specify network ports you want to make available.
5. **In the sheet's Part Name pop-up menu, click Other in the Port Name pop-up menu.**
6. **In the Description field, enter the name of the game.**

   Actually, you can enter anything you like, but specifying the game's name makes it easier for you to find this Firewall entry in the Firewall settings' Allow list the next time you want to play the game.

7. **Enter the TCP and UDP port numbers you want to make available in the TCP Port Number(s) and the UDP Port Number(s) fields.**

   Your game's documentation can tell you what numbers to enter. Some games use TCP port numbers, some use UDP port numbers, some use both. For example, Figure 9-16 shows the port numbers that the game *Quake II* requires. You can enter multiple port numbers by separating the numbers with commas, and you can specify ranges of numbers by separating the start and stop numbers with hyphens: for example, 1024-6000, 7000, 8096.

   TIP: You can find lists of popular games and the network port numbers they use on several different Internet sites, such as `www.gameconfig.co.uk`, which provides a search engine for looking up games.

8. **Click OK.**

   The new Firewall entry appears in the Allow list. From now on, you can enable and disable the ports used by this game by clicking its check box in the list.

9. **Close the Sharing window.**

Now, with the appropriate port numbers available on your Mac, you can let the gaming begin.

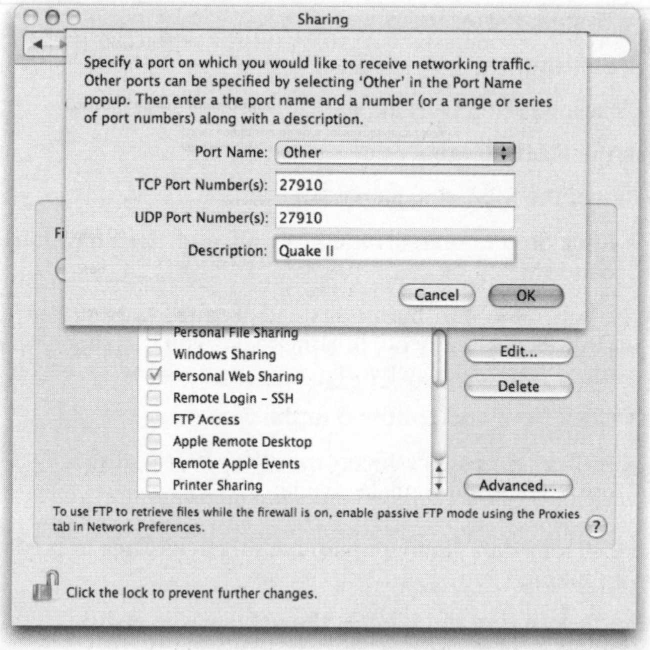
Sharing
Specify a port on which you would like to receive networking traffic. Other ports can be specified by selecting 'Other' in the Port Name popup. Then enter a the port name and a number (or a range or series of port numbers) along with a description.
Port Name: Other
TCP Port Number(s): 27910
UDP Port Number(s): 27910
Description: Quake II
Cancel
OK
Personal File Sharing
Windows Sharing
Personal Web Sharing
Remote Login - SSH
FTP Access
Apple Remote Desktop
Remote Apple Events
Printer Sharing
Edit...
Delete
Advanced...
To use FTP to retrieve files while the firewall is on, enable passive FTP mode using the Proxies tab in Network Preferences.
Click the lock to prevent further changes.

Figure 9-16: Opening ports in the Firewall for *Quake II.*

Joining Internet games

Some games, especially massively multiplayer online games, require that your Mac make contact with other players elsewhere on the Internet. When you want to play those sorts of games, your AirPort Base Station's normal setup, which ordinarily protects your AirPort network from unwanted Internet-based intrusions, can interfere. You can overcome the AirPort Base Station's protective urges in a couple of ways.

As with local network games, described in the previous section, the problem arises from port numbers. Just like your Mac's own Firewall, your base station blocks all unsolicited incoming communications from the Internet, regardless of the port number, unless you specify otherwise.

You can specify otherwise in one of two ways:

- You can establish a *default local host* in your AirPort Base Station settings and set up a Mac on the network to act as that local host. This is often called creating a *DMZ* (demilitarized zone), and it puts that Mac outside of the base station's protection, exposing all of that Mac's port numbers to Internet requests. You set up your Mac like this, for example, if you want it to serve as a game server for other network players.
- You can set up a *port mapping* list that tells the AirPort Base Station to send all communications using specific port numbers to specific local network addresses. This technique is more labor intensive, but it allows more computers to participate in the gaming and maintains the most security.

Which one of these methods you choose, or whether you need to do both, depends on the game.

Establishing a DMZ

Putting your Mac into the AirPort network's DMZ forwards all communication requests coming in from the Internet to your Mac, regardless of the port numbers they use. In a way, it's something like turning off the base station's firewall just for that Mac: Other computers on the AirPort network remain protected.

WARNING!

Because the AirPort Base Station no longer protects your Mac's network ports when you create a DMZ, you should make sure that you have enabled your Mac's own Firewall and set it to allow communications only with the port numbers that the game requires. See "Playing locally" earlier in this chapter to see how to do that.

To put your Mac into the DMZ, you have to change an AirPort Base Station setting and make a change in your Mac's network settings.

To change your AirPort Base Station settings to establish a DMZ, do the following steps:

1. **Open AirPort Admin Utility.**
2. **In the Select Base Station window, select your base station and click Configure in the window's toolbar.**
3. **Under the AirPort tab in the base station's configuration window, click Base Station Options.**

   A sheet appears with a number of tabs.
4. **Click the Ethernet Port Security tab.**
5. **Click Enable Default Host at.**

   Figure 9-17 shows what the sheet looks like after this step. Usually, the sheet assigns 10.0.1.253 as the default host IP address. Although you can change the last number of this address, you normally won't need to.

   Remember the Enable Default Host at address, because you need it to set up your Mac to use that address.

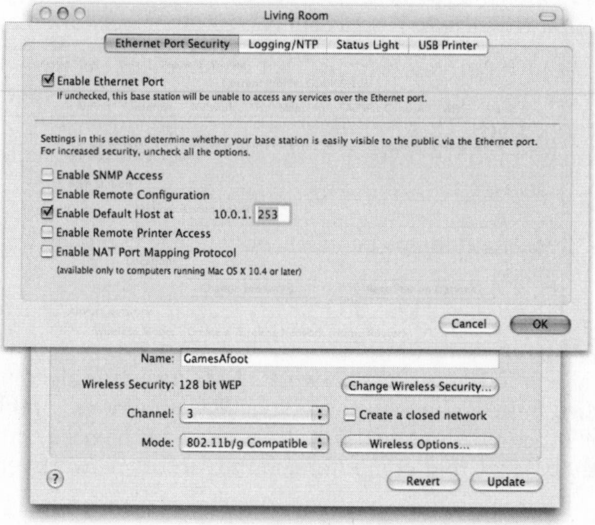

Figure 9-17: Punching a hole in the base station's firewall for one network client.

6. **Click OK.**
7. **Click Update.**

   The AirPort Base Station restarts, using the new settings.
8. **Choose AirPort Admin Utility➪Quit, or press ⌘+Q.**

You also need to set up your Mac to use the default local host address. Here's how you do that:

1. **Open System Preferences.**
2. **Click Network.**
3. **In the Network window, on the Show pop-up menu, click AirPort.**
4. **Click the TCP/IP tab.**
5. **On the Configure IPv4 pop-up menu, click Using DHCP with manual address.**
6. **In the IP Address field, enter the local host address you set up in the previous procedure.**

   Figure 9-18 depicts how the TCP/IP tab in the Network window looks after this step. 10.0.1.253 is the customary IP address for an AirPort Base Station's local default host, so chances are good that's the number you will enter here.

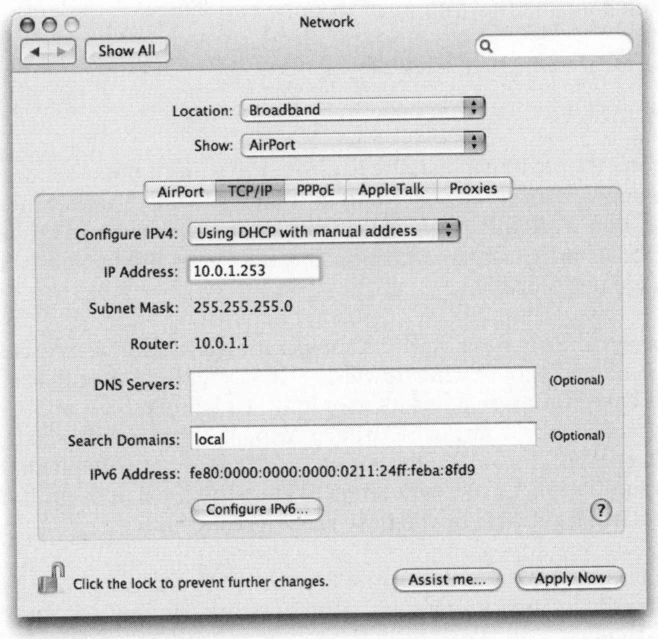

Figure 9-18: Manually assigning a network address to put your Mac in the DMZ.

7. **Click Apply Now and close the Network window.**

Now your Mac can respond to any external Internet traffic coming through your base station's Internet connection.

TIP

You can establish a new network location in your Network preferences by choosing New Location from the Location pop-up menu before you change your Mac's TCP/IP address as described above. The settings you make to your TCP/IP address apply only to the new location's configuration, and makes it much easier for you to switch back to your original network settings when you finish your gaming session.

Setting up port mapping

Port mapping tells the base station to forward all incoming Internet requests using specific port numbers to specific addresses on the local network. For example, if the AirPort Base Station receives a request that uses port number 80 — by convention, the Web server port number — it consults a list of local addresses that should receive that request and forwards the request to those addresses. Optionally, that list can also tell the base station to change the port number when it forwards the request. For example, the list can say to send all requests using port 80 to port 8080 on the receiving network clients.

You need to set up port mapping for each game player on your AirPort network, which means that you need to know two things:

- The local address of each player
- The port numbers that need to be forwarded

The second item you can usually get from the game's documentation itself, or from an online source, as described in "Playing locally" earlier in this chapter. The first item is slightly trickier, though: The DHCP server in the base station assigns local addresses *dynamically.* The addresses are not permanent, and can change periodically.

Fortunately, AirPort Base Stations usually assign local IP addresses starting with 10.0.1.2 and progressing upward — 10.0.1.3, 10.0.1.4, and so forth — until the base station reaches its user limit, which is ten for an AirPort Express and 50 for an AirPort Extreme. Also, although the base station can change a computer's local address, on most home networks it seldom does: Each computer on the network tends to keep the local address it has received until it reboots or the base station restarts.

TIP

You can find out your Mac's current IP address by opening your Network preferences in System Preferences. The Network window's Status display, which is the first thing you see in the Network window by default, lists the IP address along with the physical port that supplies your Internet connection.

After you have the list of IP addresses to which you want to apply port mapping and the list of ports you want mapped, set up your AirPort Base Station like this:

1. **Open AirPort Admin Utility.**
2. **In the Select Base Station window, select your base station and click Configure in the window's toolbar.**
3. **Click the Port Mapping tab.**
4. **Click Add.**

   A sheet descends with fields you can use to specify ports and network addresses as shown in Figure 9-19.

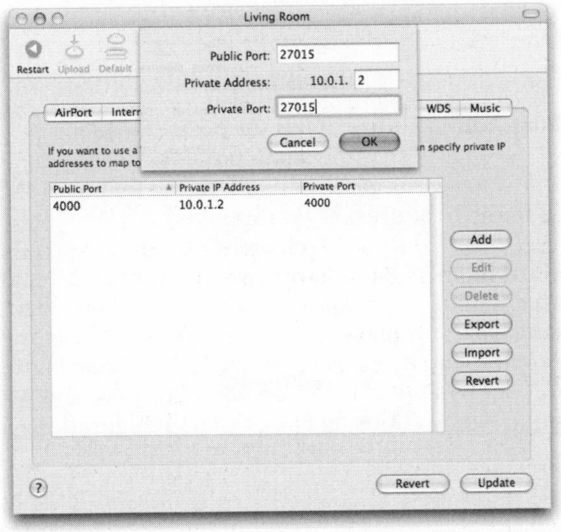

Figure 9-19: Mapping ports to allow Internet gaming access on an AirPort network.

5. **Enter the port number you want to map in both the Public Port and the Private Port fields.**

   Unfortunately, you can specify only single port numbers — no sequences and no ranges — in the base station's Port Mapping interface.

6. **Enter the final part of one of the game player computers' IP addresses in the Private Address field.**
7. **Click OK.**
8. **Repeat Steps 4 to 7 for each computer and each network port you want to forward.**

   TIP

   You can use a text editor to create a list and then import it into the Port Mapping tab in order to save time. The list must be saved as plain text — TextEdit has a Make Plain Text item in its Format menu — and contains one line of text for each port mapping entry. Each entry consists of the public port number, a tab, the local IP address, a tab, the private port number, and a return character. For example, the following shows a list of entries for three addresses that you want to receive information on port 4000:

```
4000        10.0.1.2    4000
4000        10.0.1.3    4000
4000        10.0.1.4    4000
```

You can use the Port Mapping tab's Export button to create and save port mapping text files from previously entered settings, and you can use the Import button to apply previously saved or manually created settings.

9. **Click Update.**

   The AirPort Base Station restarts, using the new settings.

10. **Choose AirPort Admin Utility➪Quit, or press ⌘+Q.**

Your base station now forwards game network traffic that uses the port numbers you specified to the IP addresses you supplied.

If you have an AirPort Express base station, you can create a separate game-playing profile that contains your game settings, so you can switch between them and your usual base station settings. See Chapter 4 for more about AirPort Express profiles. You can also save any AirPort Base Station's settings in AirPort Admin Utility by choosing File➪Save A Copy As, and you can open a saved configuration file by choosing File➪Open Configuration File.

Part IV
Taking Care of Business

The 5th Wave By Rich Tennant

©RICHTENNANT
I don't know nothin' about your computer doohickey, missy. But I do know that Virgil and I just come across proof of an alien presence right here in these woods.

In this part . . .

Corporate enterprise networks and Macs? Isn't that like dogs dancing with cats? It doesn't have to be that way.

In this part, you can learn how to make friends with your network administrator and help that poor benighted soul bring your Mac into the corporate fold. This part also includes tips and techniques when you have to work on the road or from home.

Chapter 10

Fitting In at the Office

In This Chapter

- Learning a new language
- Peeking at the Directory Access utility
- Saying the secret word
- Tunneling into the office
- Making friends with Windows

When it comes to getting down to business, Macs are, and really have always been, a minority platform. Let's face it: For most business-oriented computing environments, whether private sector, governmental, or nonprofit, it's a Windows world. Once upon a time the conventional wisdom was that, "No one ever got fired for buying IBM." Today, if you substitute "Windows" for "IBM," the wisdom is just as true — and just as conventional.

It's not that Macs can't fit into the business world: In most cases, they can fit in just fine. But most business organizations employing more than a few dozen people have information technology (IT) departments that — for various reasons, some reasonable and some arbitrary — have standardized on Windows and its related technologies. More often than not, IT workers in business environments tend to know very little about Macs and, of that little, a good part of it is out-of-date, or simply erroneous.

Though you and I both know the many benefits of using a Mac, to an IT staff worker in a business environment, the Mac simply means another set of problems to solve, another field of expertise to master, another group of users to support. Consequently, if your place of business allows you to use your platform of choice at all, you may have to bring your own support with you.

Fortunately, when it comes to networking, you have an ally: More than ever before, your Mac understands and embraces the Windows way of working, and it usually fits quite easily into a Windows network environment, whether wired or wireless — sometimes even more easily than Windows itself. When you know how your Mac talks to the Windows-oriented networks that you'll almost certainly encounter, you can fit right in, while still standing apart.

Working with Network Administrators

Network administrators have a thankless job: When things go well, no one notices, and when something goes wrong, everyone complains. They often issue seemingly bizarre announcements that no one can understand, and enact seemingly arbitrary policies that sometimes interfere with what you want to do. But they are just like you and me: ordinary people trying to get their jobs done. Unfortunately, their jobs involve keeping complex dynamic systems running in an environment over which they have surprisingly little control, and supporting users who have very little understanding of just how complicated the whole mess can be. Day after day they deal with an onslaught of continuingly escalating Internet threats, spend huge amounts of time patching myriad user systems to fix third-party software flaws, all while maintaining an ever-changing collection of user accounts on mail servers, file servers, database servers, and application servers. It's no wonder if the job sometimes makes them come off as rigid, insensitive, and just a wee bit testy.

Working successfully with a network administrator comes down to clearly expressing what you need and why you need it, accepting that sometimes you can't get exactly what you want, and remembering that both you and the network administrator are collaborators and not combatants.

TIP

July 28th is International System Administrator Appreciation Day. No fooling. It couldn't hurt to remember that, and to send your network administrator a thank you on that day. For more information, see `www.sysadminday.com`.

Justifying your need for wireless Mac access

As you may have figured out, Mac use in the working world is more the exception than the rule. Wireless Mac use on a business network is even more exceptional.

In terms of Macs and network support, you can divide businesses into four categories:

- **No Macs allowed:** Not only does the network staff not support Macs, they prohibit their use. Unless you work at a managerial level on the IT staff yourself, you can't do anything about this, so don't even try.
- **Benign neglect:** The network staff does not support Macs, and may not know anything about them, but will let you use one on the network if you can make a very good case for it, and if you can configure your Mac to meet certain of their requirements. The burden of support, which includes understanding and fulfilling the requirements the network staff

gives you, falls entirely upon your head. You will have to provide a very strong business case to get a wireless Mac approved for network use in this environment, and even a strong case may not be enough.

- **Foster child:** The business does allow Macs on the network, and does provide some support for them, but the network staff treats Macs as exceptions rather than the rule, and may not provide a full set of network services to your Mac. You will need to provide some support for yourself. Getting approval to use a wireless Mac in this environment also requires a strong business case, but has a reasonable chance of succeeding.
- **Equal partner:** The network staff knows and understands Macs, and provides full support for them on the network. Existing network policies cover how and in which circumstances a wireless Mac may connect to the network. Find out about such policies and see if they cover your situation; if they do, you're set.

If your place of business falls into any but the first category in the previous list, and you have managed to get a Mac as your computer, you need to justify the wireless connection. The best justifications, of course, are those that can clearly show one or more of the following:

- **Increased efficiency:** Workplace efficiency is one of the most difficult things to measure. However, if you use a laptop as your regular work computer, and you frequently attend meetings or meet with clients offsite with your laptop, you can make the case that wireless connectivity allows you to be more responsive and flexible.
- **Cost savings:** This requires some research on your part. Providing physical network ports in an office incurs cabling costs, and it may be that connecting your Mac to a wireless access point costs less than running network cables to your work area. If nearby co-workers can also use the access point, so much the better. To make this argument effectively, it helps to make friends with the network administrator, who probably knows what the cabling costs are.
- **Reduced network traffic:** A wireless network, such as an AirPort network, keeps local network traffic, such as the communication between your computer and your printer, off the main network and on the subnet that it establishes. If you share an office with several other workers and collaborate with them over the network, a local wireless router like an AirPort Base Station can reduce congestion on the rest of the business's network. Again, you should work with your network administrator to find out if this is the case where you work.

REMEMBER

Don't be surprised if your network administrator seems unenthusiastic at first about your request for wireless connectivity. Network administrators prefer wired networks because they can tell where each wired connection goes, and because they can monitor the status of each wired connection, using a well-established and understood set of tools. Wireless networks, on the other hand, create a less predictable area of connectivity, and require a different set of monitoring and management tools. It's up to you to discover your administrator's areas of concern and to work slowly and patiently to address them.

Getting the network information you need

Unless your place of business provides Mac support as a matter of course, the information you need in order to set up your Mac likely will come to you using terminology and procedures tailored to Windows users' needs, not yours. In many cases you can adapt that information for your purposes, but when you can't, you can ask for what you need using terminology that is neither Mac-specific nor Windows-specific.

Obtaining TCP/IP network setup information

These days, both Mac and Windows networks use TCP/IP, the network protocols that underlie the Internet, in order to establish and obtain network addresses. Although your Mac's Network preferences window, available through System Preferences, looks very different from its analogous configuration setup in Windows, most of the same TCP/IP control options are available for either.

Your network administrator can supply the information you need to set these options, such as whether you should set your IP address through DHCP or manually, and, if so, what your router and subnet settings should be. If your Mac is joining the business's wireless network directly, chances are your Mac's network setup will be very much like the one you use for joining an AirPort network. If you have received approval to connect an AirPort base station to a business's network, on the other hand, you'll use the AirPort Admin Utility to set the appropriate TCP/IP settings for the base station. Chapter 4 describes how to set up an AirPort base station over a LAN.

TIP

Even in heavily Windows-oriented businesses, many network administrators tend to have some familiarity with Linux as well as Windows. Because both Linux and Mac OS X share a common Unix heritage, you may be able to get useful configuration information by asking your network administrator, "How would a Linux user do this?"

Setting up Active Directory and related network services

Many modern Windows networks employ Active Directory, a Windows-centric database system used to manage user accounts on a corporate network and to set network privileges. To use any network resources on a network managed by Active Directory, you need to *authenticate* — that is, gain access to the network as a recognized user.

Fortunately, Mac OS X provides Active Directory compatibility so that you can join such networks. However, configuring your Mac to communicate with Active Directory is not something you can easily do by yourself — in fact, it's not something most Windows users can do by themselves, either. Setting up Active Directory access usually requires that a network administrator configure the machine appropriately.

Windows network administrators configure Active Directory on Windows computers every day. Not many Windows network administrators, however, know that a Mac can also authenticate to an Active Directory, and, even if they do know, they may not know how to set up Active Directory on a Mac. You can help them by showing them where the Mac keeps its Active Directory configuration tools. The Directory Access utility resides in your Mac's Utilities folder, as shown in Figure 10-1. It provides the configuration controls your network administrator needs in order to meld your Mac into the Active Directory database that manages the business's network.

Figure 10-1: The Utilities folder contains Directory Access, your key to Active Directory.

The Directory Access utility does much more than allow you to configure Active Directory — and much less. The program, in fact, is merely a shell that hosts a variety of network authentication and communication plug-ins that conform to Apple's Open Directory architecture. The Open Directory architecture makes it possible for your Mac to use a variety of network services, and to add new network services by adding the appropriate plug-ins to Directory Access. Because Directory Access uses plug-ins in order to do its work, the list of services that it makes available depends on the plug-ins it currently has installed.

When you open Directory Access, the utility presents its current list of network services. By default, Directory Access does not allow you to change anything until you click the padlock icon and enter your Mac's admin password. After you've done that, you can select network services to enable, disable, and configure, as shown in Figure 10-2.

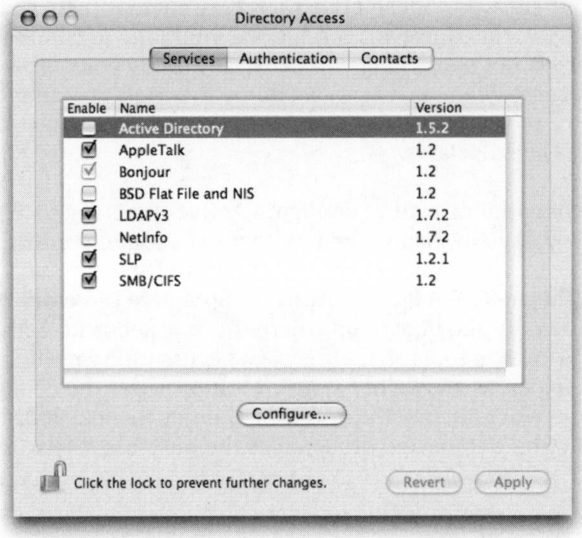

Figure 10-2: Use Directory Access to make Active Directory active.

Out of the box, Mac OS X does not enable Active Directory, and that's a good thing, because, as I've said, getting it properly set up for a particular Active Directory network usually requires the gentle touch of a network administrator.

Although the details can vary considerably, depending on the service being configured, the general configuration process for any of the services shown in Directory Access is quite simple:

1. **Open Directory Access.**
2. **Click the padlock and enter an admin username and password.**

   You should be prepared to give your network administrator these pieces of information, or be on hand to enter them yourself when your administrator comes to configure your Mac.
3. **Click the check box for the network service you want to enable.**
4. **Click Configure to make the appropriate settings in a sheet that appears.**

   Many of the configuration sheets Directory Access presents have both simple and advanced versions, the latter of which you can access by clicking the Show Advanced Options button. Most network administrators will want to see the advanced options. Figure 10-3 shows the configuration sheet that Directory Access presents when you select Active Directory in the Services list, click Configure, and then click Show Advanced Options.
5. **Click OK to activate those settings.**

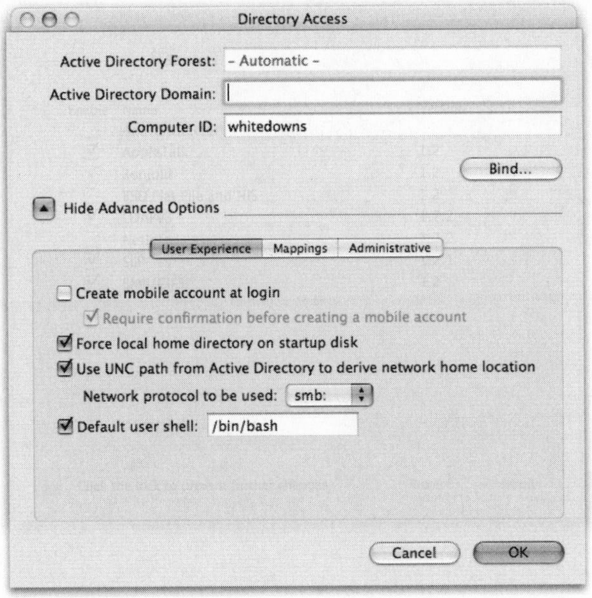

Figure 10-3: The advanced Active Directory configuration options you have available.

Unless you are a networking professional, or your network administrator has given you a clear set of instructions to follow, you should not modify any of the configuration settings that Directory Access provides. Improperly configuring network services can make your Mac unable to communicate properly over your network.

In addition to providing a plug-in for configuring Active Directory, Directory Access provides plug-ins for configuring two other network services often found on Windows networks: *LDAP* (*Lightweight Directory Access Protocol*) and *SMB/CIFS* (*Server Message Block/Common Internet File System*). These two services can either work in conjunction with Active Directory or independently from it. As with Active Directory, your network administrator, rather than you, will use Directory Access to configure these services.

LDAP provides a simple but flexible protocol that various network directory services can use to both search and update network directory databases, such as the databases that Active Directory uses to manage a network's file-sharing services, authorized users, access controls, and so on. Apple ships Mac OS X with LDAP enabled because Mac OS X's own directory services use it. Figure 10-4 shows Directory Access's LDAP configuration sheet, with its manual configuration options revealed. Your network administrator may need to change or add to the available LDAP configurations.

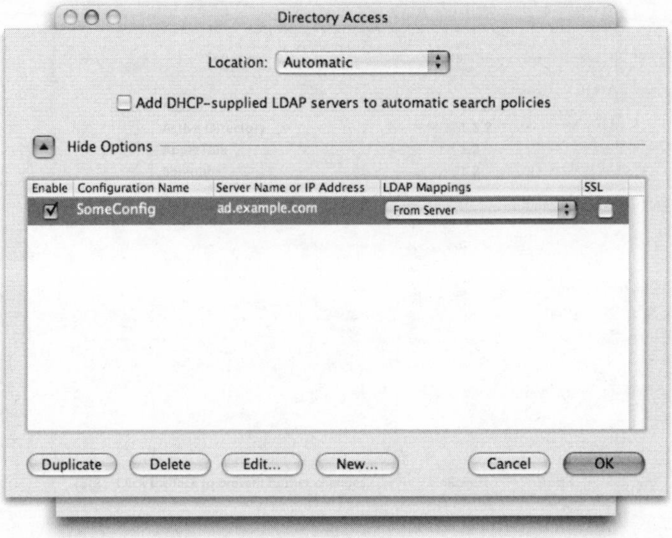

Figure 10-4: Directory Access also provides LDAP configuration settings.

Windows networks use SMB/CIFS to establish and manage communication between network devices and resources, such as file servers, printers, and similar resources. In the Windows world, it serves roughly the same purpose as Apple's *AFP* (*Apple Filing Protocol*) and Bonjour do for Macintosh networks. Mac OS X comes with SMB/CIFS enabled so that you can take advantage of Windows file sharing on a mixed local network, as described later in this chapter.

Figure 10-5 shows the simple configuration sheet that Directory Access provides for SMB/CIFS. Your network administrator may want to set the configuration for this service, or may give you instructions for configuring it, as it is a relatively simple service to configure.

TIP

Although the configuration options provided by Directory Access's plug-ins may sound familiar to your network administrator, they may look different enough from their Windows equivalents to make your network administrator feel less than comfortable using them. You can help alleviate any lingering stress by pointing your administrator to Mike Bombich's "Setting up basic authentication to an Active Directory Server" page at `www.bombich.com/mactips/activedir.html`. This page provides a Windows network administrator with a very valuable set of procedures for fitting a Mac into a Windows Active Directory network and for testing the connection.

Specifying authentication options

In addition to the Services tab, the Directory Access window provides another tab your network administrator may find useful: Authentication. Your administrator uses the controls under this tab to specify where your Mac obtains authentication information for joining the business network.

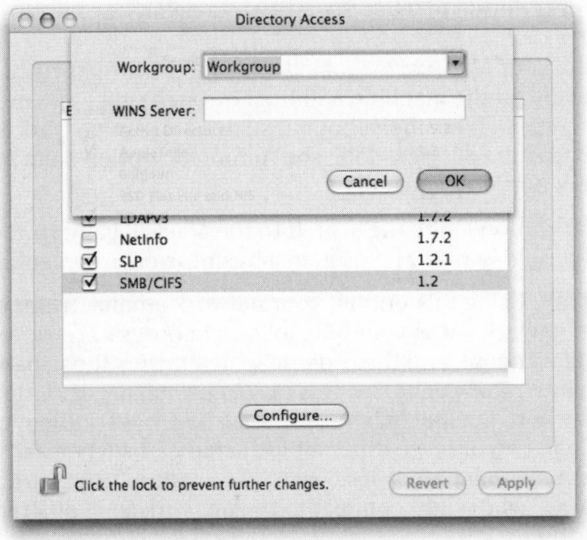

Figure 10-5: SMB/CIFS handle basic Windows file-sharing services.

Figure 10-6 shows the panel under the Authentication tab of Directory Access, as well as the available items on that panel's Search pop-up menu. Your administrator chooses one of these items to limit or expand how your Mac looks for authentication information when it joins the network.

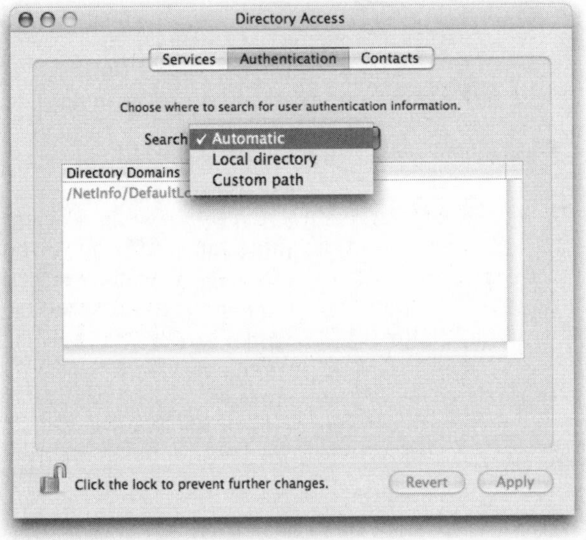

Figure 10-6: Choosing where your Mac looks for authentication information.

Here's a brief description of what these choices mean:

- **Automatic:** Your Mac starts looking through the directory information stored locally on the machine and then through other network directory sources to which it has been bound, such as an Active Directory setup created under the Services tab. The Automatic option suits mobile computer users particularly well.
- **Local:** This option limits the search to the Mac's locally stored information, and probably won't be used in a business environment.
- **Custom path:** Using this option, your network administrator can specify the exact network paths your Mac follows in order to retrieve its authentication information, and the order in which it tries those paths. A network administrator might find this option useful, particularly in instances where the Mac might otherwise find valid authentication information along two or more different available paths when searching automatically. A custom path is less useful for a laptop user, particularly a wireless user, who might connect to the network from different locations at various times.

When your administrator chooses the Custom Path item from the Search pop-up menu, an Add and a Remove button appear in the Authentication panel, as shown in Figure 10-7. Clicking the Add button presents a sheet from which an administrator can choose from the available network search paths. By clicking the Remove button, an administrator can exclude paths from being searched for authentication information, thus reducing network load.

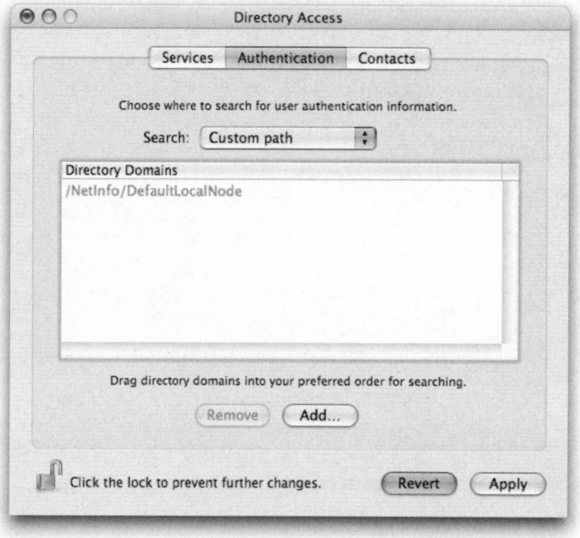

Figure 10-7: Add or remove a custom search path for finding authentication information.

TIP

As a favor for your administrator, you can download Apple's Open Directory administrative guide, a PDF file that explains almost everything a network administrator might want to know about how Macs participate in various networks. Although the title, "Mac OS X Server Open Directory Administration," implies that the document covers only Apple's Mac OS X Server software, it covers much more than that. You can find this highly useful, and highly technical, document at `http://images.apple.com/server/pdfs/Open_Directory.pdf`.

Becoming secure

Business networks require stringent security practices. In fact, much of the resistance that business network administrators tend to exhibit towards Macs and towards wireless networks arises from security concerns. You can overcome such resistance by working with your network administrator to make sure that your Mac and, if you have one, your AirPort base station, adhere to the business's security requirements. The good news is that both Macs and AirPort base stations provide powerful and sophisticated security options that can meet or exceed the security requirements of most business networks.

Choosing a practical password

Very often — and, in the opinion of some security experts, all too often — authentication systems for business networks rely upon a user password in order to grant access. After a network administrator has established your network account, whether through Active Directory or some other method, you usually end up using a password to access that account.

Passwords have certain advantages as an authentication method:

- You can easily create them.
- You can easily change them.
- You always have them with you.

But using passwords for authentication also has drawbacks:

- Someone who knows you can guess your personalized passwords.
- Anyone can pretend to be you if they have your password.
- You can forget passwords and find yourself unable to access your data.
- You can't easily tell if someone has stolen your password.

Because of the advantages of using passwords as an authentication method, the majority of networks continue to use them, rather than more complex and more expensive methods. But, because of the drawbacks, network administrators usually establish specific rules for both creating and for changing passwords that reduce the risk of using them as a means for authentication.

Though the specific rules for creating and using passwords vary from business to business, they usually involve one or more of the following recommendations in some form:

- **Passwords must not be too short:** The longer the password is, the harder it is to guess by chance. For example, there are more than 17,000 different 3-letter passwords that use only the basic English 26 alphabetic characters. This might seem like a lot until you realize that a malicious computer program that tries to crack passwords can go through all of those possibilities in a very short time. On the other hand, you can create over 200 billion possible passwords using 8 characters. Many administrators reject passwords shorter than 8, 10, or even 12 characters.
- **Passwords must never be shared:** A secret shared by two people seldom remains a secret. Furthermore, a shared password defeats the entire purpose of authentication. Many businesses consider sharing passwords to be grounds for disciplinary action, or even dismissal.
- **Passwords must not be too personal:** If you use something easy to remember because it is personal, such as your spouse's name, your pet's name, or your street address for your password, someone who knows you can guess what that password is. And not everyone who knows you is necessarily trustworthy.
- **Passwords should not appear in the dictionary:** Even long, complex, and obscure words should be avoided because the hackers' password-cracking tools often employ dictionary searches. Avoid the names of literary characters: Hackers can read, too.
- **Passwords must be changed regularly:** The longer you use a password, the more likely it is that someone can crack it. Most network administrators require that you change your network password every few months, and they often prohibit you from basing new passwords on passwords you have used before.

Even if your network administrator does not impose a strict password policy, you should try to follow these recommendations; after all, it's your network account that's at stake. To help you, Apple provides the Password Assistant, a tool that can create new passwords to meet various security requirements, and that can even evaluate how secure your current passwords are.

REMEMBER

You can find the Password Assistant lurking inside your Mac's Keychain Access utility, a program that manages the passwords stored on your Mac's keychain.

cf. Pogue, Mac OS X: Tiger, pp. 471-475

Here's how to bring up the Password Assistant:

1. **Open Keychain Access.**

   You can find the Keychain Access program located in your Utilities folder. When you open the program, a window showing you all of the items contained on your keychain appears.

2. **Choose File⇨New Password Item, or press ⌘+N.**

   A sheet appears containing fields you can fill out to create a new password, as shown in Figure 10-8.

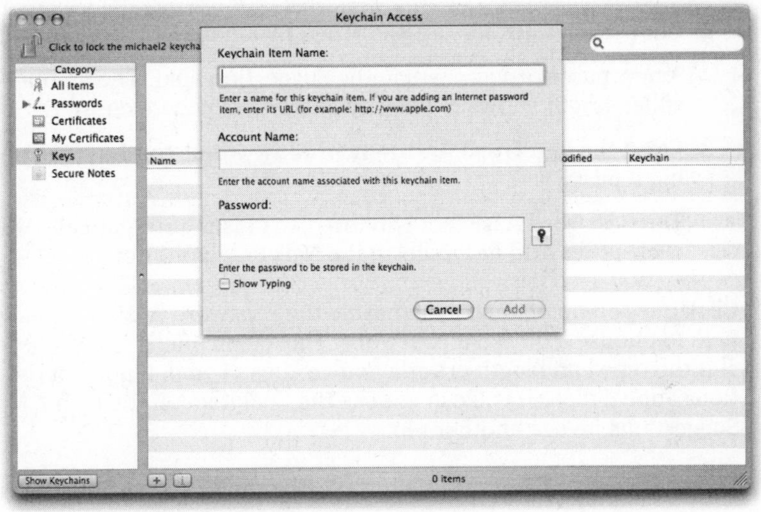

Figure 10-8: You create new passwords to store on your keychain with Keychain Access.

3. **Click the key icon to the right of the Password field.**

   This button invokes the Password Assistant, depicted in Figure 10-9.

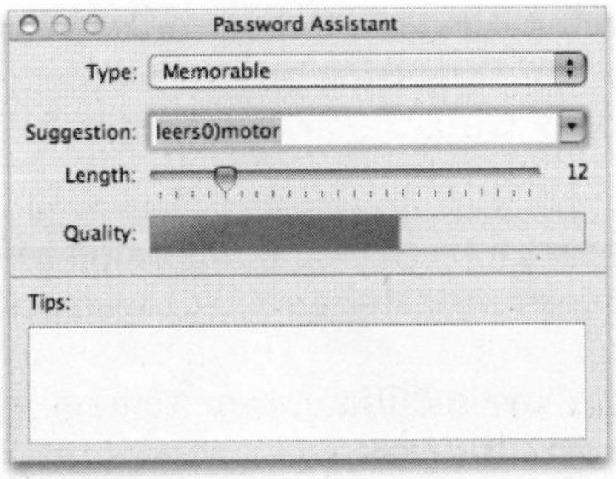

Figure 10-9: The Password Assistant both generates and evaluates passwords for you.

After you finish using the Password Assistant, you can click Cancel in the new password sheet if you don't want to save the password on your keychain. Of course, you can save the password if you want to: A password doesn't have to be associated with any particular application or function to be stored on your keychain.

Here's how to use the Password Assistant:

1. **Choose the type of password you want from the Type pop-up menu.**

   The password appears in the Suggestion field. Whenever the password in the Suggestion field changes, the new password also appears in the New Password Item sheet's Password field.

2. **Choose a length for the password by adjusting the Length slider.**

   A new password appears in the Suggestion field. The number to the slider's right shows how many characters you've chosen.

3. **Close the Password Assistant window when you have finished choosing a password.**

   The Password Assistant window also closes automatically when you click either Add or Cancel in the New Password Item sheet.

Click the pop-up menu button beside the Password Assistant's Suggestion field to choose a different password. This menu offers ten passwords that suit the type and length criteria you've chosen, and, if none of the suggestions appeal to you, you can choose the final item on the menu, More Suggestions, to see another ten.

The Password Assistant's Quality indicator shows how secure the password is:

- A short red bar indicates a weak password.
- A medium-length yellow bar indicates a somewhat secure password.
- A long green bar indicates a strong password.

Here are the types of passwords you can choose from the Type pop-up menu:

- **Memorable:** This option provides passwords that you can easily remember but that will foil most password-guessing programs. Memorable passwords usually consist of two words separated by numbers and punctuation, such as `rabbit)32basin`.
- **Letters & Numbers:** This provides a password comprising random combinations of both upper- and lowercase alphabetic characters and numeric characters, such as `iHkG5XXhmfcO`. These passwords can't easily be guessed, but you may have a difficult time remembering one.

- **Numbers Only:** These passwords contain numeric characters only. The Tips field always recommends that you mix other characters in to make the password more secure. However, some programs may require numeric passwords.
- **Random:** Use this option for the most unguessable, uncrackable password. This option provides passwords consisting of letters, numbers, and punctuation, such as `i]{?XL_\m89M`.
- **FIPS-181 compliant:** The Password Assistant uses a process that follows the Federal Information Processing Standard 181 to create passwords. Businesses working under U.S. government contracts often have to use such passwords.

The Type menu has one other item: Manual. Choose Manual, enter a password of your own devising in the Suggestion field, and the Quality indicator shows how strong the password is. In addition, the Tips field offers suggestions for improving the password's security.

TIP

You can enter a password in the Suggestion field at any time — you don't even have to choose Manual from the Type pop-up menu.

Connecting an AirPort base station to the business network

Setting up an AirPort base station in a business environment requires close coordination with your network administrator. For one thing, you have to work with your network administrator to configure your base station to use the security protocols required by the network. Furthermore, because your AirPort base station creates a sub-network connected to the main network, you need to work with your network administrator to ensure that the network addresses that the base station provides fit properly into the business network's address allocation policy.

REMEMBER

Chapter 4 covers how to set up a home network, and the same AirPort Admin Utility configuration controls described in that chapter come into play in a business environment as well. However, the actual settings made with the AirPort Admin Utility for a business network will be different, and most likely specified or directly set, by your network administrator. To find out more about how to use the AirPort Admin Utility, refer to Chapter 4.

First, you need to connect the AirPort base station to the network, and that means that you have to specify which method your AirPort base station uses to connect to the business network. You also need to specify how the base station obtains its network address. You perform both of these tasks using the configuration settings under the AirPort Admin Utility's Internet tab, as shown in Figure 10-10. For the majority of business networks, you'll connect the AirPort base station to the network by way of an Ethernet cable. Your network administrator should tell you how to configure the available settings under this tab.

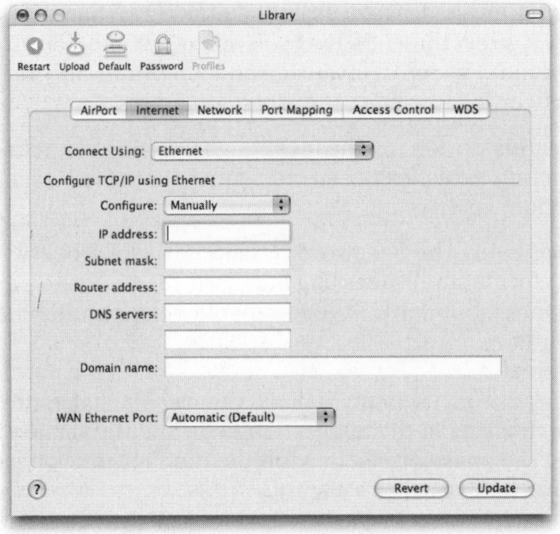

Figure 10-10: Your network administrator should provide the Internet connection settings.

After you've connected the AirPort base station, you specify how it supplies network addresses to the users of the AirPort network. You use AirPort Admin Utility's Network tab, shown in Figure 10-11, for this. Once again, your business's network administrator should have a set of guidelines for how you set the controls and fields under the Network configuration tab. In particular, the administrator can tell you whether you should specify an LDAP server, and, if so, which one you should use. Managed network environments often use LDAP servers to provide the information that helps automatically configure network clients.

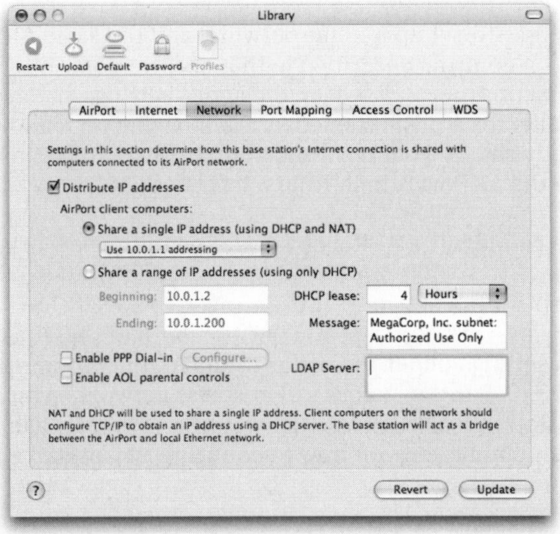

Figure 10-11: Your network administrator tells you how to distribute network addresses.

Finally, you need to specify the base station's security settings. You access these settings under the AirPort Admin Utility's AirPort tab: Click the Change Wireless Security button to see a sheet with the various security options your base station can use. Large business networks tend to use devices known as *RADIUS servers* for authenticating network users.

Although RADIUS stands for *Remote Authentication Dial-In User Service*, you don't need a dial-in connection to use a RADIUS server. Modern RADIUS servers can authenticate users connecting to secure networks through a variety of means, including wireless connections.

Figure 10-12 shows the security configuration sheet as it appears when you choose WPA2 Enterprise from the Wireless Security pop-up menu. The enterprise version of *WPA (Wireless Protected Access)* security is designed to work with a business network's RADIUS servers. On this sheet you specify the addresses of the RADIUS server(s) that your AirPort network's clients use when authenticating, and the password — also known as a *shared secret* — they use when contacting the RADIUS server. Once again, your network administrator should tell you how to configure this sheet to work with your business network.

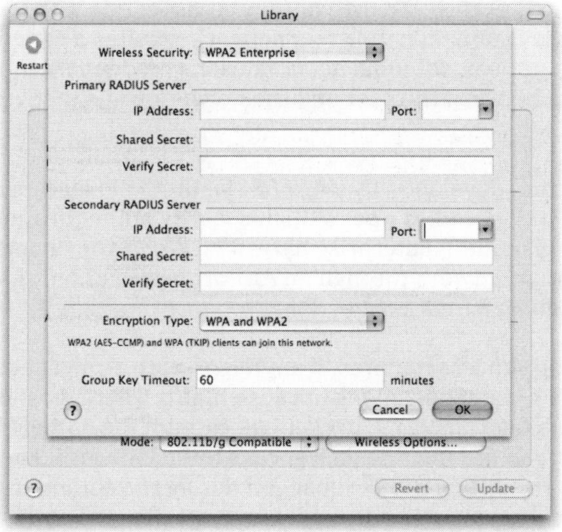

Figure 10-12: WPA2 Enterprise security requires an external RADIUS server.

After you've configured the AirPort base station, you and other users of your network are not quite ready to connect. When authenticating with a RADIUS server, each user also needs to establish that authentication using a secure wireless encryption technology supported by the *802.1X* protocol. You can find out how to establish an 802.1X encrypted connection in the next section of this chapter.

Getting Access

Walk into any corporate headquarters, and you'll likely see security guards, politely but relentlessly checking visitors' IDs before letting them get beyond the lobby or reception area. Enterprises of almost any sort, whether commercial, governmental, or institutional, don't like strangers wandering through offices, rifling through file cabinets, and otherwise making free with the tools and records that keep the operation going. Virtual security guards stand at an enterprise's network doors as well, demanding that your virtual self present proper credentials before you can enter the virtual premises. These virtual guards employ various complex and powerful software technologies to establish a potential network user's bona fides before establishing a network connection.

Establishing an 802.1X connection

As computer networks become more essential to an enterprise's operation, and as the attacks hackers use to gain unauthorized access to them become more sophisticated, verifying a user's identity simply by checking lists of plain text passwords against the user accounts associated with them just doesn't cut it. Today, authenticating to a network requires a complex set of protocols both to verify a potential network user's credentials and to encrypt the transmissions between the user and the network while that verification takes place.

The 802.1X standard, developed by the *IEEE* (Institute of Electronic and Electrical Engineers), specifies a set of high-security authentication protocols commonly used on modern enterprise networks. When you connect your Mac to an enterprise network through an AirPort base station, chances are you'll do it through an 802.1X connection.

The authentication process requires, if anything, more stringent security than other network communications because that is where most network attacks by hackers take place: If a hacker can capture the authentication messages, the hacker can use them to gain access to the network. The 802.1X protocol secures the information exchanged during the authentication process to keep electronic eavesdroppers from stealing the authentication information.

An 802.1X connection consists of three participants, which interact as follows:

- **The supplicant:** This is your Mac, which sends out the request for a connection, properly packaged in one of the forms that the 802.1X protocol supports. 802.1X supports multiple authentication methods, either singly or combined.

- **The authenticator:** This receives the authentication request from the supplicant, and forwards it along to an authentication server. An AirPort base station, set up to use WPA2 Enterprise security as described in the previous section "Connecting an AirPort base station to the business network," is an authenticator.
- **The authentication server:** This participant receives the supplicant's request from the authenticator, and then sends a series of identity challenges back to the authenticator, which, in turn, packages them for secure transmission outside of the network and sends them on to the supplicant. After the supplicant answers all the challenges successfully, the authentication server instructs the authenticator to provide the supplicant with an appropriate level of network access.

You have to configure your Mac to use an 802.1X connection for the first time, but, after you have successfully established the connection, you can re-establish the connection easily with no additional configuration work: Your Mac's AirPort and network software retain the 802.1X configuration you use with each authenticator, such as your AirPort base station, and will use that configuration the next time you connect through that authenticator.

You use the Internet Connect application to create and configure your 802.1X configuration like this:

1. **Make sure of these two things:**
   - **You are in range of an AirPort base station that provides a connection to your enterprise network.**
   - **Your AirPort card is turned on.**

   The final step in this series requires you to connect to the network so you can exchange the necessary configuration information with the network's authentication server.
2. **Open the Internet Connect application.**

   You can find this application in your Applications folder. However, if you have your AirPort status menu showing on your menu bar, you can choose Open Internet Connect from that menu.
3. **Do one of the following:**
   - **Click the 802.1X icon on the Internet Connect window's toolbar.**
   - **If you don't see the 802.1X icon on the Internet Connect window's toolbar, choose File⇨New 802.1X Connection.**
   - **Press ⌘+Shift+X.**

     This menu command becomes disabled when you already have the 802.1X icon visible on the toolbar.

The 802.1X connection pane appears in the window. Figure 10-13 shows the items available in this pane.

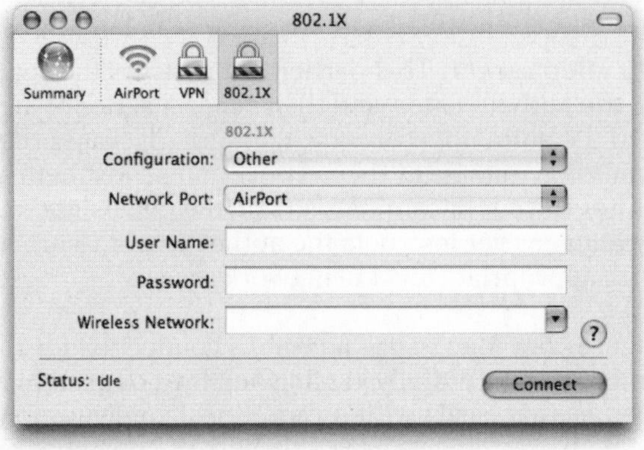

Figure 10-13: Use the 802.1X connection pane to authenticate to enterprise networks.

WARNING!

Although you can remove toolbar icons in most Mac applications — including Internet Connect — by ⌘+dragging them off the toolbar, do not remove the 802.1X icon from the Internet Connect toolbar because you won't easily be able to get it back: The New 802.1X Connection item on the File menu remains disabled even when you remove the icon.

4. **Choose Edit Configuration from the Configuration pop-up menu.**

   A sheet appears, as shown in Figure 10-14. You use this sheet to create a new 802.1X configuration by clicking +. You can also delete a configuration you no longer need by selecting it and pressing –, and you can modify a configuration by selecting it and changing the settings shown in the right-hand side of the sheet.

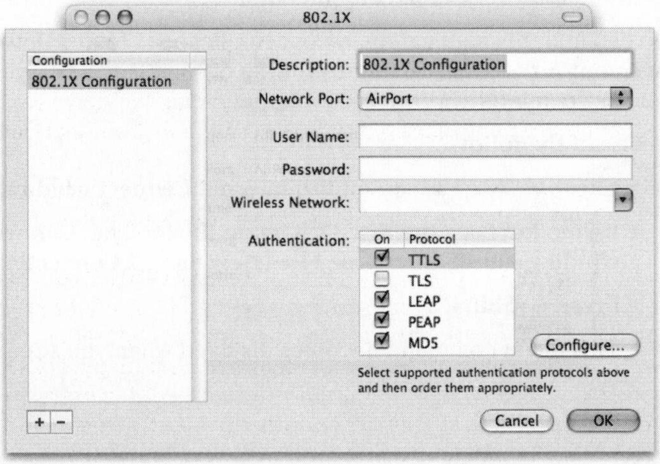

Figure 10-14: You can add, delete, and modify 802.1X configurations with this sheet.

Each 802.1X configuration contains the following items:

- **Description:** Enter a name for the configuration in this field. This configuration name appears as one of the items in the Configuration pop-up menu in Internet Connect's 802.1X pane, so enter a name that reminds you what the configuration is for.
- **Network Port:** You choose the network port the configuration uses to send and receive its authentication messages from this pop-up menu. The menu shows all currently active network ports.
- **User Name:** Enter the username or account name for your enterprise network account in this field.
- **Password:** Enter the password you use to authenticate to the enterprise network.
- **Wireless Network:** When you choose AirPort from the Network Port pop-up menu, as shown in Figure 10-14, you can then choose an available wireless network from the Wireless Network pop-up menu. You can also enter a wireless network name manually in the Wireless Network field so that you can use *closed networks* — that is, networks that don't broadcast their network name.
- **Authentication protocols:** Use this list to enable or disable the authentication protocols your enterprise network's authentication server uses. You can also arrange the order in which the configuration applies the protocols by clicking and dragging a protocol's name up or down the list. You can even configure any additional settings the authentication protocol requires by selecting the setting and clicking Configure. Your network administrator should provide a list of appropriate protocols and their settings.

5. **Make your configuration settings in the sheet and click OK.**

   The username, password, and other settings for the configuration now appear in Internet Connect's 802.1X pane.

6. **Click Connect.**

   Depending on your enterprise network's authentication server and the protocols it uses, your Mac and the server may exchange various digital certificates that your Mac will use the next time it authenticates.

After you've connected, you can close Internet Connect. You don't need the Internet Connect application open to maintain the connection.

TIP

You can choose Internet Connect's Export Configurations item, found on the File menu, to save an 802.1X configuration, and you can choose the File menu's Import Configuration item to use a stored configuration. Network administrators often employ the export feature to set up an 802.1X configuration on one Mac, save it, and then provide it to other Mac users so they don't have to create or modify 802.1X configurations directly. You might want to export your 802.1X configuration to make a backup copy of it.

Using a VPN connection

When you make a wireless connection with your enterprise's network through a base station connected directly to that network, your AirPort card and the base station encrypt the wireless portion of the connection so that no eavesdroppers can listen in on the transmissions and steal business secrets. The wired portion of the enterprise network may not have, nor require, any additional security. However, connecting to an enterprise network from a remote location, such as a broadband connection in a hotel room, or your own cable modem Internet connection at home, is another matter. Most enterprises don't trust the security of the public Internet, and with good reason: Hackers have a far better chance of stealing a communication from outside of the enterprise's well-guarded digital walls than inside them. *Virtual Private Networks* (*VPN*s) provide a way to guard the communications traveling over the Internet between your remote location and the enterprise network.

A VPN connection protects your network transactions with a remote enterprise network not by making your transmissions impossible to intercept, but by making such interceptions useless. When you establish a VPN connection, your Mac and the remote network encrypt both the data and the addressing information associated with it, and then package that encrypted information inside of a standard TCP/IP packet. The encrypted information doesn't even have to adhere to the Internet's TCP/IP standard: The encrypted information can conform to other, perhaps proprietary, network protocols. Network professionals call this procedure *tunneling*: In a sense, your communications with the remote enterprise network tunnel through the public Internet and pop up, safe and sound, inside the secure settings of the enterprise, where they can be decrypted and sent along to their eventual destination.

Using the Mac OS X VPN software

Various technology vendors have created VPN systems, some proprietary and some more open. Mac OS X supports two widely used open VPN standards, *PPTP* and *L2TP over IPSec*.

In case you're curious, here's what those two complex, acronym-laden terms mean:

- **PPTP:** Developed by a multi-vendor consortium headed by Microsoft, and submitted as an international standard, the *Point-to-Point Tunneling Protocol* can use 40-bit or 128-bit encryption, the latter providing a reasonably strong level of security.
- **L2TP over IPSec:** This stands for *Level-2 Tunnel Protocol over IP Security*, and consists of two technologies. L2TP is a tunneling protocol based upon both the Layer 2 Forwarding protocol from Cisco Systems, and Microsoft's PPTP. IPSec is a new and still-evolving Internet standard that specifies both encryption protocols and key exchanges. This last item provides the ability for a connection to change its encryption keys on

the fly, so that a hacker who intercepts the transmission has to crack multiple encryption keys before deciphering the complete message. L2TP over IPSec provides very strong security.

You don't have to understand how these standards differ in order to use them, and you won't have to choose between them, in any case: Which one you use depends on which one your enterprise's network supports.

You can set up either sort of VPN connection using your Mac's Internet Connect application. Each time you choose New VPN Connection from Internet Connect's File menu, you see the sheet shown in Figure 10-15, from which you select the type of VPN connection you want to create. After you select the type of VPN connection, a toolbar icon appears in the Internet connect toolbar for that connection. You can have VPN icons on the toolbar for each VPN you may use: Companies and institutions often have more than one VPN, with each one dedicated to a particular purpose or set of users.

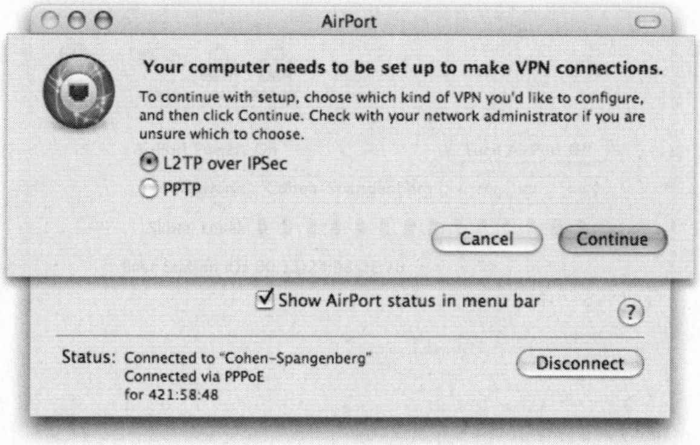

Figure 10-15: Internet Connect prompts you to choose between VPN connection types.

The Internet Connect panels for both L2TP over IPSec and PPTP connections look very similar. Figure 10-16 shows a newly created PPTP connection, but the same fields and pop-up menu appear for a newly created L2TP connection as well.

The fields available in the VPN connection panel vary depending on the connection's configuration, which you can change by choosing Edit Configurations from the Configuration pop-up menu. In fact, you can set up multiple configurations for each VPN connection you create, and choose the one you want to use from the Configuration pop-up menu.

Figure 10-17 shows the configuration sheet that appears for an L2TP VPN connection when you choose Edit Configurations from the Configuration pop-up menu. The configuration sheet for a PPTP connection is similar. Both configuration sheets have settings for account names and server names, and both provide authentication options relevant to the type of connection. You set the configuration options, using the settings provided by your network administrator: You won't be able to figure them out on your own — after all, if you could, VPN wouldn't provide much security, would it?

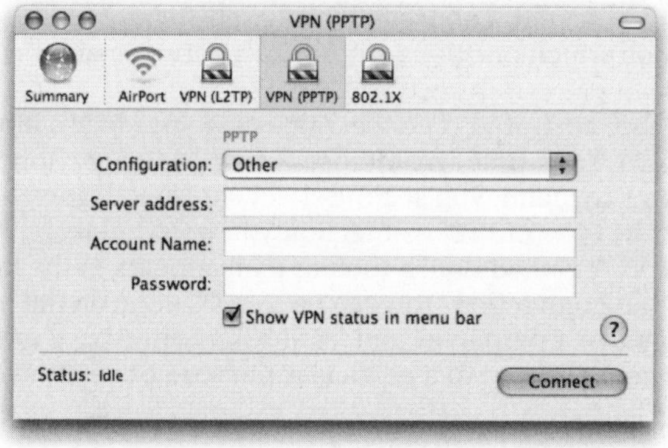

Figure 10-16: A VPN connection panel in Internet Connect — you can have more than one.

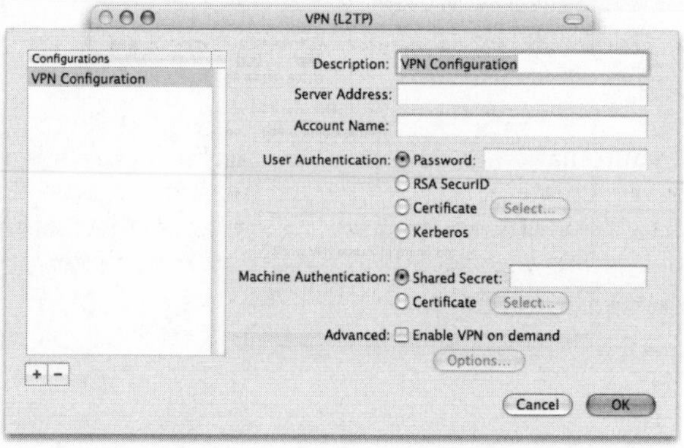

Figure 10-17: Configuring an L2TP virtual private network connection.

TIP: Like the 802.1X authentication configurations described earlier in this chapter, a network administrator can create and save configuration files for VPN connections, which can then be given to VPN users to import into Internet Connect.

Your Mac treats VPN connections like separate network ports, similar to your Ethernet, modem, or AirPort network ports. A VPN connection you create appears in your Network preferences Network Port Configurations panel, as shown in Figure 10-18. Like any other network port, you can enable or disable the VPN port, and drag it up or down the network port list to change the order in which your Mac attempts to make network connections. You can also delete VPN connections you no longer need.

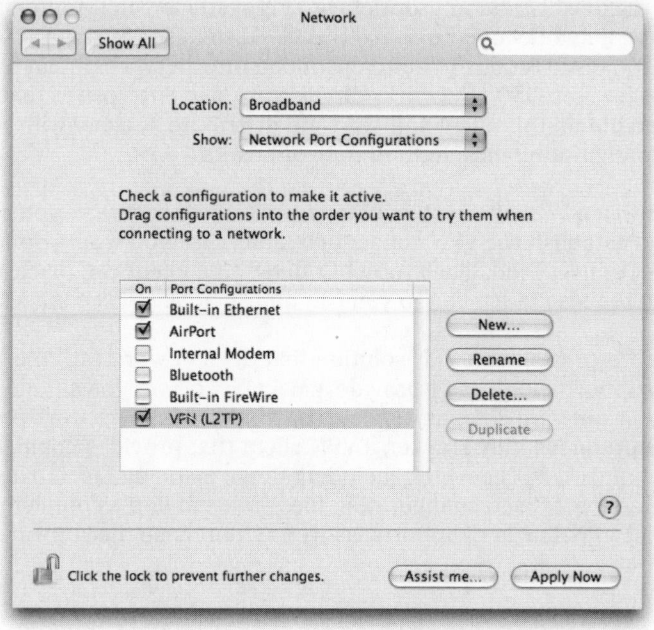

Figure 10-18: Your Mac treats a VPN connection like any other network port.

Internet Connect automatically checks the Show VPN Status in Menu Bar check box when you create a new VPN connection. You can use the Connect item on this status menu, shown in Figure 10-19, to quickly establish a VPN connection. The Connect item uses the first available VPN connection listed in your Network preferences' port configurations list.

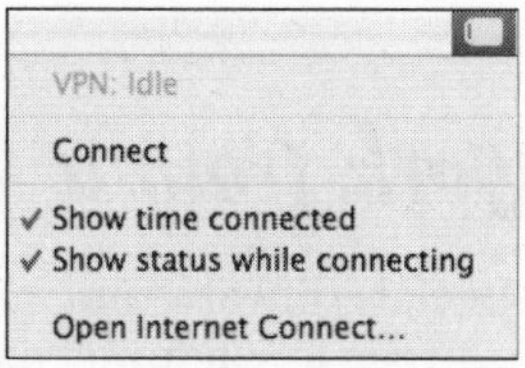

Figure 10-19: Use the VPN Status menu to establish a VPN connection quickly.

Glancing at other VPN solutions

As mentioned at the start of the previous section, the PPTP and the L2TP over IPSec VPN capabilities that Mac OS X provides are not the only way to play the VPN game. Various technology companies have developed their own VPN solutions, often coupling their own VPN devices with client software for the various operating systems the devices support.

Among the most common proprietary VPN systems you'll encounter on enterprise networks are those manufactured by Cisco Systems. To use a Cisco VPN, Cisco requires you to download or otherwise obtain Cisco's client software for Mac OS X. Network administrators in enterprises that use Cisco VPNs can obtain this client software and distribute it, along with the appropriate configuration files, to their network users.

To use a Cisco VPN after you configure their client software, you run the software and establish the VPN connection, much like you would do with Internet Connect. And, like Internet Connect, the client can place a status menu on the Mac menu bar so you can quickly make VPN connections.

If your enterprise uses a VPN solution that doesn't work with the Mac's built-in VPN support and doesn't provide a MacOS X client, you should check out Equinux, a software company specializing in Mac OS X network products. Equinux produces VPN Tracker, a VPN client that provides compatibility with several hundred VPN devices, including those made by Cisco: The software can even import Cisco configuration files. You can find a complete list of the VPNs that VPN Tracker supports, as well as purchase the software, from Equinux at `www.equinux.com`.

Collaborating with Co-Workers

You've made the case for connecting your Mac to your enterprise's network, you've sugar-talked your network administrator into helping you configure your Mac for authentication, and you've even wrangled a VPN account so you can connect from home or the road — you must have had a reason for going through all this, right? Whatever your reasons, they almost certainly had something to do with working with others in your enterprise. And that means sharing files and other information with them over the network.

Viewing a Windows network from a Mac

Gone are the days when you had to install special software, and maybe even purchase a special network adapter, in order to see a Windows-based network on a Macintosh. Mac OS X has that capability built right in, and it is enabled by default.

A software service called SMB/CIFS provides the key that unlocks the Windows network kingdom, as mentioned earlier in this chapter in "Setting up Active Directory and related network services." If you use Directory Access's default configuration for SMB/CIFS, your Mac can show you any Windows computer that has shared its files with the other computers in a local Windows workgroup called Workgroup — which, not surprisingly, is Window's default name for the local workgroup. As also discussed in "Setting up Active Directory and related network services," you should ask your network administrator what settings to make to Directory Access's SMB/CIFS configuration so you can see the file servers available to the workgroup in which you participate.

And what do I mean by, "Your Mac can show you any Windows computer?" Just this: When you open the Finder's Network window, which you can do by choosing Network from the Finder's Go menu, you see an icon that represents a file-sharing Windows computer. That icon looks just like the icon the Finder uses when it shows a file-sharing Mac in the Network window: All servers look alike to the Finder.

That is, they look alike until you open them. When you open a Windows server, you see an authentication dialog like the one shown in Figure 10-20. Unlike the dialog you see when connecting with a file-sharing Mac, this one proudly announces its Windows-ishness by telling you that you're authenticating to an SMB/CIFS File System, and it provides a space to let you enter a workgroup name, a username, and a password.

Figure 10-20: SMB/CIFS lets you use files shared by a computer running Windows.

The authentication dialog's title, by the way, is redundant: the "FS" in CIFS stands for "File System," which means that the heading at the top of the dialog actually expands to say "Server Message Block/Common Internet File System File System Authentication."

If you don't see the server you want in the Finder's Network window, you can enter the server address manually like this:

1. **Choose Go➪Connect to Server.**

   This command brings up the window shown in Figure 10-21, which has a Server Address field in which you can type the address of the server. Windows server addresses begin with `smb://` instead of the `afp://` used for Mac server addresses.

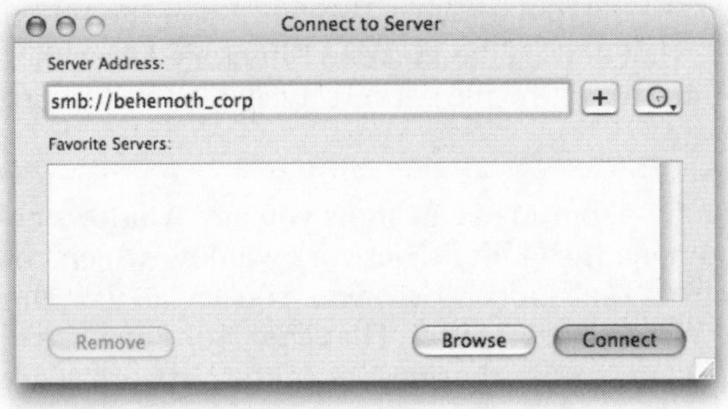

Figure 10-21: Enter a server address by hand.

2. **Click Connect.**

   You see a progress window as your Mac looks for the Windows server on the network. When your Mac makes contact with the server, you authenticate as previously shown in Figure 10-20.

REMEMBER

You don't always need to authenticate to the servers you see displayed in the Finder's Network window. When you authenticate to Active Directory, as described earlier in this chapter, the Active Directory system looks up the servers on the network to which you have access and makes them available to you. Thus, the servers you see listed in your Finder's Network window may consist of servers to which you are already authenticated, as well as some to which you must authenticate separately. What's more, on a large enterprise network, you may see dozens of servers listed depending on the size of your workgroup, the network privileges you have been granted, and the configuration of the network.

Sharing files with Windows users

In most enterprise networks, you share files with Windows users by storing the files on a server to which both you and they have access. In fact, one of the reasons that Windows networks have workgroups is to provide a way for a network administrator to allocate shared server space among different groups of co-workers. Nonetheless, you may occasionally find that you need to share files directly from your Mac to a Windows co-worker. Mac OS X provides a way for you to do that, too.

To share files with a Windows user, you need to turn on Windows file sharing, which you do like this:

1. **Open System Preferences.**
2. **Click Sharing.**
3. **In the Sharing window, click Services.**
4. **In the Service list, click Windows Sharing.**

The first time you turn on Windows Sharing, the Sharing window presents a small alert icon like the one shown in Figure 10-22. As the fine print in the Sharing window explains, you have to choose at least one user account on your Mac that will have its files shared: Unlike Mac sharing, Windows sharing does not provide guest access. You have to share all the files in an account's Home directory.

If you plan to share files from your Mac with Windows Sharing, create an account that you will only use for Windows sharing purposes, and put the files you want to share in that user account's Home directory.

To pick which accounts to share with Windows Sharing, perform the following steps:

1. **Click Enable Accounts.**

   You see a sheet listing all the user accounts on your Mac, like the one shown in Figure 10-23.

2. **Click the check box beside an account that you wish to share.**

   A dialog, similar to the one in Figure 10-24, prompts you to enter the account's password.

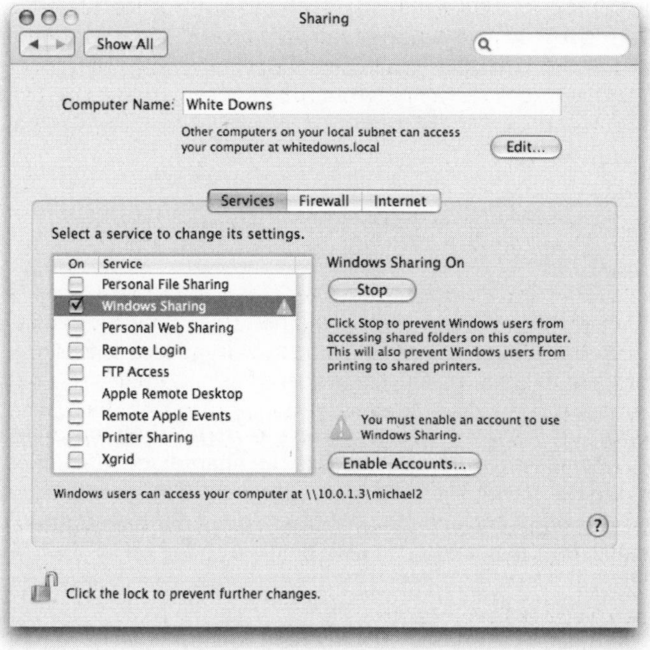

Figure 10-22: When you turn on Windows Sharing, you have to pick a user account to share.

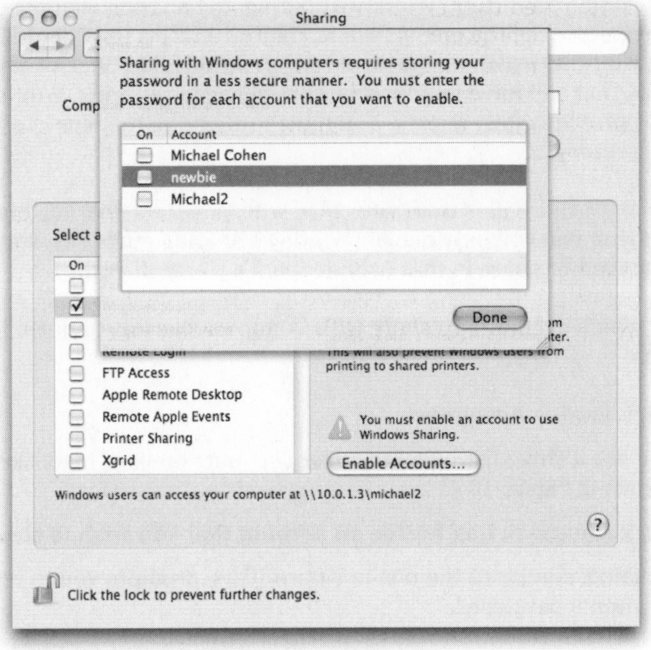

Figure 10-23: Pick an account to share from this sheet.

Figure 10-24: You have to enter the password of the account you want to share.

WARNING!

In order for Windows network users — or, for that matter, Mac users on the Windows network — to access the shared account's files, you must provide them with the account's password. This presents a security risk. Make sure that the account you share does not have Admin privileges, because that password can also be used to log on to your Mac directly.

3. **Enter the account's password and click OK.**

   A check mark appears by the account name.

4. **Click Done.**

   The sheet retracts, and the warning icon disappears from the Sharing window. In addition, the Enable Accounts button changes to Accounts.

You can add shared user accounts by clicking the Accounts button in the Sharing window again and clicking the check box by the account's name in the sheet. You can also disable Windows Sharing for an account by clicking the check box by the shared account's name. You have to specify the user account's password when you disable Windows Sharing, just like you do when you share it.

If you seek more control over which folders your Mac shares, you should look into Michael Horn's donation-ware program SharePoints. With SharePoints, you can pick the folders you want to share rather than sharing all the folders in a user account. You can also create user groups, and give members of different groups access to different folders. You can download SharePoints from `www.hornware.com`.

Chapter 11

Hitting the Road

In This Chapter

- Assembling a travel kit
- Getting WiFi room service
- Sipping coffee and sending e-mail
- Assigning your AirPort Express multiple identities
- Putting a PDA into play
- Using the modem in your pocket

Read the ads, the articles, and the brochures: Wireless technology is all about portability, mobility, and freedom. Or so the device vendors and the advertisers and the technology writers — like me — would have you believe. But, like so much else, the devil lurks in the details when it comes to wireless technology and the bright beckoning dream of untethered freedom. Yes, wireless technology really can enable you to roam free and far from the close confines of your office, but freedom requires work.

You need to plan what to pack. You need to know where you're going and what you'll find when you get there. You need to know how to use the collection of gear you've acquired for your travels. And, most importantly, you need to decide why you're packing up all the gear in the first place.

And all that's what this chapter will help you do. Let the journey begin.

Getting Ready to Go

They say that getting there is half the fun. Maybe that was true once upon a time, but these days getting there more often means long lines, security checks, no in-flight meal, and seats designed for a rather smaller species than the one to which you happen to belong. *Being* there is where the fun is, and it is a whole lot more fun if you manage to bring what you need, to leave what you don't need, and to confirm your reservations ahead of time.

Packing your bags

I know some people who just toss a few items into a battered piece of luggage, whip out a credit card, and jet off to parts unknown at the drop of a metaphorical hat. I'm not one of those people: Even for a two-day business trip I can spend hours deciding what goes with me and what stays behind. Digital technology may have changed what I pack, but it hasn't made planning what to pack any easier for me.

TIP

I find it helpful to divide the gear I plan to take into two categories: little-bag items and big-bag items. The little bag is the bag I carry with me, and the big bag is the luggage that I reluctantly hand over to the baggage-check folk after glaring suspiciously at them for a short, tense moment. In fact, I actually have *two* little bags: a little-little bag, which carries just my basic gear and which I use for short flights and short stays, and a somewhat more spacious little bag, with room for a few other non-computer items I might want to carry with me when I take a longer flight or plan a longer stay. This bigger little bag might also contain a book to read, a snack, a water bottle, a toothbrush, and even a change of underwear in case my real-big bag is lost.

I find it helpful to have a list of all the things that I might take with me, so I can winnow the list down to what I will take with me, and also decide into which bag I will pack each item.

- **Laptop carrying case:** This is usually the little bag that I mentioned. I've used a variety of computer carrying bags over the years and have learned that the best carrying bag for me is one with enough padding to protect my laptop from miscellaneous bumps and jostles; that allows me to get at the laptop quickly should a security agent want to examine it; that has enough storage pockets to hold the cords and other equipment that I absolutely must have; and that can be carried over the shoulder as well as by a handle. If you don't want to get a dedicated computer bag, you may be able to obtain a protective padded sleeve for your computer, allowing you to use a more traditional piece of carry-on luggage.

  REMEMBER

  The height, width, and length of your carrying case should add up to no more than 45 inches. Most United States airlines restrict carry-on bags to this size. Furthermore, many airlines prefer that these dimensions come in the proportions of 9 × 14 × 22 inches, because those dimensions allow the case to fit beneath most airline seats.

- **Power supply for your laptop:** Depending on your laptop, this item might be a big-bag candidate, especially if you have a heavy power supply and don't intend to spend much time working while in transit. No matter which bag you put it in, you have to have it when you get to where you're going: Batteries last only a few hours. Because I have had luggage lost too many times, I usually pack the power supply in my little bag with my laptop: I'd rather put up with the added weight to carry than carry the added weight of worry.

- **Ethernet cable:** You may not need this item while in transit, but you may not be able to get along without it when you reach your destination: if you have a broadband connection at your destination, you use this cable to connect your laptop or wireless base station. Because these cables are light and easy to store, putting one in your little bag is not a bad idea. If you're the type of person who doesn't mind wearing a belt and suspenders, you can put one in each bag.

  Consider purchasing a retractable Ethernet cable that stores itself in a small enclosing case when not in use. You can find them in many computer supply stores for under $20.

- **AirPort Express Base Station:** Weighing in at under 7 ounces, and taking up only 12 cubic inches in your bag, you want to bring this with you in one bag or another if you think there's even a remote chance of connecting to broadband at your destination. You might think that you won't need to roam wirelessly around a small hotel room, but I've found that in-room Ethernet ports are sometimes placed in oddly inconvenient places, like behind the TV or under the bed. If you travel with a companion who also has a WiFi-enabled computer, the base station becomes even more useful.
- **Back-up battery:** If you plan to work in transit, you should consider carrying a charged back-up laptop battery in your little bag if you can handle the extra weight: the battery adds about a pound that you'll have to tote around.
- **Power strip:** Electrical outlets in most commercial lodgings tend to be hard to find and inconveniently placed. Packing a lightweight power strip in your big bag will empower your collection of devices.
- **Phone cable:** Sometimes, a plain old modem connection may be the only online connection you have available to you. I pack a six-foot retractable phone cable in my little bag when I travel, and it has paid off on more than one occasion.
- **Replacement software:** Rarely, although once is too often when it happens to you, software can crash in such a way that it gets corrupted, and the only thing to do is to reinstall it. You may want to bring the installation CDs of your most important software with you if you have the space in your big bag. Be sure to bring your software registration numbers, too.
- **Back-up drive:** I always travel with a small FireWire portable drive in my big bag. Before I leave, I back up my most important files to it, and, when I'm feeling especially paranoid, I back up my most important software to it, too. Few things feel more frustrating than to find yourself hundreds of miles from home with a computer you can't use because you accidentally deleted the one file you really need. Make sure you bring the necessary cables and power supply for the drive as well.

TIP

The latest generation of USB flash memory drives can provide several gigabytes of storage. Depending on the size of the files upon which you usually work, one of these drives can serve as an excellent, very lightweight back-up drive, which can easily fit in your little bag.

- **Portable printer:** Lightweight inkjet printers can weigh about five or six pounds, and sometimes even more when you add in a battery, so you probably want to pack this in your big bag if you bring it at all. Though it may seem a luxury, having a printer can pay off quickly when you have to print directions to a tourist attraction, or your daily conference schedule, or handouts for a meeting.
- **Cell phone:** As discussed later in this chapter in "Going online with your Bluetooth Phone," you can use certain cell phones as wireless modems, depending on your service plan and the phone's capabilities. With a Bluetooth-enabled phone, you can go online wherever you can get a phone signal, and you won't even need to take the phone out of your pocket.

TIP

Rather than bring your phone charger, you can get a USB phone charging cable so you can charge the phone's battery directly from your Mac laptop's USB port. These devices cost around $20 or so.

- **PDA:** I know what you're thinking — why do I need my PDA when I have my laptop computer with me? See "Lightening the load with a PDA" later in this chapter for some answers. When I bring a PDA, it usually goes in my little bag, in an outer pocket where I can get at it easily.

Your list might be shorter or longer than this one, depending on the other stuff you might take, such as digital cameras or iPods. To help me plan which of the items I should pack for any given trip, I ask myself these questions:

- **Why am I going?**

  My answer to this question usually falls into one of three categories: business, pleasure, or a combination of the two. I tend to pack different digital gear for a business trip than I do for a tropical vacation. For the latter, I may want my digital camera and my iPod speakers more than for the former.

- **What do I plan to do when I get there?**

  Though related to the previous question, this is not the same question as the previous one in different words. For example, I might take a business trip to attend a conference, for which I would require one assortment of portable digital equipment, or I might be traveling in order to work directly with a business associate on a particular project, which could require that I take a different equipment assortment. Similarly, for a vacation trip, I might want to take different stuff to Disney World than I would for a stay in a vacation cottage in Maine.

- **How much do I want to carry?**

  The answer to this question helps me decide what goes into the little bag that I mentioned earlier, and how I answer it depends on the individual trip I'm taking. If I know I'll be taking several connecting flights, and trotting from one airport terminal to another, I know I'll want to carry much less stuff than if I'm taking a car trip somewhere. In addition, how I answer this question also may depend on how I answer the next question.

- **What can I afford to do without for a few days or even longer?**

  This question, as well as the previous one, helps me decide which items are little-bag candidates and which items are big-bag candidates. What goes in the little bag are those things that I absolutely can't do without, such as my laptop computer itself: If I lose that, most of the other equipment ends up being dead weight. What goes in the big bag is stuff that I can do without in the event that my luggage is stolen, or is lost, or somehow falls into the grasp of an enraged silverback gorilla who decides to batter it against the tarmac a few dozen times. Such items might include blank CDs or a spare battery or a portable inkjet printer.

Planning your access

As Buckaroo Banzai pointed out, "Wherever you go, there you are." What he didn't point out is that you will want Internet access when you get there. Fortunately, if you do a little planning, you usually can get some sort of access wherever it is that you are going.

Using in-room broadband

Hotels or similar commercial lodgings often provide in-room broadband access. Most lodgings advertise broadband access on their list of amenities, and if not, you can often find that information from the lodging's Web site. Of course, if the place you plan to stay doesn't have a Web site, chances are less promising that it has broadband access available, but it still might. When in doubt, call and ask.

When checking out a lodging's in-room Internet service, you should inquire specifically about a few things:

- **Charges:** Most lodgings charge for in-room broadband service. Find out what that charge is, and whether it's a flat rate per day or week, or whether it is a minute-by-minute charge.
- **Restrictions:** Some lodgings provide unfettered Internet access, and others may employ firewalls or some other kind of network configuration that could block non-Web access, including things like access to your ISP's e-mail server. In addition, some establishments may restrict you from attaching your AirPort Express to their network and sharing your signal.

- **Wireless access availability:** Some lodgings actually supply wireless access themselves. If yours does, you should ask them if they support Macs; and if the person answering your question seems unsure, ask if they require any special software to use their service. If the answer is no, chances are that your Mac will work just fine with their service.

Bringing your own service

If your destination does not provide broadband service, you may have to fall back to the older, more traditional, much slower modem that's built into your laptop and forgo the advantages of wireless access — at least while you're in your room.

TIP

If you are willing to pack an AirPort Extreme instead of an AirPort Express, you can share a modem connection wirelessly. Consider this option only if you really require both modem access and a wireless connection to your computer, because an AirPort Extreme both costs more and weighs more than an AirPort Express.

Check with your current *Internet Service Provider (ISP)* to see if it provides dial-up access as well. Some DSL providers provide a dial-up account as part of their service, or provide it for a small additional monthly fee. Cable providers are less likely to do so, but you should check with your provider just in case. If your ISP does provide dial-up access, and if that ISP has a local access number at your travel location, you're in business.

If you travel regularly to places where you know you will need modem access, you can also get an account from national services like America Online that provide a certain number of hours per month for less than they charge for unlimited accounts. For a one- or two-week vacation in a resort area, you may also have some luck finding a local ISP that offers short-term dial-in accounts. You may not find such limited accounts practical for intensive Web browsing, but they usually suffice for sending and receiving e-mail.

WARNING!

Some commercial lodgings do not have standard phone jacks in guest rooms, and others may have jacks that look like a standard phone jack but that connect to something else, and that something else might damage your computer. Before you connect any of your devices to a phone jack in your room, ask the establishment if you can do so safely. You may also want to inquire what the lodging charges for local calls: Some charge far more than others for local calls, and may even charge by the minute.

Communicating in Public

If you could see with WiFi eyes, you'd behold a landscape of WiFi hotspots, floating cloudlike around businesses, schools, and homes. You can use a good number of those hotspots to get Internet access even if you have no Internet service where you're lodging.

Finding public hotspots

More and more often, businesses where people congregate, such as restaurants, coffee shops, and bookstores, have begun to offer free wireless Internet access as a way to attract customers. You may also find WiFi access in other public venues, such as libraries. Some municipalities and community groups have even begun to offer WiFi access in places like city parks. In addition, many businesses offer WiFi access for a small fee. But how do you find such havens of wireless bliss?

One answer to that question, if you are a Mac OS X 10.4 user, comes in the form of a simple Dashboard widget: the JiWire WiFi Hotspot Finder, available for free download from `http://www.jiwire.com/macosxtiger.htm`. If you're running Mac OS X 10.3, JiWire also provides a downloadable hotspot directory. And if that doesn't work for you, you can consult JiWire's main Web site at `www.jiwire.com` to consult its online directory of over 70,000 hotspots. And if *that* doesn't work for you, you can try `www.wifi411.com` to consult its directory of over 35,000 hotspots.

If, like me, you choose to download and uncompress the widget from JiWire, you can install it like any other widget with a simple double-click: Mac OS X knows what to do from there, as shown in Figure 11-1.

Figure 11-1: Mac OS X 10.4, a.k.a. Tiger, knows how to install Dashboard widgets.

After installing, the JiWire widget immediately checks for available wireless networks within range of your computer and, as Figure 11-2 illustrates, shows these networks to you, along with their signal strengths, broadcast channels, and, indicated by a small key icon, whether or not they require passwords. You can even join a network by clicking its name in the widget. These features may help when you finally arrive in the vicinity of a hotspot, which the widget's other main function helps you do.

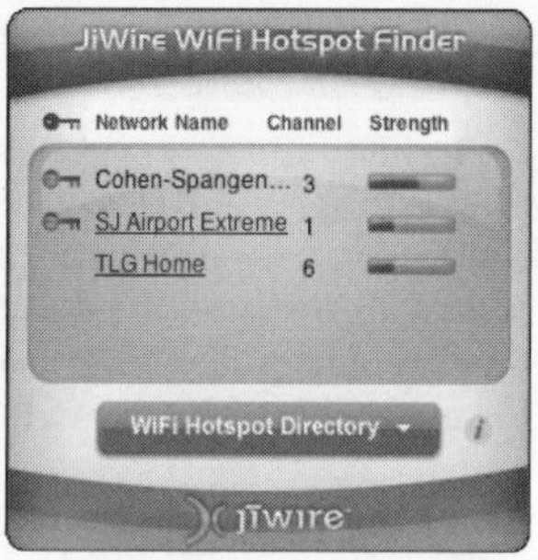

Figure 11-2: You can see and join WiFi networks with the JiWire Dashboard widget.

To find WiFi hotspots with the widget, follow these steps:

1. **Click WiFi Hotspot Directory.**

   A location drawer slides out of the bottom of the widget, as shown in Figure 11-3.

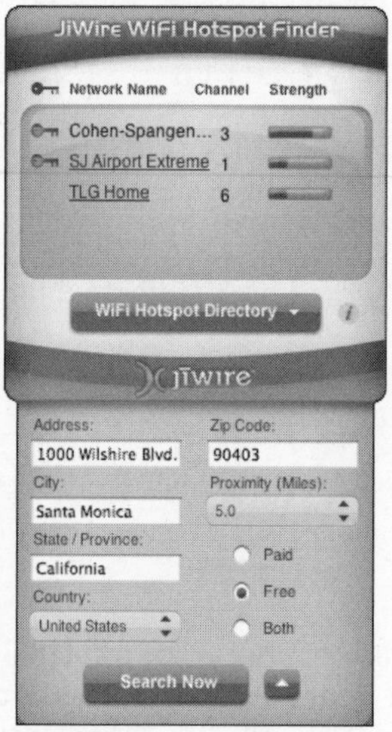

Figure 11-3: Tell the widget where you are so it can find nearby hotspots.

2. **Enter the address or your current or planned location in the appropriate fields.**

   TIP: To use the widget's hotspot finder function, you first must be connected to the Internet. Therefore, you might want to use the widget before you leave on your trip in order to see a list of places at your destination that provide WiFi access, which you can then use when you arrive.

3. **Choose a distance from the Proximity menu.**

This menu focuses the search for nearby hotspots. You can restrict the search for hotspots to those within a mile of the location, or expand the search area all the way up to a 100-mile radius.

4. **Click Free, Paid, or Both to limit the kinds of hotspots shown.**

   Some businesses, as mentioned earlier, provide free hotspots that anyone can use. Others require that you pay a fee, or that you have an account with the WiFi provider in order to use the hotspot.

5. **Click Search Now.**

   The Dashboard widget swivels around and shows you a list of hotspots arranged by increasing distance from the location that you entered in Step 2, as shown in Figure 11-4. You can click the + beside any hotspot to have the widget open your Web browser and show you more details about that hotspot from JiWire's Web site, and you can click the Map link to see a map in your Web browser displaying the hotspot's location.

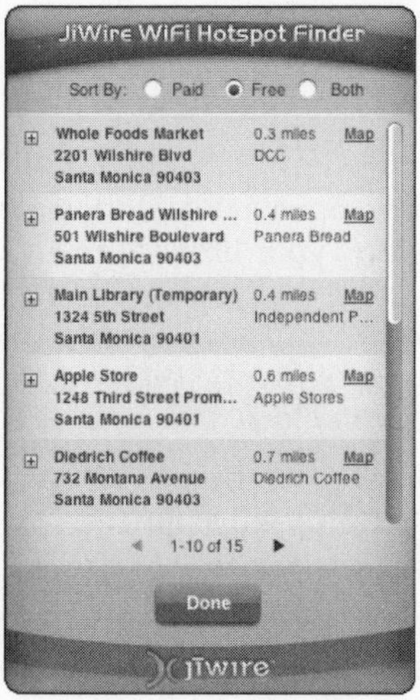

Figure 11-4: The widget provides a list of local hotspots you can use, arranged by distance.

Using public hotspots

Armed with a list of nearby public WiFi hotspots, you march on over to the nearest one, sit down, open your Mac laptop — and then what? Why, you connect to the wireless network, just as you would with your own AirPort network.

Let me show you what I mean. Today I walked a few blocks to a neighborhood bakery/cafe that offers free WiFi access. When I got there, I sat down, opened my iBook, and clicked the AirPort status menu. As I hoped would happen, the name of the cafe's network appeared on my AirPort menu, as shown in Figure 11-5. I chose the network's name from the menu and that was that: My AirPort status icon showed that I was connected.

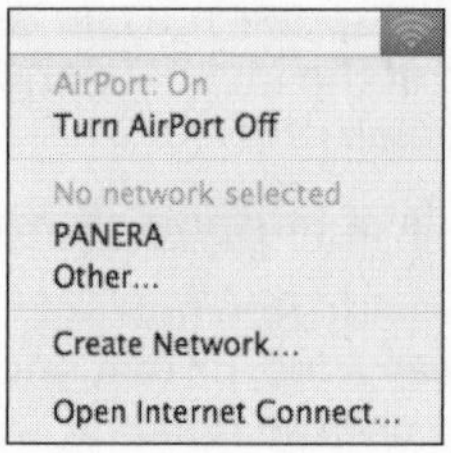

Figure 11-5: See a network, choose a network.

Of course, it's one thing to join an open wireless network, and another to get Internet access through it. Most public hotspots, whether free ones like that shown in Figure 11-5, or fee based, let you join the local wireless network quite simply. But the hotspot's providers usually put in an extra hurdle that you must leap in order to make your way to the rest of the Internet. This hurdle is the authentication step, and leaping it usually requires that you use your Web browser to provide authentication information.

Although the details vary from hotspot to hotspot, here's the general authentication process:

1. **Open your Web browser.**

   If you have configured your browser to open a default home page, you won't see it. Instead, the hotspot diverts your browser to an authentication page. If, on the other hand, you have configured your browser to open to a blank page like I have, simply go to any Web page for which you have a bookmark to have the hotspot divert you to its authentication page. The authentication page for my local bakery/cafe's hotspot is shown in Figure 11-6.

2. **Fill out the items that the authentication page requires.**

   For my neighborhood bakery/cafe hotspot, this step simply involves clicking the Go Online button shown in Figure 11-6. For a hotspot that imposes a fee, you will fill out a form of some sort. The items on the form, of course, can vary widely, depending on the hotspot provider.

   Hotspot operators often refer to this authentication process as the *UAM (Universal Access Method).* When the hotspot employs the UAM to redirect your browser to its authentication page, that page comes from a Web server connected to the hotspot. The information you send to that server when you fill out the authentication page uses *Transport Layer Security (TLS),* which protects the information you send from other users of the hotspot. You'll appreciate this if you need to provide credit card information to use the hotspot.

After you submit the authentication form with proper information, you have Internet access as long as you stay connected to the network. You usually need to authenticate to the hotspot with your Web browser before you can use any other Internet application, such as your e-mail program.

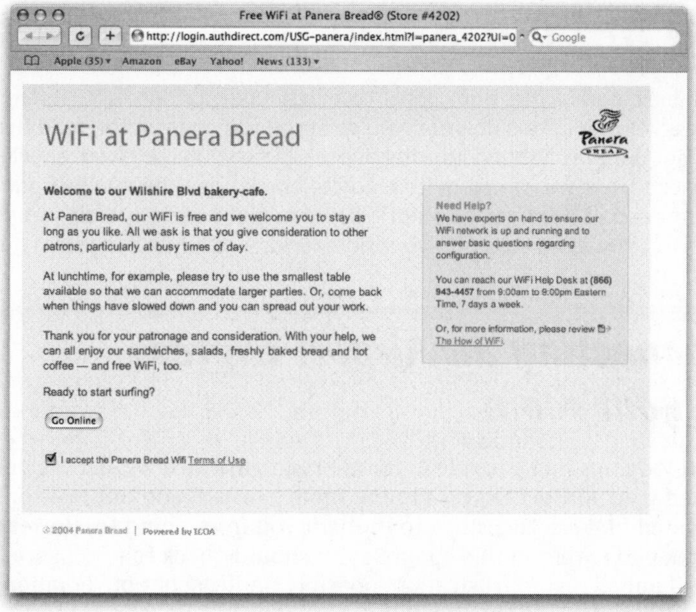

Figure 11-6: A very simple hotspot authentication page.

Appropriating bandwidth

Not all accessible, unprotected wireless networks that you may stumble upon are intended for public use. Sad to say, lots of people in the past few years have purchased wireless base stations and set them up without giving a thought to security.

For example, from where I sit right now, my AirPort status menu displays three wireless networks in my vicinity other than my own network, and only one of them has a password protecting it. With one click I can join either of the other two unprotected networks and use its Internet connection. There's nothing to stop me: The law, for now at least, seems to provide no clear guidelines.

I can't tell you whether you should use any unprotected WiFi access point that you happen to stumble across. Technologically, nothing prevents you from doing so. Legally, you may or may not be in the clear, depending on the jurisdiction in which you happen to find yourself, the particular circumstances in which you obtained access, and what you did with that access.

I don't want to sound like Jiminy Cricket here, but I ask you to consider this: If you happened to own a wireless network that you either forgot to secure or that you didn't know how to secure, and you discovered that other people were using both it and its Internet connection, a connection for which you were paying, how would *you* feel about that?

Working in Your Room

With luck, and having done some research as described in "Planning your access" earlier in this chapter, you've found lodgings on the road that come with in-room broadband Internet access. Now you've arrived, checked in, and unpacked, but you're not quite ready to crack open the minibar, grab a bag of over-priced salted nuts, and settle down to some serious Web surfing and e-mail yet. First you need to set up your home-away-from-home wireless network.

Connecting an Airport Express in your room

Most lodgings that provide in-room broadband Internet access permit you to attach your AirPort Express to the room's broadband outlet — or, at least, they don't take active steps to prohibit you from doing so. However, as I've mentioned earlier in this chapter, you should check to make sure that you can, if only to avoid frustration, possibly unpleasant conversations with lodging staff, and unanticipated charges on your bill. So let's assume that you can use your AirPort Express with the lodging's network.

What you want to do is take advantage of the Express's configuration profile capability to create a configuration for your lodging's connection. That way, when you get back home, you can simply switch the AirPort Express to your home profile.

Although, just as with public hotspots, individual in-room network configurations can vary significantly, here's how to set up a new AirPort Express profile for use with many, if not most, of them:

1. **Plug in your AirPort Express base station, but do not connect it to the room's Ethernet port.**

   If you connect your base station when you turn it on, it will try to connect using its current profile, which might confuse the base station or the lodging's network.

2. **Open AirPort Admin Utility.**

3. **In the Select Base Station window, double-click your AirPort Express base station's name.**

   You may have to wait a few moments until the base station finishes its start-up process before its name appears in the window.

4. **In the base station configuration window's toolbar, click Profiles.**

   A sheet appears listing the current set of configuration profiles stored in the AirPort Express.

5. **Click Add and type a name for the new profile.**

   A new profile appears in the list, enabled and ready to be named, as shown in Figure 11-7. When created, a new profile takes on the settings of the previously enabled profile.

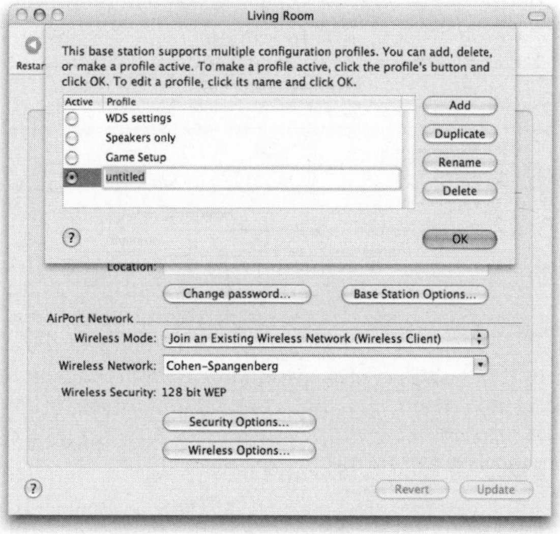

Figure 11-7: Create a new profile and name it.

6. **Click OK.**

   The sheet retracts and the name of the newly created profile appears below the configuration window's toolbar, as shown in Figure 11-8. You can now change the profile's settings to work with your lodging's network connection.

7. **Choose Create a Wireless Network (Home Router) from the Wireless Mode pop-up menu.**

   You may not be at home, but you are at your home away from home.

8. **In the AirPort Network section of the window, in the Name field, enter a name that identifies your in-room wireless network.**

   If you perform the next step, make sure you give the network a name that you can easily remember and type.

9. **Optionally, click the Create a Closed Network check box, and then click OK in the sheet that appears.**

   The confirmation sheet, shown in Figure 11-9, warns you of the consequences of creating a closed network: A closed network keeps others, such as guests staying in adjacent rooms, from detecting the presence of your network. However, it also requires you to enter the network's name by hand every time you want to join it, which explains why you gave the network a memorable and eminently enter-able name in the previous step.

Figure 11-8: Now that you have a new profile, you can change its settings.

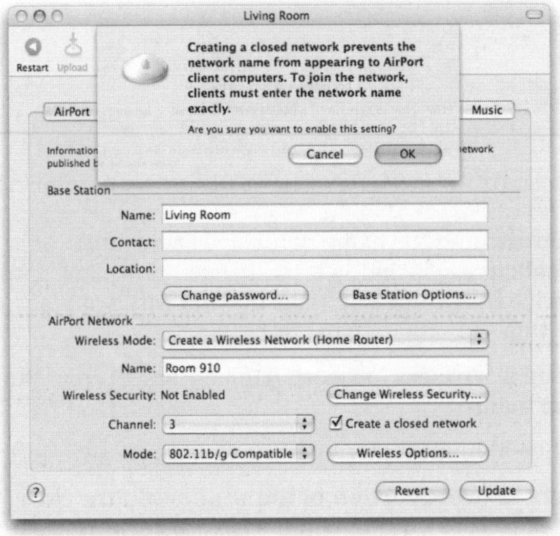

Figure 11-9: Creating a closed network protects your privacy, but it has consequences.

WARNING!

Do not forget the name of your closed network, because you will also need to type that name to administer the base station when you want to change profiles or otherwise adjust the AirPort Express's configuration at some future time.

10. **Click Change Wireless Security.**

11. **In the sheet that appears, make the security settings that seem most prudent to you and then click OK.**

Figure 11-10 shows the security settings I tend to use when I travel. You may choose different ones, depending on the type of AirPort card in your Mac and your own level of comfort. Chapter 4 describes in more detail the security settings you can make.

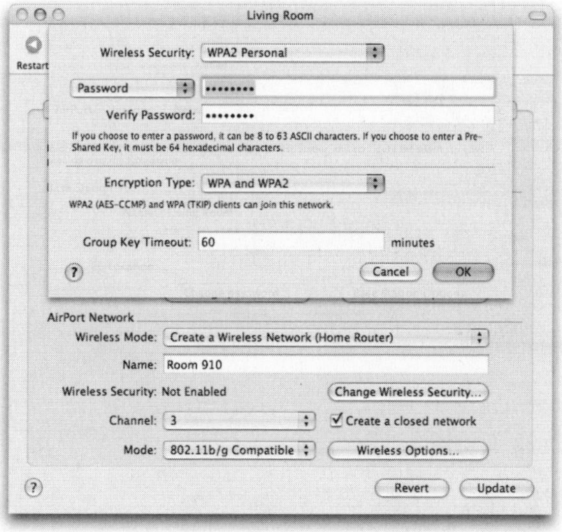

Figure 11-10: In unfamiliar surroundings, enabling network security is not a bad idea.

12. **Click the Network tab and make the appropriate settings in the configuration panel that appears.**

    In most cases, the settings shown in Figure 11-11 work. Your room may provide an instruction sheet for using your in-room network connection that specifies different settings, and, if so, you should follow those instructions instead.

13. **Click the Internet tab and make the appropriate settings in the configuration panel that appears.**

    Like the previous step, the settings shown in Figure 11-12 work for the majority of in-room network connections, but you should, of course, rely on the instructions that your lodgings may provide to you instead of the ones shown here.

14. **Click Update.**

    Your AirPort Express resets itself to use the new configuration. You can quit the AirPort Admin Utility.

After the AirPort Express resets, you need to join the network you just configured.

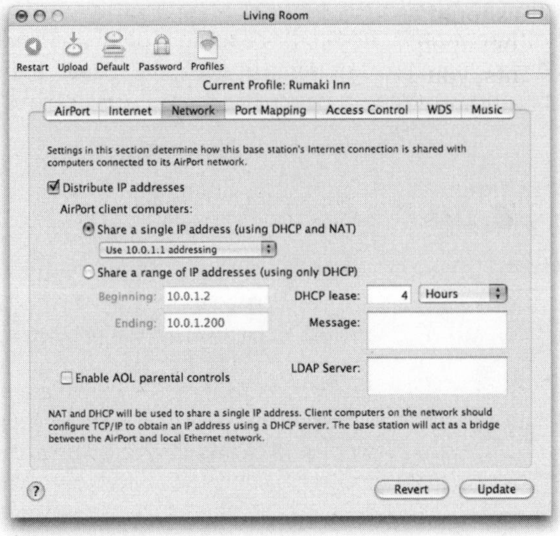

Figure 11-11: These settings work for most establishments, but you should check first.

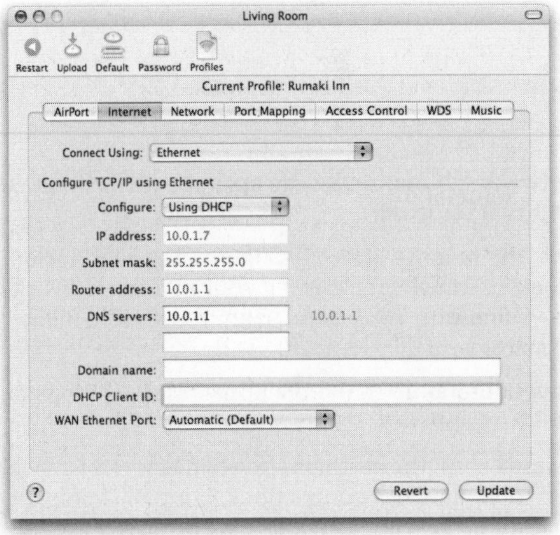

Figure 11-12: The settings shown here apply to many in-room network connections, too.

TIP: If you have created a closed network, as described in Step 9, you can join the closed network by choosing Other from your AirPort status menu or from Internet Connect's Network pop-up window and entering the network's name and password in the dialog that appears.

Finally, if the place that you're staying doesn't provide a cable, connect the AirPort Express to the in-room network connection with the Ethernet cable you packed — see "Packing your bags," earlier in this chapter.

Your AirPort Express should join the network provided by your lodgings. To test out the connection, open your Web browser and try to load a Web page. In many establishments, you'll see an authentication page much like you would at a wireless hotspot, as described earlier in "Using public hotspots."

Being secure on a foreign network

When you connect to a network provided by your lodgings, you have little idea, and no say, about how the network handles security, so you need to make sure that you provide yourself with as much security as you can.

In the previous section, I listed a couple of settings you can make to your base station's configuration that shield access to your in-room wireless network from others: that is, creating a closed network and setting an appropriate level of wireless security. In addition, applying the network settings shown earlier in Figure 11-11 effectively provides a firewall between your Mac and the rest of the Internet.

However, you should take other security precautions as well:

- **Use your Mac's firewall:** Your Mac has a built-in firewall you can enable that restricts incoming network connections. You can enable the firewall with the Sharing preferences in System Preferences. Chapter 9 describes the firewall in more detail.
- **Avoid making financial transactions online:** Chances are, you are no less safe making online purchases or similar transactions over your lodgings' network than you are over your connection at home, especially if the Web pages used for those transactions use secure encryption — any URL that begins with `https://` employs such security. However, I like to err on the side of better-safe-than-sorry whenever I can.
- **Use a VPN to connect to your place of business:** Your employer probably requires this no matter where you connect from if a VPN is provided at all, but in the event that you do have a choice, use the VPN.

Above all, keep in mind that you don't know who your neighbors are, nor what equipment and software they may have at their disposal. You don't have to stay up nights worrying and fretting, just try to exercise some caution in your online pursuits while you're on the road.

Meanwhile, crack open that minibar and have a cold one. You've earned it.

Taking Advantage of Portable Gadgets

This book is called *AirPort and Mac Wireless Networks For Dummies*, but, as you may have already seen elsewhere in these pages, your Mac is not the only wireless player on your digital team. Other wireless devices have contributions they can make to your increasingly digital lifestyle, both working with and sometimes substituting for your Mac.

Lightening the load with a PDA

When you spend a day out of town on business, perhaps dashing from meeting to meeting or prowling the endless aisles of a crowded and noisy tradeshow floor, you may not want to schlep an expensive, not-quite-lightweight-enough notebook computer with you. In some cases, a *PDA (Portable Digital Assistant)* with wireless capability can take on many of the tasks you might otherwise relegate to your Mac.

The latest generation of PDAs have grown far more sophisticated and powerful than their earliest forebears, boasting larger and brighter color screens, vastly increased storage capacities, more sophisticated handwriting recognition, and increasingly improved interoperability with your laptop or desktop Mac. They have truly begun to earn the sobriquet of Digital Assistant.

Can you make do with a PDA instead of a Mac when you travel? It all depends on what you need the PDA to do. Here's what a modern PDA, such as the PalmOne LifeDrive that I have sitting beside me right now, can do rather well:

- **Appointment calendar:** The PDA won't run iCal, but there isn't a PDA made that doesn't have its own appointment calendar software. Apple's iSync software can synchronize your iCal calendars with the PDA's calendar software.
- **Address book:** The PDA won't run Apple's Address Book application either, but it will have its own contact manager software. Once again, iSync can synchronize the contact lists in Address Book with the PDA's contact lists.
- **Meeting notes:** Most PDAs have handwriting recognition of some sort that you can use to take meeting notes. When you get back to your lodgings or return home, you can transfer the note files to your computer for cleanup — which, if your handwriting is anything like mine, you'll probably have to do: I think I once heard my PDA's handwriting recognition software cry.
- **Photo storage:** High-capacity PDAs, such as the LifeDrive with its 4GB internal disk drive, have enough room to allow you to store photos from a digital camera for later transfer to your computer.

- **E-mail retrieval:** A PDA with wireless network capability can use a hotspot or other wireless network connection to access your e-mail account and let you read your e-mail on the run. You can also respond to e-mail if you feel comfortable using the handwriting recognition software or the on-screen keyboard.
- **Light Web browsing:** Even the best, full-color PDA displays don't provide enough screen area to show most Web pages to their full advantage, but you can usually read most Web pages adequately enough on your PDA to have a satisfying browsing experience while sipping a cappuccino at your local wireless hotspot.
- **Voice recording:** Many PDAs have a built-in microphone so you can record short voice notes, and those PDAs with sufficient storage capacity may let you record entire conversations or meetings. You can transfer the audio files to your computer later for editing or transcription.
- **Entertainment:** Need some light reading for your flight? Handheld PDAs can hold quite a few books without increasing your carry-on weight by an ounce — `www.pdasupport.com/eBookSites.htm` provides a list of sites that offer e-books you can download. And although a PDA is not an iPod, that doesn't mean they can't store and play MP3 files. What's more, a high-capacity PDA with a color screen, like the one I'm currently using, can even store an entire feature film and show it on its color display.

Chances are, you won't want to use a PDA for things like extensive word processing, image editing, or complex database or spreadsheet entry: After all, its a Portable Digital *Assistant*, not a substitute for a fully equipped desktop or notebook Mac. But PDAs do work well for accessing information of various kinds, they can handle light data-entry tasks, and, being assistants, they are designed to exchange information easily with your computer.

When you use your wireless PDA with your AirPort network, it needs to authenticate just like your Mac does, which means that you have to enter the AirPort network password on your PDA the first time you connect to the network. If you use WPA Personal or WPA2 Personal security for your network, the PDA can use the same password that your Mac does. However, if your network uses WEP security, the PDA will have to use what Apple calls the *equivalent network password*, just like the wireless music players described in Chapter 7 have to use. This password consists of a string of either 8 or 26 letters and digits. When you configure your AirPort Base Station with the AirPort Admin Utility, you can obtain this password by clicking the Password icon on the toolbar. Figure 11-13 shows the sheet that appears with the equivalent network password.

Depending on the type and the capability of the PDA you have, you may or may not be able to synchronize appointments and contacts wirelessly using Bluetooth or WiFi. Strange as it may seem, in some cases PDAs that have Bluetooth and WiFi capability still require a physical connection, such as a USB cable, to exchange information with your Mac. Before you invest in a PDA as a notebook substitute, check with the manufacturer to see if the PDA supports wireless Mac synchronization. You don't necessarily need wireless synchronization, but it does mean one less cable to carry around.

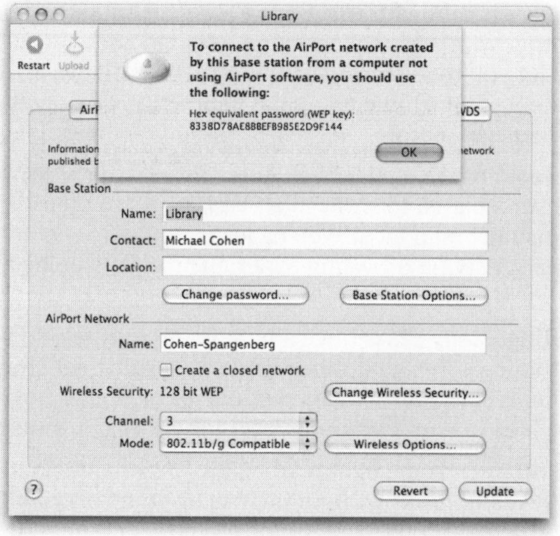

Figure 11-13: An AirPort network password rendered as a string of hexadecimal digits.

Going online with your Bluetooth phone

Sometimes you can find yourself in a location where you have no WiFi hotspots in the vicinity, the phone system doesn't allow modem hookups, and the Internet seems tantalizingly close and yet seemingly unreachable. If you have a cell phone with Bluetooth capability, however, you may be able to turn it into a low-speed but usable wireless modem for your Mac.

Understanding Bluetooth

If you don't know what Bluetooth is, you're not alone: lots of people don't know what this wireless technology is, or what it's doing on their Macs. Here's a brief introduction to Bluetooth if you haven't already bitten into it.

In the late 1990s, a group of cell phone manufacturers and other digital device makers, lead by Sweden's Ericsson, developed Bluetooth as a common technology that would allow their products to exchange information wirelessly with each other over short distances, rather than requiring them to be connected with cables. Bluetooth has several characteristics that appeal to the makers of cell phones and small electronic devices:

- It uses encryption to make the information exchanged between devices be secure from interception and tampering.
- It consumes very little electrical power.
- It costs relatively little to include in a digital device.

Bluetooth is not a replacement for a wireless network technology like AirPort: Bluetooth devices typically offer a maximum broadcast range that's about one-fifth the range of an AirPort network, and Bluetooth devices exchange information more slowly than devices on an AirPort Extreme network do. However, Bluetooth's range and speed are quite sufficient for devices like wireless keyboards, telephone headsets, and portable printers, to name just three of the many kinds of devices that have begun to incorporate Bluetooth.

Like any digital technology, Bluetooth comes with its own collection of special terms and concepts. The most important ones you need to know are *discovery* and *pairing*:

- **Discovery:** Before a Bluetooth device can connect to another Bluetooth device, it first needs to *discover* that other device. It does this by transmitting a request for either a particular kind of a device ("Hey, are there any Bluetooth printers out there?") or for a specific device that it knows about ("Hey, Michael's cell phone, are you out there?") and then listening for a response.

  You do not need to make a Bluetooth device discoverable in order to use it. Generally, the device making a request doesn't need to be discoverable, only the devices that respond to the request.

- **Pairing:** The first time a Bluetooth device discovers another Bluetooth device, the two devices usually *pair*, a process that allows the two devices to exchange information securely.

  To establish the secure connection, one device sends a code, often called a *PIN* or a *passkey*, to the other device, and the other device sends a similar or identical code back. Usually, the pairing process requires a user to enter the passkey on the device's keyboard or touch-pad, if it has one.

  After the two devices have paired, at least one of them, such as your Mac, records that the other device exists, and stores certain identifying information about the paired device, so that the two devices do not need to be paired again.

Both discovery and pairing are described in the next section, "Pairing your Mac with a Bluetooth phone."

Many of the most recently released Macintosh models have Bluetooth built in as a standard feature. Owners of older Macs without built-in Bluetooth can attach a small, inexpensive Bluetooth transmitter and receiver, often called a *dongle*, to a USB port. Mac OS X can work with most commercially available Bluetooth USB dongles.

Putting Bluetooth to work on your Mac comes down to three simple steps:

1. **Turn your Mac's Bluetooth adapter on.**

   If Bluetooth is built into your Mac, you turn it on using your System Preferences, like this:

   1. **Open System Preferences.**
   2. **Click the Bluetooth icon to see the Bluetooth preferences window.** You can find the icon in the Hardware group of icons in the System Preferences window.
   3. **Click the Settings tab in the Bluetooth preferences window.**
   4. **Click Turn Bluetooth On.** The button label changes to Turn Bluetooth Off when you do this.

   TIP

   You can close the Bluetooth preferences window once you've performed this last step, or you can do yourself a favor and click the Show Bluetooth status in menu bar check box before you close the window. You can use the Bluetooth status menu that appears on your menu bar the next time you want to turn Bluetooth on or off, or whenever you want to connect to a Bluetooth device or to set up a Bluetooth device, saving yourself another trip into System Preferences.

   If you have a Bluetooth dongle, you may not need to visit the Bluetooth preferences window or the Bluetooth status menu: simply plugging the dongle into a USB port is usually all you need to do to turn Bluetooth on.

2. **Set up your Bluetooth device.** You usually only need to do this if you've never used the device with your Mac before. The Set Up New Device button, located under the Devices tab in the Bluetooth preferences window, brings up your Mac's Bluetooth Setup Assistant to guide you through the setup process, which varies depending on the type of device you're setting up. If you have the Bluetooth status menu on your menu bar, you can choose its Set up Bluetooth Device command to bring up the Bluetooth Setup Assistant.

3. **Use the Bluetooth device with your Mac.**

The next section describes how to set up a Bluetooth phone so your Mac can use it.

Pairing your Mac with a Bluetooth phone

You use the Bluetooth Setup Assistant in Mac OS X to pair your Mac with a Bluetooth phone — or with any other Bluetooth device for that matter. This assistant walks you through discovering the device, pairing your Mac with the device, and setting up the various tasks your Mac can perform with that device. To use the Bluetooth Setup Assistant, of course, you must have Bluetooth active on your Mac: The previous section, "Understanding Bluetooth," explains that process.

To bring up the Bluetooth Setup Assistant and pair your Mac with your cell phone, follow these steps:

1. **Turn on the phone and make sure it is discoverable and ready for pairing.**

   You should consult the phone's documentation to find out how to activate Bluetooth on your phone and to make the phone discoverable, as the procedure varies among phone models and brands. On my phone, I have to navigate through several menus — Connect ➪ Bluetooth ➪ Discoverable — but, unless you have the same kind of phone as me, you'll probably follow a somewhat different path.

   Your phone documentation should also tell you how to give your phone a Bluetooth name. If you like, you can give your phone a distinctive name, which will appear in Step 4 below when your Mac discovers your phone. Otherwise, your Mac will refer to your phone by the default name assigned to it by the phone's manufacturer.

2. **Perform one of the following procedures.**

   If you have the Bluetooth status menu visible on your menu bar:

   1. **Click the Bluetooth status menu.**
   2. **Click Set up Bluetooth Device.**

   If you *don't* have the Bluetooth status menu visible on your menu bar:

   1. **Open System Preferences.**
   2. **Click the Bluetooth icon to see the Bluetooth preferences window.**

      You can find the icon in the Hardware group of icons in the System Preferences window.

   3. **Click the Devices tab in the Bluetooth preferences window.**
   4. **Click Set Up New Device.**

   The Bluetooth Setup Assistant window appears.

3. **Click Continue.**

   The Bluetooth Setup Assistant presents a list Bluetooth device types from which you can select the type you want to set up.

4. **Click Mobile Phone and then click Continue.**

   The Assistant searches for your phone and, after ten seconds or so if all goes well, the phone appears in the Assistant's Mobile Phone list.

5. **Click your phone in the Assistant's phone list, and then click Continue.**

   The Assistant displays a passkey, and tells you that you need to enter the passkey into your cell phone. In most cases, your cell phone display also presents a message asking you if you want to pair with the device — your Mac — that just contacted it.

6. **Respond to the prompts on your cell phone display and enter the passkey into your cell phone when directed.**

   When the pairing completes, the Assistant presents a checklist of mobile phone services. By default, the Assistant puts check marks by all the available services.

7. **Uncheck all the options in the checklist.**

   The first two check boxes tell the Assistant that you intend to synchronize the mobile phone's contacts lists and calendar events with your Mac, which you don't want to do right now. The third check box, Access the Internet with your phone's data connection, is what you eventually want to do, but not right now, because doing so requires you to have some information about your Internet service provider and to choose the type of connection you want to establish. The next section, "Finding your phone's capabilities," discusses how to choose the type of connection, and the section "Setting up the wireless phone connection" later in this chapter gives instructions on how to set up the connection.

8. **Click Continue.**

   The Assistant finishes setting up the mobile phone and displays a congratulatory screen.

Finding your phone's capabilities

Nothing beats the cell phone industry when it comes to making something as seemingly simple as a cell phone brain-bogglingly complicated. The variety of features and service plans offered, and the welter of terms used to describe those features and plans, can discourage all but the most persistent comparison shopper. But take heart: With the help of the next few pages, you can figure out if your phone can provide your traveling Mac with a wireless modem.

Mac OS X's Bluetooth modem capabilities support two kinds of wireless modem connection:

- **Dial-up:** This method literally uses your cell phone to dial the same number as your Mac would, using a direct connection from its modem port. The Bluetooth phone acts as a modem, albeit a slow one: Typical connection speeds rarely go higher than 14 kbps, roughly one-third the speed of your Mac's built-in modem. At those speeds, you can read e-mail and browse some Web sites if they aren't too data intensive. Cell phone providers treat these kinds of connections like voice calls, charging you by the minute.
- **Direct digital high speed:** Many cell phone service providers provide a direct digital connection to the Internet through their own service. Though the providers tend to call these connections "high-speed," they only seem so compared to the dial-up cell phone connections just described, providing connection speeds in the 40 to 56 kbps range, roughly equivalent to your Mac's built-in modem port.

You'll find two kinds of direct digital connection available from the various cell phone providers, both of which Mac OS X supports. Which connection method you use depends on your provider's cell phone network, and keeping these methods straight in your head requires that you have a high tolerance for acronyms.

Ready? Here's a taste of the alphanumeric soup served up by the cell phone industry:

- **1xRTT:** Providers operating networks that use the CDMA wireless phone protocol, such as Verizon and Sprint, offer 1xRTT connections.

  CDMA (Code Division Multiple Access) is a wireless protocol that provides multiple digital connections on a single frequency by tagging each chunk of data with a digital code. The kind of code used for this tagging allows some very clever mathematics to extract the data belonging to each individual connection from the stream of combined connections.

- **GPRS:** Providers that operate networks using the *GSM (Global System for Mobile Communications)* wireless protocol offer *GPRS (General Packet Radio Service)* data connections. Such providers include T-Mobile, Cingular, and AT&T.

  GPRS provides multiple digital connections on a single frequency by slicing the transmissions into discreet time chunks, using a method called *TDMA (Time Division Multiple Access).*

For either type of direct digital connection, you need to purchase that service from your mobile provider, often as a separate package in addition to your regular mobile-voice service. These direct digital service plans charge by the amount of data transmitted and received while your phone is connected, rather than by the amount of time the phone spends connected.

In addition to determining the type of connection you can establish, you also need to determine whether your cell phone can function as a modem at all.

The built-in features ordinarily available on particular cell phone models may be disabled at the service provider's discretion. For example, although your phone's manufacturer may say that the phone has Bluetooth-modem capability, the service provider may have disabled that feature.

If you are not sure about your phone's features, your Mac's Bluetooth System Preferences can help you determine if your Bluetooth phone offers modem capabilities.

Before you perform the following steps, you first need to have paired your Mac with your Bluetooth phone, a process described in the previous section, "Pairing your Mac with a Bluetooth phone."

To see which cell phone features your Mac has detected, follow these steps:

1. **Activate Bluetooth on your phone and make the phone discoverable.**

   Consult your cell phone's manual if you don't know, or can't remember, how to accomplish this.

2. **Activate Bluetooth on your Mac.**

   You can find instructions for doing this in "Understanding Bluetooth" earlier in this chapter.

3. **Open System Preferences and click the Bluetooth icon in the System Preferences window.**

4. **In the Bluetooth preferences window, click the Devices tab.**

5. **In the Bluetooth Devices list, click your cell phone's name.**

   You saw your cell phone's name when you first paired your phone with your Mac, as described earlier in "Pairing your Mac with a Bluetooth phone." When you click the phone's name in the list, the Bluetooth features appear in the Bluetooth Devices list, itemized under the Device Services heading as shown in Figure 11-14. The Dial-up Networking feature is the one that provides modem capability.

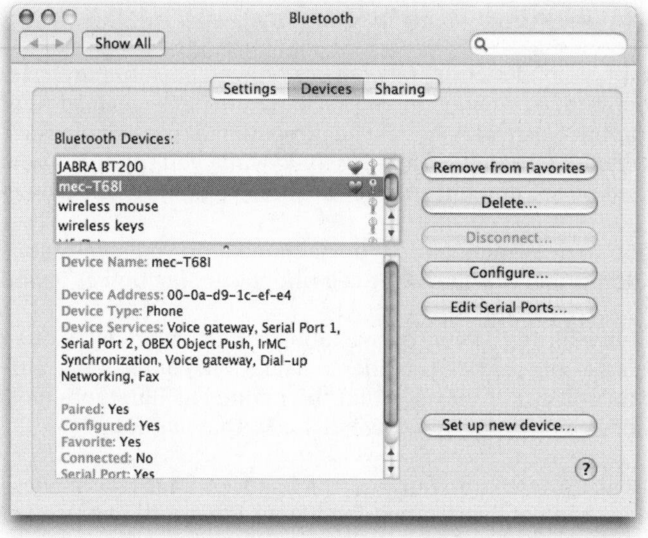

Figure 11-14: Check your phone's device features in the Bluetooth System preferences.

Setting up the wireless phone connection

If your phone can perform crazy Bluetooth modem tricks, and you have decided what kind of Internet connection you will use with the phone, you can configure your Mac and your phone to join forces in your quest for Internet access.

TIP

Before you begin this connection odyssey, you may want to create a new network location dedicated to your Bluetooth connection experiments. That way, you can avoid messing up your current network settings. You can find out how network locations work and how to create them in Chapter 6.

Assuming you have previously paired your Bluetooth phone with your Mac — and if you haven't, see "Pairing your Mac with a Bluetooth phone" in this chapter — you can follow the steps below to set up your Mac to use your Bluetooth phone as a modem:

1. **Activate Bluetooth on your phone and make the phone discoverable.**

   Consult the documentation that came with your phone to find out how to do this.

2. **Activate Bluetooth on your Mac.**

   "Understanding Bluetooth" in this chapter tells you how this is done.

3. **Open System Preferences and click the Bluetooth icon in the System Preferences window.**

4. **In the Bluetooth preferences window, click the Devices tab.**

5. **In the Bluetooth Devices list, click your cell phone's name.**

   Figure 11-14 in the section immediately preceding this one shows you how your Bluetooth preferences window should look — if you take into account that the figure shows *my* Bluetooth devices and not yours, of course.

6. **Click Configure.**

   The Bluetooth Setup Assistant swings into action, as shown in Figure 11-15. The Assistant asks you for information to set up the Bluetooth modem.

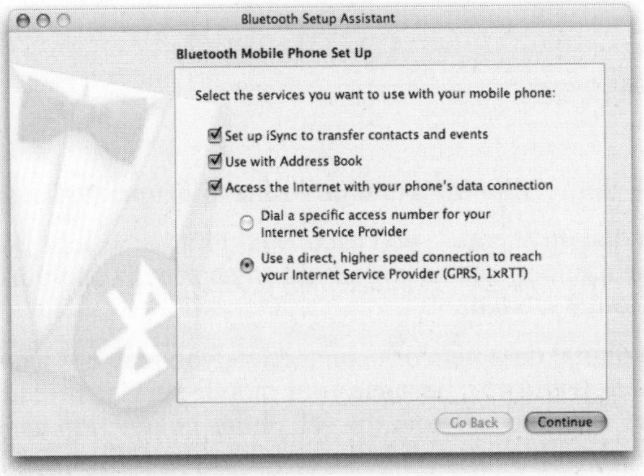

Figure 11-15: The Bluetooth Setup Assistant handles the phone configuration for you.

7. **Check the Access the Internet with your phone's data connection option.**

   You may notice that Set up iSync and Use with Address Book are both checked on this screen every time you run the Bluetooth Setup Assistant. Unless you also want to have Apple's iSync program move your contact information to your cell phone, I suggest you uncheck these two options.

8. **Click the radio button beside the connection method you want to use.**

   Click the first choice — Dial a Specific Access Number for Your Internet Service Provider — if you want to call an existing provider over an ordinary voice connection. The other choice uses your cell phone service provider's digital data network, to which you must subscribe in order to use the service.

9. **Click Continue.**

   The Assistant shows you a panel like the one in Figure 11-16 if you chose a dial-up connection. The panel for a digital data network connection displays slightly different text, and changes the Phone Number field to a GPRS CID String field, but is otherwise identical.

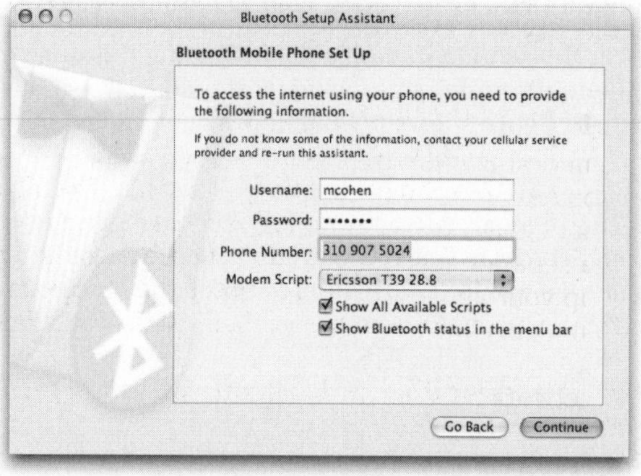

Figure 11-16: Configuring a dial-up connection for a Bluetooth phone.

10. **Enter your connection information in the appropriate fields.**

    - For a dial-up account, you usually enter a name, a password, and a phone number, just as you would when setting up your Mac to use its internal modem.
    - For a digital data network connection, you may not need to use a name or password, because your mobile service provider already knows who you are from the cell phone connection itself. You need to check with your mobile service provider for the GPRS CID string you should use. They usually look something like this: `*99*1234567#`.

11. **Choose a modem connection script from the Modem Script pop-up menu.**

    A modem connection script contains the commands that a particular modem, such as the one built into your cell phone, can respond to. Your Mac uses the modem script you select to tell the modem to connect to the Internet. Different cell phone models often require different modem scripts, and picking the right one from the set of scripts that Apple provides can be a hit-or-miss process. Your best bet is to do some online research to find out what other folk who have your cell phone use.

    TIP: You can find additional modem scripts, along with information about using them and other useful tips, at `www.taniwha.org.uk`. After you download the script, put it in the Modem Scripts folder inside your Mac's main Library folder. Preferably, you do that before you run through the steps here.

12. **Click Continue.**

    The Setup Assistant makes the necessary configuration changes and presents you with a congratulatory message.

After you have finished these steps, you can skip ahead and follow the steps in "Making the Bluetooth modem connection" to make your first connection. Or you can read the following section first.

Customizing your network location for Bluetooth

TIP: Before you make your first Bluetooth modem connection, you may want to visit your Network preferences and clean things up a little. The following procedure is optional, but I highly recommend it if you have taken my earlier advice and created a separate Bluetooth network location. In the following procedure, you set up your Bluetooth location settings to make sure your Mac uses the Bluetooth modem connection rather than another network port.

To customize your network location for Bluetooth use, follow these steps:

1. **Open System Preferences and click Network.**
2. **From the Show menu, choose Network Port Configurations.**

   Depending on how you have set up your Mac, you may need to click the lock icon in the bottom left of the Network preferences window and then enter an Admin password before you can make any changes to your Mac's network preferences.

3. **In the Port Configurations list, uncheck all the ports except the Bluetooth port.**

   You may also want to drag the Bluetooth port to the top of the list. When you finish, your Network preferences window should look similar to Figure 11-17.

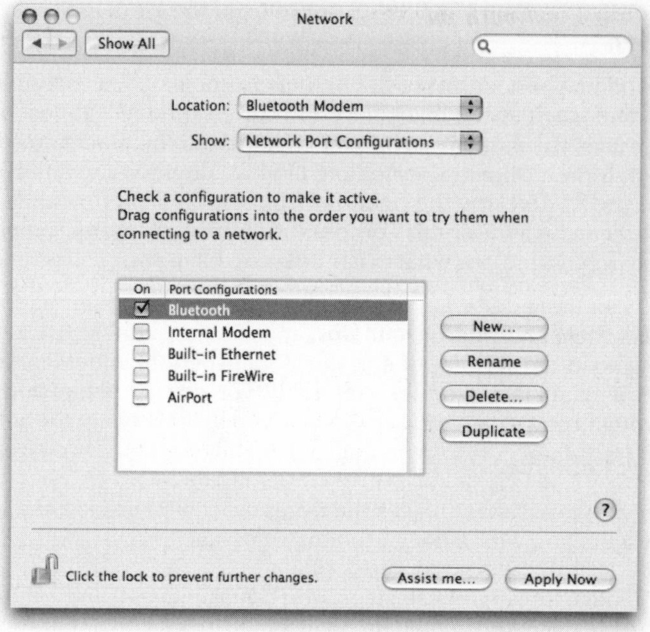

Figure 11-17: Make your Bluetooth phone your primary network connection.

4. **Click Apply Now.**
5. **From the Show menu, choose Bluetooth.**

   The Network preferences window shows the Bluetooth connection settings specified with the Bluetooth Setup Assistant, as described in "Setting up the wireless phone connection." Double-check them to make sure they are correct. If not, use the Bluetooth Setup Assistant again to make the changes required.

   If you need to switch modem scripts, you can do so by clicking the Bluetooth Modem tab in this preference window. You can choose from the same list of modem scripts the Bluetooth Setup Assistant offered by clicking the Modem script pop-up menu available under this tab.
6. **If you made any setting changes in the previous step, click Apply Now.**
7. **Close the Network preferences window.**

With your Bluetooth modem set up and your network preferences fine-tuned for Bluetooth modem connectivity, you can try out this miracle of modern science.

Making the Bluetooth modem connection

Although it does take a bit of work to set up your Bluetooth cell phone to work with your Mac, as described in the last few sections, you only have to go through those procedures once — unless you change phones or Internet Service Providers. Actually making the Bluetooth modem connection itself is quite simple.

To establish the Bluetooth modem connection, follow these steps:

1. **Activate Bluetooth on your phone and make the phone discoverable.**
2. **Activate Bluetooth on your Mac.**
3. **Open the Internet Connect application.**

   If you've performed the steps described in "Setting up the wireless phone connection," you should see a Bluetooth icon on Internet Connect's toolbar, as shown in Figure 11-18.

Figure 11-18: Internet Connect grows a Bluetooth icon when you set up a Bluetooth modem.

4. **Click the Bluetooth icon.**
5. **Click Connect.**

   If all goes well, your Mac and your phone make the connection, though depending on the connection method, and whether the mobile network is congested or not, it may take a few dozen seconds or so. After Internet Connect makes the connection, the application's window should look similar to the one shown in Figure 11-19.

TIP

After you've established that your connection works, you can click the check box that puts a small status menu, which includes a Connect item, on your menu bar. You can use that menu the next time you want to open a Bluetooth connection without having to launch Internet Connect. Of course, you still need to remember to have Bluetooth on, your phone on, and to make your phone discoverable, but it's always nice to get rid of one step.

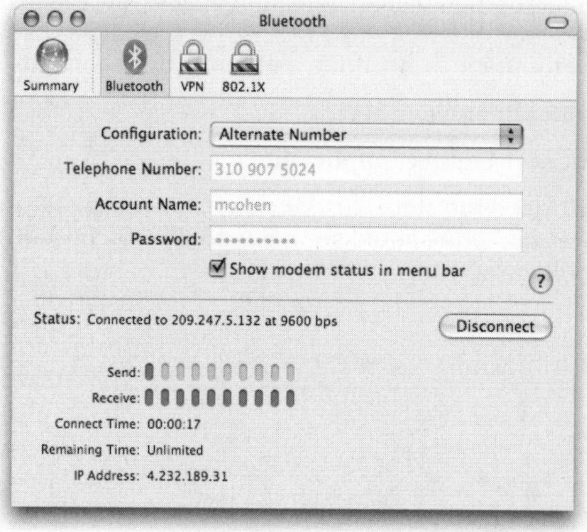

Figure 11-19: Success! Your Bluetooth phone has put your Mac on the Internet.

Part V
The Part of Tens

The 5th Wave By Rich Tennant

"Yes, it's wireless; and yes, it weighs less than a pound; and yes, it has multiuser functionality... but it's a stapler."

In this part . . .

Four chapters. Ten items in each one. Bite sized nuggets of wireless information, a tasty dessert platter of factoids and tips.

Chapter 12

Ten Troubleshooting Tips

In This Chapter

- Using preventive measures
- Employing diagnostic measures
- Jiggling and poking

In theory, there's no difference between theory and practice, but in practice, there is. Want an example? In theory, after you have set up your AirPort network, it should operate flawlessly from that point onward, but in practice, it may not: Signals sometimes vanish, the Internet connection may go away, your connections can slow to a crawl. These things may not happen frequently, and some of them might never happen to you. If they do, though, you can often restore things to working order with a minimum of fuss and expense.

Preventing Problems

You don't put on a seat belt because you plan to have a car accident. You put one on just in case you have a car accident. Here are a couple of AirPort seat belts you can strap on.

Create a separate admin account

When you set up your Mac, Mac OS X creates a user account for you with administrative privileges, and, if you're like most Mac users, you use that account every day. Over time, as you install and remove software and otherwise work and play, the preferences and settings associated with that account can get written and rewritten and modified and, occasionally, corrupted — including settings that may affect your Mac's network connections.

You can use the Accounts preference in System Preferences to create a second user account and to give that account administrative privileges. Don't use this account for day-to-day work, though: Keep it as your fallback/troubleshooting account. When things begin to go wonky in your main account, switch to the fallback account and see if the problems persist. If they don't, it means that the problem relates to your main account's settings and not the Mac itself. You may then be able to fix your main account by deleting from your main account's Library folder and preference files or cache files that might have become corrupted. At the very worst, you can move your documents and other files to your fallback account and continue working until you have a chance to fix the problem.

Save your AirPort configuration

It may not have disk drives, a keyboard, or a screen, but your AirPort Base Station is actually a computer, albeit a specialized one. And, like any computer, the software running it can sometimes have bugs that could cause it to stop working. When you finish setting up your AirPort base station, use the Save a Copy As item on the AirPort Admin Utility's File menu to save a copy of the base station's configuration. Then, if you ever encounter problems that require you to completely reset the base station, you can double-click that file to reload the configuration back into your base station.

Diagnosing Problems

Half of fixing a problem is figuring out just what went wrong in the first place. Before you cart your equipment to the Genius Bar at the closest Apple Store, you can try a few things to acquire the information that may help you out of your difficulties. Use these few figurative stethoscopes to listen to your network's metaphorical heartbeat.

Check your System logs

You may not know this, but your Mac keeps a plain text log of messages that get sent from your system software every time something unusual or notable happens. You can see these messages with the Console utility in your Utilities folder: Open the Console application and click File⇨Open System Log. Each message includes the date and time that it was sent. Although most of the messages may seem like veritable gobbledygook, you may spot some that pertain to your network connection or that mention AirPort. Even if you can't understand them, perhaps the genius at the Genius Bar, or a knowledgeable friend or colleague, can spot the clue that unravels your AirPort malfunction mystery.

Keep a trouble log

The worst problems to diagnose and fix are intermittent ones. If you have occasional AirPort signal dropouts or similar anomalies, you should write down what you were doing, what happened, what time of day it was, and anything else you can think of. Sometimes you may spot a pattern that helps you fix things. For example, if the problems you encounter happen only at certain times of the day, your network signal may be experiencing interference from something your neighbors are doing, such as their making dinner in the microwave oven that just happens to be on the other side of the apartment wall between your AirPort Base Station and their kitchen.

Try another Mac

The fault, dear Brutus, may lie not in our stars but in our Macs — in other words, you may have a hardware or software problem in your Mac that causes the network error rather than a problem in the AirPort Base Station or your Internet connection. See if you can obtain another Mac with an AirPort card and use that with the network for a while. If the problem persists, then your Mac may be okay. If not, then you have isolated the problem to your Mac and eliminated the network itself as the problem source. This diagnostic technique works especially well in a multi-Mac household or office: Getting a second Mac for your spouse or offspring lets you rationalize these purchases as necessary diagnostic tool acquisitions.

Solving Problems

Sometimes you can solve problems with the laying on of hands, either by way of a judicious jiggle, a tentative tweak, or a stern prodding. These last few tips may fix intermittent and irritating network drop-outs.

Check the cables

Yes, sometimes faulty wires in a wireless network can cause the problem. You have at least two cables, and maybe more, that you can check:

- The cable from your base station to your Internet connection, such as a DSL modem or cable modem
- The cable from your cable or DSL modem to the wall jack
- The cable from your base station's LAN port to your Ethernet hub, as well as the cables that go from that hub to your wired devices

Try unplugging things one by one and swapping known good cables with suspect ones. With luck, a cable replacement may be all you need to get your network back up to speed.

Change the channel

I cover this in Chapter 5, but it bears repeating: Interference, either from a nearby base station or some other device, may affect reception on your network's current channel. Use the AirPort Admin Utility to change the base station's channel, and keep a log, like I mentioned earlier, so you can record the current channel setting and its affect on network performance.

Get out of the car and get back in

In other words, reboot the base station and your cable or DSL modem. You can do this by disconnecting power from them and then reconnecting the power after waiting a couple of minutes. Rebooting often fixes things when the software in your base station or high-speed modem becomes confused. You may also want to try choosing Turn AirPort Off in your AirPort Status menu to turn off the power to your Mac's AirPort card, and then choosing Turn AirPort On to restart it.

Check the antenna

Desktop Power Macs have a detachable AirPort antenna on the back of the case. If you have such a Mac, and the AirPort reception seems iffy, the problem may be antenna-related. Try detaching and reattaching the antenna.

For iBooks, PowerBooks, and iMacs with installable AirPort cards, you can check to see that the antenna cable is firmly inserted in the AirPort card: In portable Macs especially, the cable can occasionally work its way loose. See Chapter 2 for details regarding how to access the AirPort cards in various Macs.

Reset the base station completely

When all else fails, you can try completely resetting the AirPort Base Station to its factory-default state and then running the appropriate AirPort Setup Assistant. I sometimes call this the nuke-and-pave option: It means completely reconfiguring the base station (and often the network clients), but it can solve otherwise recalcitrant problems. Chapter 3 describes how to reset various base station models. Also, see the second tip in this chapter: If you have a stored copy of a known good configuration for your base station, reloading it after a reset can save you time and trouble.

Chapter 13

Ten Wireless Security Measures

In This Chapter

- Locking down the base station
- Hardening your Mac

It's a jungle out there, and the glinty-eyed beasts crawling through the underbrush do not mean you well. What *you* see as an AirPort Base Station, they see as prey; and what *you* see as priceless data, they see as fresh meat. To keep yourself safe in the jungle, remember these ten measures. The first six show you how to secure the base station, and the last four how to make your Mac safer.

As far as securing your base station remember that AirPort base stations have very powerful security options built in — which do you absolutely no good at all if you don't use them. When enabled, these options keep uninvited guests from using your network to steal Internet access or to gain access to other machines on the network. How much trouble you want to go to securing your base station and the network it creates is up to you, of course, but you should at least consider the first six tips.

As far as making your Mac safer, remember you can't always control a wireless network's security — especially one that you don't own — but you can control your own computer's security. Even when you use a public hotspot or other nonsecure network, you still want to protect your Mac from hackers who may want to rifle through your files for credit card numbers and other sensitive information. Take some Mac-hardening steps to keep your data safer by following the last four tips.

Use the Strongest Feasible Wireless Security Setting

WPA2 Personal for home networks and WPA2 Enterprise for business networks provide the strongest security, but not all AirPort base stations and cards support these. If your equipment can, use them. If not, use WEP with

128-bit encryption at the very least. Chapter 4 describes where you can find these settings. (Hint: The Change Wireless Security button in AirPort Admin Utility's configuration window is a good place to start.)

Turn Off Remote Management and Similar Options

The Base Station Options button in the AirPort pane of AirPort Admin Utility's configuration window brings up a sheet of check boxes, several of which enable access to the base station from outside the AirPort network, and all of which should remain unchecked unless you have a network administrator who requires one or more of them to be turned on.

Hide Your Network

Your AirPort base station ordinarily shouts out the digital equivalent of, "Hi there, stranger, here I am," at regular intervals so that computers looking for a base station can find it. You can make your base station more demure by checking the Create Closed Network option in the AirPort pane of AirPort Admin Utility's configuration window. Although this requires you to specify the network by name when you join it, it keeps casual strangers from attempting to enter your network uninvited.

Use Different Passwords for Administration and for Joining

You'd be surprised how many people use the same password for both the base station and for the network it creates, which makes it easy for any network user to launch AirPort Admin Utility and play havoc with the network settings. You can change both passwords in the AirPort pane of AirPort Admin Utility's configuration window:

- Click Change Password to set a password for administering the base station.
- Click Change Wireless Security to change the password that network users need in order to join the AirPort network.

Consider a Members-Only Policy

The Access Control pane of the AirPort Admin Utility's configuration window gives you a place to enter the AirPort IDs of the wireless clients allowed to join your AirPort network. Only clients on the list can access the network. You can find a Mac's AirPort ID — also known as the wireless card's *media access control* (MAC) address — in the AirPort pane of the Network preferences window in System Preferences.

pp. 102-103

p. 16
Cf. 74

TIP

Include a description for each client ID you enter so you can keep track of which users have which IDs.

Turn Your Base Station Off

No, not always — what good is a network if it's not turned on? But just because you can run a network 24/7 doesn't mean you have to. Turn off the base station during the hours that you aren't normally at home, or when you won't be using it for a day or two. Why give hackers extra hours to attempt to crack into your network?

Turn Off File Sharing

Although file sharing can be convenient in some work or home situations, that doesn't mean you should just enable it and forget it. Keep file sharing turned off until you actually need to share some files, and when you finish sharing them, turn the feature back off. This goes double for Windows file sharing. Although it does take a few clicks to get to the System Preferences' Sharing window, spend the time — don't exchange a minor inconvenience for a major security headache.

pp. 227-232

Raise Your Firewall

Your Mac has a very good firewall, and, as Chapter 6 explains, you can set it up easily and quickly in the System Preferences' Sharing window. The firewall keeps outsiders from accessing services on your Mac without your express invitation. This is a Good Thing.

pp. 157-159

Keep Your Keychain Locked

Your Mac's Keychain stores all the passwords you ordinarily use for things like e-mail, certain secure Web sites, AirPort network access, and various applications. When you unlock the Keychain, your Mac silently provides passwords as they are needed. This means anyone who accesses your Mac when it's unattended can use your stored passwords. After you use the Keychain Access application in your Utilities folder to lock your Keychain, your Mac asks for your Keychain password when it needs to get at one of the stored passwords in the Keychain. Although repeatedly having to provide a password can be tiresome, it's not as tiresome as having your data stolen.

Strengthen your passwords

pp. 285-289

Built into the Keychain Access program is Apple's Password Assistant, which, as Chapter 10 explains, can help you devise very secure, difficult-to-crack passwords. Take advantage of the Assistant to make passwords for your Mac, your network, and any of the password-requiring services that you use, such as e-mail.

Cf. Pogue, Mac OSX: Tiger, pp. 471-475

Chapter 14

Ten Obscure Facts about Wireless Technology

In This Chapter

- Spelling obscure acronyms
- Going for the distance records
- Seeing movie stars and musicians
- Counting on things

If you want to find practical information about wireless technology, may I direct you to most of the other pages in this volume? What you find here in this tiny section hidden near the back of the book is information that will do you absolutely no good — unless you find yourself in a trivia contest.

Release the factoids!

You'd Go Blind

According to Apple's specifications, the AirPort Extreme base station's transmission power is 15 dBm, which equals about 0.032 watts. A normal incandescent reading light is about 2,000 times more powerful.

Marco! Polo!

An AirPort base station, like other wireless base stations, sends out a beacon message periodically so that other wireless devices can lock onto the signal. Usually, the beacon message gets sent about ten times every second, which sounds like a lot until you realize that it still leaves enough available bandwidth for the base station to stream high-fidelity techno-emo-punk to your entertainment system.

But I Thought They Drank a Lot of Coffee There

The most unwired city in the United States in 2005, according to the annual survey from Intel on such matters, is Seattle, Washington. Mobile, Alabama, on the other hand, ranked only 72nd despite its name.

And It Includes a Mint on the Pillow

A Gartner, Inc., survey released in 2005 claims that more than 60,000 hotels around the world now offer wireless access. On the other hand, the same survey says that only 30 percent of travelers use such services.

No Thanks, I'd Rather Walk

In 2002, Casey West, Meng Wong, Michael G. Schwern, and David H. Adler established a high-speed wireless network of a different sort. The network, using 802.11b technology, allowed a Volkswagen Passat and a Hyundai Tiburon to communicate while barreling down I-70 in the American heartland. The network participants engaged in chat sessions and played MP3s, but occasionally lost contact when large tractor trailers came between the two network nodes.

I'd Like To Buy Another Vowel

Back before it was called WiFi, the industry-standard acronym for wireless network technology went by the name of DFWMAC: Distributed Foundation Wireless Media Access Control. It almost sings, doesn't it?

Pretty Fast for a Preteen

The first systems that fulfilled the requirements of the 802.11 standard were tested in 1994. They had a transmission speed of 1 megabit per second, which was pretty zippy at the time. AirPort, when it reached the marketplace in 1999, achieved speeds 11 times faster and can now reach speeds that exceed the original devices fiftyfold.

What Have You Done for Me Lately?

The original AirPort card and base station rolled out by Apple at the end of the last century used technology manufactured by Lucent Technologies, an offshoot of Bell Labs. Another offshoot of Bell Labs was an operating system known as Unix, which today provides the core of Apple's Mac OS X.

Not to be Confused with Hedley Lamarr in Blazing Saddles

In 1942, Hedy Lamarr, under her married name of Hedy Kiesler Markey, was issued patent 2,292,387 for the "Secret Communication System" that eventually gave the world Bluetooth. That same year she also appeared as Dolores Ramirez in *Tortilla Flat,* Lucienne Talbot in *Crossroads,* and Tondelayo in *White Cargo.* Lamarr's friend and co-inventor, George Antheil, the avant-garde composer, composed "Water Music for 4th of July Evening" that year.

Can You Hear Me Now?

The 150-foot range of a typical WiFi signal was exceeded by a factor of more than 4,800 in August of 2005 by Trango Broadband Wireless and Microserv Computer Technologies, which collaborated to send a 2.3 Mbps WiFi signal across a distance of 137.2 miles, eclipsing the previous record of 125 miles. The transmission consisted of an FTP link, which, contrary to vicious rumor, didn't consist of an illegal MP3 download of "Water Music for 4th of July Evening."

Chapter 15

Ten Web Resources for Wireless Networking

In This Chapter

- Understanding wireless stuff
- Getting wireless stuff
- Fixing wireless stuff

When it comes to getting information about the technologies that make the Internet work, the place to look is the Internet. Scattered throughout that vast Web-work you can find most of the stories behind Internet technologies, learn from the personal experiences of countless folk who use those technologies, and obtain innumerable tips and tools for handling those technologies — if you know where and how to look, of course.

Configurations and Cards

This category contains places where you can find out about the hardware specifications for various Mac models and AirPort-related devices, and acquire some useful setup information.

Apple.com

No surprises here: Apple's site contains an extensive set of pages that explain all things AirPort. It also contains a bunch of other stuff, too, and sometimes finding your way around Apple's site to get what you need can be frustrating. To help you worm your way through the Apple, I've listed some useful pages for finding out what your Mac and AirPort card can do:

- **Do-it-yourself parts and service:** At the lower right of the page at `http://www.apple.com/support/diy/`, Apple provides a pop-up menu that lists all kinds of Mac models. Choose your model from the list and you leap to the page in Apple's Knowledge Base that lists the installable parts for that model (such as the compatible AirPort cards) and the procedures for installing them. Many of the procedures are in several languages, so you can both learn how to install a part and get a chance to practice your language skills at the same time.
- **AppleSpec:** `http://www.info.apple.com/support/applespec.html` contains a list of links to all the individual specification pages for Apple products released after November 1997, listed by the name and date they were introduced. As an added bonus, the final link on the page takes you to the AppleSpec page for what I think of as fond memories.

Non-Apple sources of configuration information

Apple's site, though often definitive, is not exhaustive when it comes to helping you set stuff up, especially when it comes to equipment or software not produced by Apple. These sites are useful sources of non-Apple wireless information:

- **Mac OS Wireless Adapter Compatibility List:** If you need to find which non-Apple access points and wireless devices work with your Mac, `http://home.earthlink.net/~metaphyzx/Wireless.htm` offers an extensive list. Listed by manufacturer, the table on this page provides the device type, the physical connection, the drivers used, and the compatible versions of Mac OS. This is *the* place to go when you want to find a wireless PCMCIA card for your PowerBook that's running Mac OS 8.6.
- **Sharing AirPort Base Station Experiences:** Constantin von Wentzel's site at `http://www.vonwentzel.net/ABS/index.html` provides all sorts of hacks and tips for setting up base stations in ways that Apple dares not discuss. As von Wentzel says, "All information presented here is for entertainment purposes only and all projects you undertake as a result of the following pages are at your risk alone," and you may void your equipment warranties if you do some of the things you find here.

News and Articles

Wireless network technology, and the commercial environment in which it exists, has evolved very quickly: tracking all the latest news is a real challenge. These sites can tell you which company has just introduced which product, which company has bought which, and what it all may mean.

- **O'Reilly Wireless Dev Center:** This site at `http://www.oreillynet.com/wireless/` aggregates news announcements, important blog posts, and detailed articles. A complete list of the latter, with short abstracts, can be found at `http://www.oreillynet.com/pub/q/all_wireless_articles`. The site is maintained by, and aimed at, the developer community, which means it would probably have a Technical Stuff icon next to just about every item in it if it were a *For Dummies* book. The writers, however, tend to be quite articulate, and the articles are usually understandable by the non-specialist.
- **MacInTouch Wireless Guide Reader Reports:** Ric Ford's venerable MacInTouch site began collecting reports from its readers about wireless technology in 2000. The reports now span 31 parts at this writing, with each part containing several months' worth of questions, answers, and commentary. Every time Apple releases a new version of the AirPort software, this is where I go to find out what's right and wrong with it. You can find the complete list of reports at `http://www.macintouch.com/wirelesslanreader.html`.
- **WiFi Net News:** Maintained and largely written by Mac aficionado Glenn Fleishman, `wifinetnews.com` usually posts several short articles each day about new wireless products, wireless industry machinations, and events of interest to wireless users. I check it out every couple of days to see which way the wind is blowing in the wireless world.

Facts, Tips, and Downloads

You have come to the grab-bag section, ladies and gentlemen. Here are a few places I regularly poke around when I'm looking for something in particular — a factoid, a utility, or a tip about how to do something:

- **Apple AirPort Support page:** At `http://www.apple.com/support/airport/`, Apple provides links to their latest AirPort software and documentation. I go here first anytime I need an Apple utility or an installation guide. From this page you can jump to user discussions, acquire network management tools, and get to Apple's most recent support articles for AirPort issues.

- **Mac OS X Hints Networking Forum:** Click the Networking link at forums.macosxhints.com to see ongoing discussions that often contain interesting and useful tips for solving recalcitrant networking problems. I use this site's search engine when I'm looking for the answer to a specific problem, but I also just like to browse around.
- **Wikipedia:** At en.wikipedia.org, you can enter the always-updated and expanding encyclopedia created by the general public. I go here whenever I need to find out about a specific topic or technology; the articles about networking in particular can be quite detailed. I'm often astonished at how often a search for an arcane — to me — networking term brings up an article of depth and substance. The Wikipedia Dashboard widget available at http://www.apple.com/downloads/dashboard/reference/wikipedia.html provides instant access to this remarkable trove of knowledge for Mac OS X 10.4 users.

Index

• B •

• D •

• E •

• F •

• J •

• K •

• L •

• M •

• N •

• O •

• P •

• T •

• U •

• V •

• W •

• Y •

• Z •

Notes

Notes

Notes

Notes

Notes

Notes

Notes

Notes

Notes

Notes

BUSINESS, CAREERS & PERSONAL FINANCE

0-7645-5307-0

0-7645-5331-3 *†

Also available:

- Accounting For Dummies † 0-7645-5314-3
- Business Plans Kit For Dummies † 0-7645-5365-8
- Cover Letters For Dummies 0-7645-5224-4
- Frugal Living For Dummies 0-7645-5403-4
- Leadership For Dummies 0-7645-5176-0
- Managing For Dummies 0-7645-1771-6
- Marketing For Dummies 0-7645-5600-2
- Personal Finance For Dummies * 0-7645-2590-5
- Project Management For Dummies 0-7645-5283-X
- Resumes For Dummies † 0-7645-5471-9
- Selling For Dummies 0-7645-5363-1
- Small Business Kit For Dummies *† 0-7645-5093-4

HOME & BUSINESS COMPUTER BASICS

0-7645-4074-2

0-7645-3758-X

Also available:

- ACT! 6 For Dummies 0-7645-2645-6
- iLife '04 All-in-One Desk Reference For Dummies 0-7645-7347-0
- iPAQ For Dummies 0-7645-6769-1
- Mac OS X Panther Timesaving Techniques For Dummies 0-7645-5812-9
- Macs For Dummies 0-7645-5656-8
- Microsoft Money 2004 For Dummies 0-7645-4195-1
- Office 2003 All-in-One Desk Reference For Dummies 0-7645-3883-7
- Outlook 2003 For Dummies 0-7645-3759-8
- PCs For Dummies 0-7645-4074-2
- TiVo For Dummies 0-7645-6923-6
- Upgrading and Fixing PCs For Dummies 0-7645-1665-5
- Windows XP Timesaving Techniques For Dummies 0-7645-3748-2

FOOD, HOME, GARDEN, HOBBIES, MUSIC & PETS

0-7645-5295-3

0-7645-5232-5

Also available:

- Bass Guitar For Dummies 0-7645-2487-9
- Diabetes Cookbook For Dummies 0-7645-5230-9
- Gardening For Dummies * 0-7645-5130-2
- Guitar For Dummies 0-7645-5106-X
- Holiday Decorating For Dummies 0-7645-2570-0
- Home Improvement All-in-One For Dummies 0-7645-5680-0
- Knitting For Dummies 0-7645-5395-X
- Piano For Dummies 0-7645-5105-1
- Puppies For Dummies 0-7645-5255-4
- Scrapbooking For Dummies 0-7645-7208-3
- Senior Dogs For Dummies 0-7645-5818-8
- Singing For Dummies 0-7645-2475-5
- 30-Minute Meals For Dummies 0-7645-2589-1

INTERNET & DIGITAL MEDIA

0-7645-1664-7

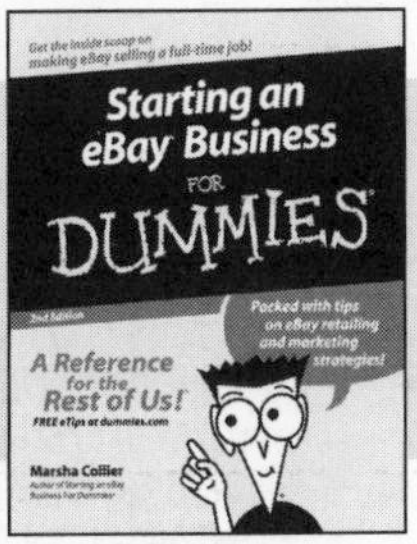

0-7645-6924-4

Also available:

- 2005 Online Shopping Directory For Dummies 0-7645-7495-7
- CD & DVD Recording For Dummies 0-7645-5956-7
- eBay For Dummies 0-7645-5654-1
- Fighting Spam For Dummies 0-7645-5965-6
- Genealogy Online For Dummies 0-7645-5964-8
- Google For Dummies 0-7645-4420-9
- Home Recording For Musicians For Dummies 0-7645-1634-5
- The Internet For Dummies 0-7645-4173-0
- iPod & iTunes For Dummies 0-7645-7772-7
- Preventing Identity Theft For Dummies 0-7645-7336-5
- Pro Tools All-in-One Desk Reference For Dummies 0-7645-5714-9
- Roxio Easy Media Creator For Dummies 0-7645-7131-1

*** Separate Canadian edition also available**

† Separate U.K. edition also available

Available wherever books are sold. For more information or to order direct: U.S. customers visit www.dummies.com or call 1-877-762-2974. U.K. customers visit www.wileyeurope.com or call 0800 243407. Canadian customers visit www.wiley.ca or call 1-800-567-4797.

WILEY

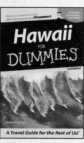

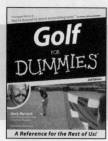

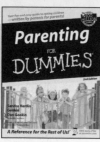

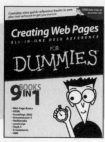

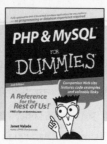

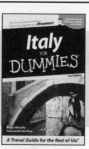

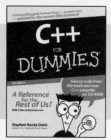

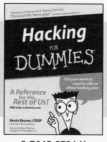

HEALTH & SELF-HELP

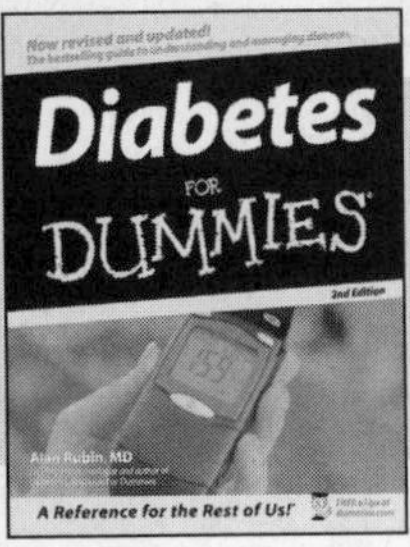

0-7645-6820-5 *†

0-7645-2566-2

Also available:

- Alzheimer's For Dummies 0-7645-3899-3
- Asthma For Dummies 0-7645-4233-8
- Controlling Cholesterol For Dummies 0-7645-5440-9
- Depression For Dummies 0-7645-3900-0
- Dieting For Dummies 0-7645-4149-8
- Fertility For Dummies 0-7645-2549-2
- Fibromyalgia For Dummies 0-7645-5441-7
- Improving Your Memory For Dummies 0-7645-5435-2
- Pregnancy For Dummies † 0-7645-4483-7
- Quitting Smoking For Dummies 0-7645-2629-4
- Relationships For Dummies 0-7645-5384-4
- Thyroid For Dummies 0-7645-5385-2

EDUCATION, HISTORY, REFERENCE & TEST PREPARATION

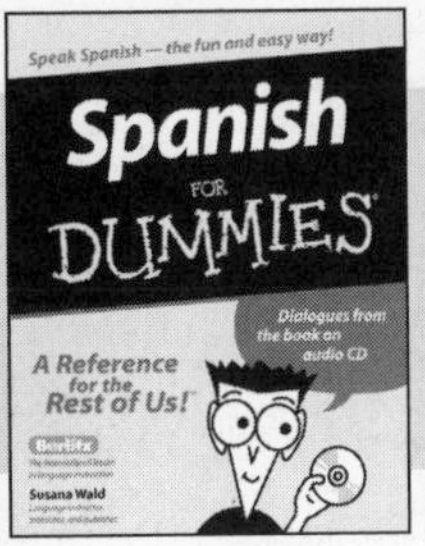

0-7645-5194-9

0-7645-4186-2

Also available:

- Algebra For Dummies 0-7645-5325-9
- British History For Dummies 0-7645-7021-8
- Calculus For Dummies 0-7645-2498-4
- English Grammar For Dummies 0-7645-5322-4
- Forensics For Dummies 0-7645-5580-4
- The GMAT For Dummies 0-7645-5251-1
- Inglés Para Dummies 0-7645-5427-1
- Italian For Dummies 0-7645-5196-5
- Latin For Dummies 0-7645-5431-X
- Lewis & Clark For Dummies 0-7645-2545-X
- Research Papers For Dummies 0-7645-5426-3
- The SAT I For Dummies 0-7645-7193-1
- Science Fair Projects For Dummies 0-7645-5460-3
- U.S. History For Dummies 0-7645-5249-X

Get smart @ dummies.com®

- **Find a full list of Dummies titles**
- **Look into loads of FREE on-site articles**
- **Sign up for FREE eTips e-mailed to you weekly**
- **See what other products carry the Dummies name**
- **Shop directly from the Dummies bookstore**
- **Enter to win new prizes every month!**

*** Separate Canadian edition also available**
† Separate U.K. edition also available

Available wherever books are sold. For more information or to order direct: U.S. customers visit www.dummies.com or call 1-877-762-2974. U.K. customers visit www.wileyeurope.com or call 0800 243407. Canadian customers visit www.wiley.ca or call 1-800-567-4797.